Dictionary of Environmentally Important Chemicals

Dictionary of Environmentally Important Chemicals

DAVID AYRES

and

DESMOND HELLIER
Queen Mary and Westfield College,
University of London, UK

BLACKIE ACADEMIC & PROFESSIONAL
An Imprint of Chapman & Hall
London · Weinheim · New York · Tokyo · Melbourne · Madras

**Published by Blackie Academic and Professional, an imprint of
Thomson Science, 2–6 Boundary Row, London SE1 8HN**

Thomson Science, 2–6 Boundary Row, London SE1 8HN, UK

Thomson Science, 115 Fifth Avenue, New York, NY 10003, USA

Thomson Science, Suite 750, 400 Market Street, Philadelphia, PA 19106, USA

Thomson Science, Pappelallee 3, 69469 Weinheim, Germany

First edition 1998

© 1998 David Ayres and Desmond Hellier

Thomson Science is a division of International Thomson Publishing   I(T)P

Typeset in 10/12 pt Times by Pure Tech India Ltd, Pondicherry
Printed by The Alden Group, Oxford

ISBN   0 7514 0256 7

A catalogue record for this book is available from the British Library

♾ Printed on acid-free text paper, manufactured in accordance with
ANSI/NISO Z39.48–1992 (Permanence of Paper).

Glossary of acronyms and abbreviations

AAS	atomic absorption spectrometry (analysis)
ACGIH	American Conference of Government Industrial Hygienists
ADI	allowable daily intake
ATSDR	Agency for Toxic Substances and Disease Registry (USA)
BATNEEC	best available technology not entailing excessive cost
BCF	biological concentration factor (see text)
BHA	butylated hydroxyanisole (chemical)
BHT	butylated hydroxytoluene (chemical)
BOD	biochemical oxygen demand (see text)
BPM	best practical means
Bq	becquerel: 1 disintegration per second (dps) = 1 becquerel (Bq)
CAS	Chemical Abstracts Service
CEC	Commission of the European Communities
CFCs	chlorofluorocarbons (chemicals)
Ci	curie
CNS	central nervous system
COD	chemical oxygen demand (see text)
COSHH	Control of Substances Hazardous to Health (UK)
DBP	di-*n*-butyl phthalate (chemical)
DDE	dichlorodiphenyldichloroethene (chemical)
DDT	dichlorodiphenyltrichloroethane (chemical)
DEHP	bis(diethyl-2-hexyl) phthalate
DFG	Deutsche Forschungsgemeinschaft (German Science Foundation)
DNA	deoxyribonucleic acid (genetic code material)
DOE	Department of the Environment (UK)
ECETOX	European Chemical Industry Technology and Toxicology Centre
ECD	electron capture detector
ECT	*Encyclopedia of Chemical Technology* (Kirk–Othmer)
EEC/CE/EC	European Community
EPA	Environmental Protection Agency (USA)
EPA	Environmental Protection Act 1990 (UK)
EU	European Union (formerly European Community)
FDA	Food and Drug Administration
FID	flame ionization detector
GC	gas chromatography (analysis)
GC/MS	GC combined with mass spectrometry (analysis)
GDP	gross domestic product
GLC	gas–liquid chromatography (analysis)
Gy	gray (100 rad)
HMIP	Her Majesty's Inspectorate of Pollution (UK)

HMSO	Her Majesty's Stationery Office
HPLC	high-performance (pressure) liquid chromatography (analysis)
HSC	Health and Safety Commission (UK)
HSE	Health and Safety Executive (UK) (particularly important is the EH series of Reports; EH40 gives the recommended Occupational Exposure Limits annually)
IAEA	International Atomic Energy Authority
IARC	International Agency for Research on Cancer
ICP-OES	inductively coupled plasma optical emission spectrometry (analysis)
ICPMS	inductively coupled plasma mass spectrometry (analysis)
IPC	Integrated Pollution Control (Europe)
IPCC	International Panel for Climate Control
IPCS	International Program for the Control of Substances
LAAPC	Local Authority Air Pollution Control
LC_{50}	lethal concentration for 50% survival
LD_{50}	lethal dose for 50% survival
LC_{Lo}	lowest published lethal concentration, in a known time
LFG	landfill gas
LOAEL	lowest observable adverse effect level (toxicology)
LOEL	lowest observable effect level (toxicology)
LPG	liquified petroleum gas
MAC	maximum allowable concentration (USA)
MAFF	Ministry of Agriculture, Fisheries and Food
MAK	Maximale Arbeitsplatz-Konzentration (maximum concentration at the workplace in the Federal Republic of Germany, set by the German Commission on Health Hazards)
μCi	microcurie $= 3.7 \times 10^4$ disintegrations s^{-1}
MEL	maximum exposure limit (toxicology)
NAAQS	National Air Quality Standards (USA)
nCi	nanocurie (nCi) $= 37$ disintegrations s^{-1} [1 curie (Ci) $= 3.7 \times 10^{10}$ disintegrations s^{-1}]
NIOSH	National Institute of Occupational Safety and Health
NOAEL	no observable adverse effects level (toxicology)
NOEL	no observable effects level (toxicology)
NRA	National Rivers Authority (UK, incorporated into the Environmental Agency, 1996)
NRPB	National Radiological Protection Board (UK)
NTP	normal temperature (273.15 K) and pressure
ODP	ozone depletion potential
OECD	Organization for Economic Cooperation and Development
OEL	occupational exposure limit (toxicology)
OSHA	Occupational Safety and Health Administration
P_{ow}	partition coefficient (see text)
PAH	polycyclic aromatic hydrocarbons (chemical)
PAN	peroxyacetyl nitrate (chemical)
PCB	polychlorobiphenyl (chemical)
PCDD	polychlorodibenzo-p-dioxin (chemical)
PCDF	polychlorodibenzofuran (chemical)

PCP	pentachlorophenol (chemical)
PEL	permitted exposure limit (toxicology)
PM_{10}	particles of diameter <10 μm
ppb	parts per billion (10^9)
ppm	parts per million
PVC	poly(vinyl chloride) (chemical)
PWR	pressurized water reactor
Rad	absorbed radiation of 0.01 Jkg^{-1}
Rem	Röntgen equivalent humans
RNA	ribonucleic acid (genetic code material)
RTECS	Registry of Toxic Effects of Chemical Substances, edited by NIOSH
SI	Statutory Instrument
STP	standard temperature and pressure
STEL	short-term exposure limit (toxicology)
$t_{1/2}$	half-life, the time for half of the original amount of a radioactive isotope to decay
TCDD	tetrachlorodibenzo-p-dioxin (chemical)
TD_{Lo}	lowest published toxic dose
TDI	tolerable daily intake (toxicology)
TEF	toxic equivalent factor (toxicology)
TEQ	toxic equivalent quantity (TEF X concentration)
TLV	threshold limit value (see text)
TWA	time weighted average (see text; TLV)
VCM	vinyl chloride monomer
Waldsterben	forest death
WDA	Waste Disposal Authority (UK)
VOC	volatile organic compound
WHO	World Health Organization

Introduction

In recent years there has been considerable media comment on the problems arising from the misuse of chemicals. This concern has been fuelled by a number of incidents which attracted international attention, examples include the release of dioxin at Sevéso, the Flixborough explosion and the deaths at Bhopal consequent on the release of methyl isocyanate.

Those with a professional interest are often concerned at media comments about public exposure to cancer-inducing chemicals, when no distinction is drawn between those which are active in experimental animals and the very few proven human carcinogens. To put the problem in perspective, one notes that there are some 70 000 synthetic chemicals now in use; the United Nations list (*Consolidated List of Products Barred, Withdrawn, Severley Restricted or not Approved by Governments*, United Nations, New York, 1991) includes 676 toxic compounds at all levels of risk, which is 1% of the whole. It needs also to be borne in mind that this list includes those of historical importance whose use is now banned or restricted, and also that many of these are pesticides which are inevitably toxic to some life forms. If one includes arsenic compounds with the element itself, then only 18 of the United Nations list are proven human carcinogens.

Entries in this book are given for some chemicals whose toxicity and/or persistence has led to their sale being discontinued. These include compounds such as DDT, aldrin and dieldrin – they show how the subject has developed and illustrate through case studies how risks come to be evaluated and controlled. The cost of substitutes often means that toxic chemicals remain in use in the poorer developing countries.

Faced with so many potential entries, the authors faced a problem of choice. It was decided to include substances listed in three of the following regulatory agencies:

American Conference of Governmental Industrial Hygienists
European Community Directives of Dangerous Substances
German Commission for Investigation of Health Hazards of Chemicals in the Work
  Area
International Agency for Research on Cancer
United States Environmental Protection Agency list of priority pollutants (1995)

The lists of substances restricted by the above agencies are collated in the technical reports of the European Chemical Industry Ecology and Toxicology Centre (Avenue Louise 250, B.63, B-1050 Brussels, Belgium).

Within this constraint, we have also covered chemicals which pose some risk to members of the general public, in the hope that the book will be of use to those writing for the press and media on environmental topics. Within reasonable limitations of space and time we have not been able to include a systematic review of major topics such as food additives, medicines and cosmetics.

The sectionalized list of reference sources will enable the reader to extend the search. Among the principal sources, we drew most extensively on *Kirk–Othmer – Encyclopedia of Chemical Technology*, the *Dictionary of Substances and Their Effects (DOSE)*, the series of reviews published by the *World Health Organization/ International Labour Office* and those from the *International Agency for Research on Cancer*

Analytical procedures fall outside our remit, but a relevant method for any of the entries can be found from this section of the reference list. Standard procedures are given in the manuals published by NIOSH, OSHA and the USEPA, while explanations of techniques are given by Fifield, F. W., and Haines, P. J., *Environmental Analytical Chemistry*, Thomson Science, London, and by Alloway, B. J., and Ayres, D. C., *Chemical Principles of Environmental Pollution*, 2nd edn, Thomson Science, London.

An important feature of risk exposure from chemicals is the volume of production and the geographical distribution of manufacturing sites. For major chemicals the annual list of the top 50 produced in the USA is published annually in June or July in *Chemical Engineering News* (American Chemical Society, Washington, DC); another valuable source is the *United Nations Statistical Yearbook*, published annually by the OECD. Production levels for the less common chemicals are not systematically reviewed but figures appear from time to time in *Chemistry and Industry* (Society of Chemical Industry, London) and *Chemical Week* (Chemical Week Associates, New York). In the UK, information about the location and nature of chemical production can be obtained by inspecting the registers held by the district offices of Her Majesty's Inspectorate of Pollution. These provide information supplied by industry under the terms of Integrated Pollution Control (IPC), which became effective in 1995 through the Environmental Protection Act of 1990; the data are available on a CD-ROM – the *Chemical Release Inventory*. Information about the location and activities at chemical plants in Europe is given in *The Directory of Chemical Producers, Vols 1 and 2, Western Europe*, SRI International, Menlo Park, CA.

In the broader sphere, information about chemical products can be obtained from the European Chemical Industry Council (CEFIC); the EC publishes data relating to the control of risks on a CD-ROM named IUCLID, obtainable from the European Chemicals Bureau, Existing Chemicals, T.P. 280, I-21020 Ispra (VA), Italy. The Chemical Abstracts Service of the American Chemical Society (Washington, DC) covers regulatory information on the CHEMLIST file, which is searchable for a fee (CAS Client Services, fax 614 447 3747; telephone 614 447 3870).

Owing to the variety of compounds listed, it was not possible to restrict each entry to a standard form. However a general order of the material was followed, as shown in the following extract:

1,3-BUTADIENE CAS No.[1] 106-99-0

LD_{50} rat 3.2 $g\,kg^{-1}$ [2] LC_{50} (2 h) inhalation[3] mouse 270 $g\,m^{-3}$ [4]

$CH_2=CH-CH=CH_2$ C_4H_6 $M = 54$.

Biethylene, divinyl[5];   b.p. $-4.5°$; v.p. 760 Torr at b.p., 1840 Torr[6] at $21°$. Highly flammable, flash point $-70°$; bioconcentration factor 1.9, accumulation unlikely.

(1) The Chemical Abstracts Service system number to enable a more extensive search.

(2) The oral LD_{50} for the rat is given as a general basis for comparing the acute toxicities of different substances.

(3) LC_{50} values relating to the toxicity to mammals on inhalation or to fish in freshwater are given where relevant.

(4) Where the equivalent unit of concentration is not given it may be obtained from the relation milligrams/cubic metre $(mg\,m^{-3}) = \dfrac{\text{molecular weight}}{25.45} \times$ parts per million (ppm)

(5) Some alternative names are given but where the substance is sold for commercial use (e.g. pesticides) these may be many and the reader should refer to a dictionary of trade names and pseudonyms, e.g. Ash, M. and Ash, I. (1994) *Gardner's Chemical Pseudonyms and Trade Names*, 10th edn, Gower, Aldershot; also *The Pesticide Index*, 3rd edn British Crop Protection Council/Royal Society of Chemistry, Cambridge, 1995.

(6) Pressure is usually expressed in Torr; 1 Torr = 1/760th of normal atmospheric pressure (760 mm of mercury) = 133.3 Pa or 0.133 kPa.

All temperatures are in degrees Celsius, vapour densities are relative to that of air and densities are the mean over the temperature range 4–20°.

These data will typically be followed by some or all of the following:

Historical and synthetic details	Potential risks on reaction or storage
Production and use	Sources and human exposure
Examples of pollution incidents	Metabolism
Further examples of acute or chronic toxicity	Evidence of mutagenic or carcinogenic activity
Occupational limits (ACGIH, OSHA)	References to sources, or for further reading

Other terms used in the initial summary include $\log P_{ow}$; (see Partition Coefficient); others explained in the main text are Biological Concentration Factor, Biochemical Oxygen Demand, Chemical Oxygen Demand, Henry's Law and Threshold Limit Value.

SI prefixes used in the text are:

10	deca (da)	10^{-1}	deci (d)
10^2	hecto (h)	10^{-2}	centi (c)
10^3	kilo (k)	10^{-3}	milli (m)
10^6	mega (M)	10^{-6}	micro (μ)
10^9	giga (G)	10^{-9}	nano (n)
10^{12}	tera (T)	10^{-12}	pico (p)

Following the main entries are a list of further reading and an Appendix giving an explanation of medical terms used in the text.

ACETALDEHYDE CAS No 75-07-0
LD_{50} oral rat 1930 mg kg^{-1}; carcinogenic in animals LC_{50} fish ca 35 µg l^{-1}.

$$CH_3—C=O \qquad\qquad\qquad C_2H_4O \quad M = 44.0$$
$$\mid$$
$$H$$

Ethanal: m.p. − 123°; b.p. 20°; sp.gr. (20°) 0.78. Soluble in water, ethanol; very flammable, irritates the eyes and respiratory system.

Occurs naturally as a volatile component of cotton leaves and blossoms, also in fruits, spices, oak and tobacco leaves. Produced by the liquid-phase oxidation of ethylene, catalysed by palladium and copper chlorides. Production in the USA was 281×10^3 t and in Western Europe 706×10^3 t (1982). Annual emissions in the USA were then 5056 t and the major emitters (tonnes) were coffee roasting 4411, acetic acid manufacture 1460, production of vinyl acetate 1095, ethanol manufacture 58, acrylonitrile manufacture 52.

Uses. As an intermediate for acetic acid (61%), for ethyl acetate (19%), for peracetic acid (8%) and in the manufacture of paraldehyde, perfumes and flavourings, plastics, aniline dyes, synthetic rubber.

Acetaldehyde is released on the combustion of plastics, polycarbonate and polyurethane foam. Emissions from woodburning stoves fall in the range of 9–710 mg kg^{-1}. Smoking tobacco produces 980 µg/cigarette, marijuana 1200 µg/cigarette. Based upon adult inhalation of 22 m^3 day^{-1} intake is estimated to be 1.7 µg kg^{-1} body weight day^{-1}. Assuming a daily liquid intake of 1.4 l for a body weight of 64 kg then at a level in drinking water of 0.1 µg l^{-1} the further intake of acetaldehyde is 0.002 µg l^{-1}.

Undergoes aerial oxidation in the presence of iron or cobalt salts to give acetic acid. Reacts with ammonia to give crystalline 1-aminoethanol, m.p. 97°.

Acetaldehyde is present in small amounts in alcoholic beverages.

A tolerable concentration towards humans in air is 2 mg m^{-3}(2000 µg m^{-3}). The ACGIH have set the STEL at 25 ppm (45 mg m^{-3}) and classed acetaldehyde as a group A3 carcinogen.

See also under **Aldehydes**.

Further reading
Environ. Health Criteria, World Health Organization, Geneva, 1995, Vol. 167.

ACETIC ACID CAS No. 64–19–7
LD_{50} oral rat 3310 mg kg^{-1}; LC_{50} fish 75–88 mg l^{-1}

$$CH_3—C=O \qquad\qquad\qquad C_2H_4O_2 \quad M = 60.0$$
$$\mid$$
$$OH$$

Ethanoic acid; m.p. 17°; b.p. 118°; sp.gr. 1.05; v.p. 100 Torr 27°. Miscible with water, ethanol and benzene. Inflammable and burns the skin.

A normal metabolite of animals and plants, it occurs naturally in vegetation, e.g. juniper oil. Produced industrially by the Co- or Mn-catalysed oxidation of acetaldehyde and by the liquid-phase oxidation of low-b.p. petroleum fractions.

In 1994 production in the USA was 3.82×10^9 lb, placing it 34th in order of productivity.

The principal industrial uses are the production of acetic anhydride and vinyl acetate for the manufacture of paints and adhesives. The esters are important solvents. The thallium salt is used in ore flotation and was formerly applied to treat ringworm. Its use is now prohibited in cosmetic products.

The sources to air include fish processing, emissions from chemical plants, lacquers, wines, wood distillation and landfill leachate. Acetic acid was detected at $45\,\mathrm{mg\,l^{-1}}$, with propionic acid at $24\,\mathrm{mg\,l^{-1}}$, in groundwater near a wood preserving factory (Goerlitz et al., 1985). It is the major volatile from brown sugar and is used as a food additive in vinegar.

The concentration of acetic acid in the river Lee in the UK was generally $< 0.1\,\mathrm{ppb}$, but in 1980 it was found in seawater near Belgium at 12–240 ppb. A survey in 1984 in the USA of landfill wells revealed levels in the range 0.66–4.60 ppb. It also occurs in rainout.

All sugar-containing saps or juices are transformed biochemically to dilute acetic acid – vinegar. Acetic acid is exhaled in human breath at a level of about $70\,\mathrm{mg\,day^{-1}}$ and a concentration of $20\,\mathrm{mg\,m^{-3}}$ can be reached in confined spaces; the human intake from air is in the range 2–$32\,\mathrm{mg\,day^{-1}}$.

A vapour concentration of 800–1200 ppm is intolerable after 3 min, but long-term exposure at 60 ppm has no significant effect; a level of 10 ppm is regarded as non-irritant under industrial conditions. This level has been set as the TLV (TWA) by the ACGIH.

Reference
Goelitz, D.F., Troutman, D.E., Godsy, E.M. and Franks, B.J.(1985) Migration of wood-preserving chemicals in contaminated groundwater in a sand aquifer at Pensacola, Fl. *Environ. Sci. Technol.*, **19**, 955–961.

Further reading
Howard, P.H. (1991) *Handbook of Environmental Fate and Exposure Data for Organic Chemicals*, Vol. **2**, *Solvents*, Lewis, Chelsea, MI, pp. 1–8.

ACETIC ANHYDRIDE CAS No. 108-24-7
$\mathrm{LD_{50}}$ oral rat $1780\,\mathrm{mg\,kg^{-1}}$

$C_4H_6O_3 \quad M = 102.1$

M.p. $-73°$; b.p. $139°$; dens. 1.082; v.p. 3 Torr (20°); odour threshold 0.13 ppm. Inflammable, lachrymatory and burns the skin, reacts slowly with water to give acetic acid. Miscible with ethanol, acetone, chloroform and benzene.

Produced by the oxidation of acetaldehyde or from ketene and acetic acid:

$$CH_2\!\!=\!\!C\!\!=\!\!O \overset{H^+}{} \quad \longrightarrow \quad CH_3\!\!-\!\!C\!\!-\!\!O\!\!-\!\!C\!\!-\!\!CH_3$$

$$\underset{O}{O\!\!-\!\!O\!\!-\!\!C\!\!-\!\!CH_3}$$

It may also be obtained from coal by the Eastman process via synthesis gas along with methanol, acetic acid and methyl acetate.

Acetic anhydride is widely used as an acetylating agent, e.g. for cellulose acetate, for flavouring esters and to solubilize animal fats. It is also used in the preparation of the herbicides metolachlor and alachlor and also for nitration with nitric acid.

In December 1992 there was a substantial release of acetic anhydride together with sulphuric acid, sodium hydroxide, methanol and vinyl acetate when a freight train was derailed near Winnipeg. The spread of the components was limited by the wind chill temperature of $-75°$.

The USA TLV and the UK short term limit for exposure are both 5 ppm ($21\,mg\,m^{-3}$).

Further reading
Prager, J.C.(1995) *Environmental Contaminant Reference Databook*, Van Nostrand, New York, Vol. 1, pp. 73–76.

ACETONE CAS No. 67-64-1
LD_{50} oral rat, dog 4–11 mg kg^{-1}; LC_{50} (24 h) goldfish 5 g l^{-1}

$$CH_3\!\!-\!\!C\!\!\underset{CH_3}{\overset{O}{\diagup}}$$

$C_3H_6O \quad M = 58$

2-Propanone, dimethyl ketone; m.p. $-94°$; b.p. $56°$; sp.gr. $(25°)$ 0.79; v.p. 23 Torr $(25°)$; $logP_{ow} - 0.24$.

Produced either by the dehydrogenation of 2-propanol or during the conversion of cumene into phenol:

$$Me\!\!-\!\!CH \xrightarrow[\text{or } O_2]{\text{air}} Me\!\!-\!\!C\!\!-\!\!O.OH \xrightarrow[\text{acid}]{60-100°} OH \; + \; \underset{Me}{\overset{Me}{\diagdown}}C\!\!=\!\!O$$

In 1990 the world production of acetone was 3×10^6 t year^{-1}. It is used in the manufacture of bisphenol A and higher ketones and as an industrial solvent for fats, oils, plastics and pharmaceuticals, also in paints and varnishes. Contact with the skin can cause dermatitis through extraction of protective fat. It is used as an extractant for animal residues and plants and as a carrier for cylinder-stored acetylene.

Acetone occurs in cigarette smoke at 1100 ppm and in car exhaust in the range 2–14 ppm. It is a microcompound in blood and urine and occurs as a product of abnormal metabolism in diabetics, when GC with a Porapak Q column will detect $1.1\,mg\,l^{-1}$ ($18.9\,nmol\,l^{-1}$) in unconcentrated breath (Phillips,1987).

The ACGIH and the UK exposure limit (8 h TWA) is 750 ppm ($1780\,mg\,m^{-3}$). The STEL is 1500 ppm ($3560\,mg\,m^{-3}$).

Reference
Phillips, M. and Greenberg, J. (1987) Detection of endogenous acetone in normal human breath. *J. Chromatogr.*, **422**, 235–238.

Further reading
Howard, P.H. (1991) *Handbook of Environmental Fate and Exposure Data for Organic Compounds*, Lewis, Chelsea, MI, Vol. 2, pp. 9–18.

ACETONITRILE CAS No 75-05-8

LD_{50} oral rat $3800\,mg\,kg^{-1}$; LC_{50} fish 1000–$1850\,mg\,l^{-1}$ No evidence of carcinogenicity in animals

$$CH_3\!-\!C \equiv N \qquad\qquad C_2H_3N \quad M = 41.1$$

Methyl cyanide; b.p. 81°; dens. 0.79; v.p. 74 Torr (20°); log P_{ow} −0.34. Miscible with water, ethanol, methanol, ethyl acetate, etc., insoluble in petrol fractions. Acetonitrile (84%) forms an azeotrope with water (16%) with b.p. 76°.

Acetonitrile is produced naturally by the burning of wood and other vegetation.

It has a high dipole moment of 3.84 D and will dissolve inorganic salts such as silver nitrate, lithium nitrate and magnesium bromide. It is important as a solvent for pharmaceutical synthesis and in HPLC analysis.

Acetonitrile is obtained as a byproduct from the synthesis of acrylonitrile from propylene in the Sohio process which produces $0.035\,kg\,kg^{-1}$ of acrylonitrile; small amounts occur in coal tar. It is principally used as an extractant for fatty acids and vegetable oils, also for the extraction of unsaturated petroleum hydrocarbons from the saturated components and as a solvent for spinning and moulding of plastics.

It is inflammable and toxic by ingestion and skin contact and is anaerobically metabolized in the intestine to cyanide and thiocyanate; humans exposed to a high concentration may suffer headache,nausea and vomiting from the slow production of cyanide. Cytochrome P-450 monooxygenase converts acetonitrile to cyanohydrin:

$$CH_3CN \longrightarrow HOCH_2CN \longrightarrow HCHO + CN^-$$

which dissociates to release cyanide ion, this in turn conjugates with thiosulphate and is then excreted in urine as thiocyanate (CNS^-).

Acetonitrile undergoes photodegradation. It is very slowly hydrolysed in water at pH 10.0. The wastewater from metal plating contains high concentrations of cyanides and nitriles which can be utilized by microorganisms such as *Pseudomonas putida* which convert cyanides and acetonitrile to carbon dioxide and ammonia. A major mechanism for removal of acetonitrile is the reaction with hydroxyl radicals,which abstract H atoms; the half- life is 535 days compared with the reaction with ozone with a half- life of 860 days.

It is emitted from the thermal degradation of sewage sludge together with benzene, toluene, ethylbenzene and acrylonitrile.

The ionized complex $H^+(CH_3CN)_m H_2O_n$ is detectable at an altitude of 35 km. Levels in air range between 2 and 7 ppbv with urban levels at the upper end of the range; in the vicinity of bush fires levels of 35 ppb of acetonitrile may be reached. A sample of polyurethane foam (10 mg) burnt at 800° emitted hydrogen cyanide (26 µg) and acetonitrile (21 µg).

A group of workers were heavily exposed to the vapour when painting the interior of a storage tank with paint containing 30% of acetonitrile and thinned with the same solvent. No immediate effects were seen but after vomiting and spitting blood in the night one man died the following day from respiratory failure attributed to the slow release of hydrogen cyanide. Other similar cases of human fatalities have been recorded. Chronic exposure recorded in workers in electroplating works at air levels in the range 7–11 ppm during 5–10 years resulted in headache, weakness, changes in taste and vomiting. The onset of symptoms is delayed until the release of cyanide becomes significant.

GLC with N–P detection can detect cyanides in blood at a concentration of $50\,ng\,ml^{-1}$.

Occupational exposure limits have been set:

	TWA	STEL
In the UK	40 ppm	60 ppm
In the USA	40 ppm	60 ppm

Further reading

Howard, P.H. (ed.) (1993) *Handbook of Environmental Fate and Exposure Data for Organic Compounds*, Lewis, Chelsea, MI, Vol. 4, pp. 1–8.

Nielsen, I.R. and Howe, P.D. (1995) *Environmental Hazard Assessment: Acetonitrile*, DOE, Watford.

ACROLEIN CAS No. 107-02-8

LD_{50} oral rat 46 mg m^{-3}; Teratogenic in rats and mice; Very toxic to fish LC_{50} 0.06–0.25 mg l^{-1}

$$CH_2=CH-\underset{\underset{\displaystyle H}{|}}{C}=O \qquad\qquad C_3H_4O \quad M = 56$$

Acrylaldehyde, 2-propenal; b.p. 53°, v.p. 29.3 kPa (220 Torr) (20°); sp.gr. (20°) 0.843. It has an absorption maximum at 315 nm, $\epsilon = 26\,l\,mol^{-1}\,cm^{-1}$. Flash point – 18°, highly reactive and very soluble in water in which the half-life is 3–7 h, also soluble in ethanol and diethyl ether.

It is formed by the photooxidation of VOCs, especially 1,3-butadiene, and reacts further with ozone and hydroxyl radicals to give CO, HCHO, ketene and glycolaldehyde. The detection limit by GC–MS is 0.1 µg m^{-3}; HPLC has been used to detect it in car exhaust at 1.4 µg m^{-3}.

World production is by the catalytic oxidation of propene 59 000 t (1975) but much more is produced for direct conversion into acrylic acid and its esters (242 000 t in the USA, 1983).

Commercial acrolein is 95% pure with 3% of water plus propanal and acetone. It is used for the production of acrylic acid and as a biocide, at $6-10\,mg\,l^{-1}$, to control algae, molluscs and plants in recirculating water systems, irrigation channels, etc.

It occurs naturally in oakwood, tomatoes, wines and the degradation of amino acids when food is heated – roasting coffee gives levels in air up to $0.59\,mg\,m^{-3}$, close to car exhausts the level may reach $6\,mg\,m^{-3}$ and in production plants the range is $0.1-8.2\,mg\,m^{-3}$. Smoke from an overheated frier contributed to the deaths of two boys from respiratory failure. Smoking tobacco produces levels indoors of 3-228 µg/cigarette, with acetaldehyde and other aldehydes in the range 82-1255 µg/cigarette.

Outdoor levels of 32 µg m^{-3} were recorded in Japan, those in rainout are generally low but $0.04\,\mu g\,l^{-1}$ has been reported.

Odour perception is at $0.07\,ng\,m^{-3}$, conjunctival irritation follows at $1\,mg\,m^{-3}$ and severe eye irritation at $3\,mg\,m^{-3}$. When ingested, but not when inhaled, acrolein reacts with sulphydryl groups in the body and with the amino groups of proteins:

$$GSH + CH_2=CHCHO \longrightarrow GSCH_2\,CH_2CHO$$

while it is converted into glyceraldehyde by reaction with cytochrome P-450:

The ACGIH has set the TLV (TWA) at 0.1 ppm $(0.23\,mg\,m^{-3})$

Further reading
Environ. Health Criteria, World Health Organization, Geneva, 1991, Vol. 127.

ACRYLAMIDE CAS No. 79-06-1
LD$_{50}$ oral rat $124\,mg\,kg^{-1}$; Guinea pig $170\,mg\,kg^{-1}$; LC$_{50}$ fish $124-160\,mg\,l^{-1}$ (96 h)

$$CH_2=CH-C=O \qquad\qquad C_3H_5NO \quad M = 71.1$$
$$\overset{|}{NH_2}$$

Acrylic acid amide; m.p. 84.5°; b.p. 125°/3.33 kPa; v.p. 0.009 kPa/25°; 0.09 kPa/50°; dens. $1.122\,g\,cm^{-3}$.

Very soluble in water, acetone, ethanol, soluble in ethyl acetate, chloroform; insoluble in alkanes.

Gas chromatography with flame ionization detection responds to 0.01 ppm and HPLC to 0.1 ppm.

There is no natural source and commercial production began in 1954; production by the hydration of acrylonitrile in the USA in 1981 was 37 000 t; with sulphuric acid catalysis only 1 ppm remains unreacted, alternatively a Cu catalyst may be employed at 100°.

Polyacrylamides are used as dental fillers, grouting agents, ion exchangers and in hair sprays and textile treatment.

Acrylamide is a local skin irritant; over 60 cases of illness in workers have been reported, including contact dermatitis, disorientation, lack of concentration and hallucinations.

In the air near industrial plants levels up to $0.2\,\mu g\,m^{-3}$ have been recorded but the process water following polyacrylamide treatment of effluent was not so polluted.

In the USA the Food and Drugs Administration has set a limit of 0.2% of polyacrylamide in paper products in contact with foodstuffs. It is a suspected human carcinogen. The ACGIH classes it in group A2 with a TLV (TWA) of $0.03\,mg\,m^{-3}$.

Further reading
Environ. Health Criteria, World Health Organization, Geneva, 1985, Vol. 49.

ACRYLONITRILE CAS No. 107-13-1
LD_{50} rat $93\,mg\,kg^{-1}$; LC_{50} fish 10–$33\,mg\,kg^{-1}$; Causes cancer in animals

$CH_2 = CH - C \equiv N$ $\qquad\qquad$ C_3H_3N $\quad$ $M = 53.1$

Vinyl cyanide; m.p. $- 83°$; b.p. 79°; sp. gr. (20°) 0.81. Solubility in water $73\,g\,l^{-1}$ (20°) also soluble in acetone and benzene. Flammable and can polymerize violently if heated or exposed to pressure, peroxides or strong acids or bases. It forms explosive mixtures with air and on combustion it produces HCN, NO_x and CO.

Acrylonitrile occurs in car exhaust and in cigarette smoke, it is toxic by inhalation and by absorption through the skin and is suspected to have been the cause of prostate cancer in exposed workers. It was also linked to the higher incidence of cancers of the liver and large intestine amongst workers at a Du Pont textile fibre plant.

Acrylonitrile is in the top 50 of chemicals produced in the USA, where it is obtained by the Sohio process in which propylene, ammonia and oxygen undergo a catalysed reaction at 500° at a pressure of 50–200 kPa. This reaction gives a 98% yield of product in a single pass. US production in 1994 was 3.08×10^9 lb.

Uses include fumigation and the production of acrylic fibres by the free radical-induced polymerization, commonly by mixture with co-monomers such as methyl acrylate and vinyl acetate. Acrylic fibres are heat stable and resist chemical and biological agents; principal demand is for clothing and home furnishing.

The USA has set an exposure limit of 2 ppm ($4.3\,mg\,m^{-3}$, 8 h TWA). Acrylonitrile is a group A2 carcinogen.

Further reading
Chemical Safety Data Sheets Vol. **3**, *Corrosives and Irritants*, Royal Society of Chemistry, Cambridge, 1990, pp. 25–29.
Environ. Health Criteria, World Health Organization, Geneva, 1983, Vol. 28.

ACTINIDES
The actinides are the 14 elements from thorium (atomic number 90) to lawrencium (atomic number 103) inclusive, which follow actinium in the Periodic Table. Every known isotope of the actinide series of elements is radioactive. Almost all actinides are very toxic owing to their radioactivity and ability to accumulate in bones, liver, etc. They are described below in order of atomic number.

For **Protactinium**, **Thorium** and **Uranium**, see separate entries.

Neptunium, CAS No. 7439-99-8
Np

Neptunium is the first transuranic which does not occur naturally. It has an atomic number of 93, an atomic weight of 237.048, a specific gravity of 20.25 at 20°, a melting point of 640° and a boiling point of 3902°. Neptunium metal has a silvery appearance, and the element forms five ionic oxidation states, +2, +3, +4, +5 and +6.

At least 13 isotopes of neptunium are now recognized. Neptunium-237 and neptunium-239 are produced by the neutron bombardment of uranium in nuclear reactors. The latter isotope has a half-life of 2.35 days and decays to ^{239}Pu, the most important isotope of plutonium. ^{237}Np has a half-life of 2.14×10^6 years and is available in kilogram quantities from uranium fuel elements. Although neptunium is present in nuclear fallouts, the amounts present make it an unlikely danger to the environment.

Plutonium CAS No. 7440-07-5

Pu

Plutonium has an atomic number of 94, an atomic weight of 239, a specific gravity of 19.84 at 25°, a melting point of 641°, and a boiling point of 3232°. It exhibits oxidation numbers of +3, +4, +5, and +6. The pure metal can be obtained by reduction of the trifluoride with calcium or barium.

Plutonium exists in trace quantities in naturally occurring uranium ores. It is produced in large quantities in nuclear reactors from uranium-238:

$$^{238}U \, (n, \gamma)^{239}U \xrightarrow{\beta} {}^{239}Np \xrightarrow{\beta} {}^{239}Pu$$

Fifteen isotopes of plutonium are known. Plutonium-239 with a half-life of 24 360 years is vitally important in nuclear weapons and the industrial use of nuclear power. One pound of plutonium is equivalent to about 10^7 kW h of heat energy. Plutonium-238 with a half-life of 87.7 years is a minor constituent of nuclear weapons fallout.

Following inhalation or ingestion of plutonium it is readily absorbed by bone marrow, the half-life in bone is about 100 years. Since plutonium emits high-energy α-particles this makes it a radiological hazard to humans.

The five isotopes plutonium-238/239/240/241/242 were all released into the atmosphere from the Chernobyl accident. Plutonium and its compounds must be handled with special equipment and precautions. The maximum human body level without harm for ^{239}Pu is presently set at 0.04 μCi (0.6 μg). The acute toxicity of plutonium is the same as that of strychnine and it is, therefore, one of the most deadly poisons known

Americium CAS No. 7440-35-9

Am

Americium, a transuranic element, has an atomic number of 95, an atomic weight of 243, a specific gravity of 13.67 at 20°, a melting point of 1176° and a boiling point of 2607°. It is capable of forming compounds in the +2 to +6 oxidation states. It is formed as a result of successive neutron bombardment of plutonium isotopes:

$$^{239}Pu(n, \gamma)^{240}Pu(n, \gamma)^{241}Pu \xrightarrow{\beta} {}^{241}Am$$

Americium-241 ($t_{\frac{1}{2}} = 432$ years) is a decay product of plutonium-241. A mixture of isotopes ^{241}Am, ^{242}Am, and ^{243}Am ($t_{\frac{1}{2}} = 7.37 \times 10^3$ years) can be prepared by intense irradiation of ^{241}Am. In any nuclear disaster americium can be dispersed over a wide area. Some isotopes are used as tracers in diagnostic medicine.

Americium metal is obtained by reducing the trifluoride with barium metal at

1000°. Enters the human body by inhalation or ingestion. Deposits on the surface of the bone ($t_{\frac{1}{2}} = 100$ years) and in the liver ($t_{\frac{1}{2}} = 40$ years).

Curium CAS Registry No. 7440-51-9

Cm

Curium is a man-made element, atomic number 96. Obtained by neutron bombardment of ^{239}Pu in nuclear reactors and explosions. Curium-247 has a half-life $> 10^7$ years, ^{242}Cm and ^{244}Cm, with half-lives of 163 days and 18 years, respectively, are also detected in nuclear wastes and fallout. Maximum allowed total body burden of ^{244}Cm is 0.3 µCi because it accumulates in the bone and interferes with red blood cell synthesis. Because of the very small quantities produced, curium is not usually a radiological hazard in the environment.

Berkelium, CAS No. 7440-40-6

Bk

Berkelium, a transuranic element, atomic number 97, is produced in very small amounts in nuclear reactors and explosions by neutron bombardment of ^{239}Pu. The most stable isotope is ^{247}Bk, with a half-life of 1400 years, but ^{249}Bk with a half-life of 314 days is the usual experimental isotope. Because of the small quantities involved, this element is considered not to be a radiological hazard.

Californium CAS No. 7440-71-3

Cf

Californium, a transuranic element, atomic number 98, is made in minute amounts in nuclear reactors and explosions, by neutron bombardment of ^{239}Pu. The longest lived isotope is ^{251}Cf, with a half-life of 890 years. ^{252}Cf, which has a half-life of 2.6 years, is used in minute quantities in the treatment of cancer by radiation therapy. 1 µg of ^{252}Cf emits 170×10^6 neutrons min^{-1}. Because of the small quantities involved, this element is considered not to be a radiological hazard.

Einsteinium, CAS No. 7429-92-7

Es

Einsteinium, atomic number 99, is produced in minute amounts by neutron bombardment of plutonium in nuclear reactors and explosions. Detected in fallout following the testing of thermonuclear weapons. Eleven isotopes are recognized, varying from mass number 246 to 256. ^{252}Es has a half-life of 140 days and ^{254}Es a half-life of 276 days. ^{253}Es with a half-life of 20.5 days is the most common isotope. Because of the small quantities involved, this element is considered not to be a radiological hazard.

Fermium CAS No. 7440-72-4

Fm

Fermium, atomic number 100, is produced in very small amounts by neutron bombardment of plutonium in nuclear reactors and explosions. Detected in the fallout from the testing of thermonuclear weapons. Ten isotopes are known, varying from mass number 248 to 257. Fermium-257 is the longest lived isotope with a half-life of 80 days. Because of the small quantities involved, this element is considered not to be a radiological hazard.

Transuranic Elements 101–103 (Mendelevium, Md [CAS No. 7440-11-1], Nobelium, No [CAS No. 10028-14-5] and Lawrencium, Lr [CAS No. 22537-19-5], respectively)
These are obtained by bombardment techniques which give principally the radio-isotopes ^{258}Md, ^{259}No, and ^{260}Lr. Since they are produced experimentally in small amounts, they usually play no part in environmental considerations.

Further reading
Cotton, S. (1991) *Lanthanides and Actinides*, Macmillan, London.
Katz, J.J., Seaborg, G.T. and Morss, L.R. (eds) (1986) *The Chemistry of the Actinide Elements*, 2nd edn, Chapman & Hall, London
King, R.B. (ed.) (1994) *Encyclopedia of Inorganic Chemistry*, Wiley, Chichester, Vol. 1, pp. 3–34.
Kirk–Othmer ECT, 4th edn, Wiley, New York, 1991, Vol. 1, pp. 412–445.

ACTINIUM CAS Registry No. 7440-34-8

Ac

Actinium, a member of Group 3 of the Periodic Table, has an atomic number of 89, an atomic weight of 227, a specific gravity of 10.07 (calc.), a melting point of 1050° and a boiling point of 3200 ± 300°. Actinium has several isotopes, all of which are radioactive.

Actinium-227, a decay product of uranium-235, is a β-emitter with a half-life of 21.6 years. It is about 150 times more active than radium. Actinium-228, formed in the decay series of thorium-232, has a half-life of 6 h. Both isotopes under normal circumstances do not present any radiological hazard and are part of the natural radioactivity of thorium and uranium ores.

Further reading
Cotton, S. (1991) *Lanthanides and Actinides*, Macmillan, London.
Eisenbud, M. (1987) *Environmental Radioactivity. From Natural, Military, and Military Sources*, 3rd edn, Academic Press, New York.
Katz, J.J., Seaborg, G.T. and Morss, L.R. (eds) (1986) *The Chemistry of the Actinide Elements*, 2nd edn, Chapman & Hall, London.
King, R.B. (ed.) (1994) *Encyclopedia of Inorganic Chemistry*, Wiley, Chichester, Vol. 1, pp. 3–34.
Kirk–Othmer ECT, 4th edn, Wiley, New York, 1991, Vol. 1, pp. 412–445.
Lee, J.D. (1991) *Concise Inorganic Chemistry*, 4th edn, Chapman & Hall, London, Ch. 30, 'The Actinides.'
Merian, E. (ed.) (1991) *Metals and Their Compounds in the Environment*, VCH, Weinheim.

AFLATOXINS
These are a group of highly carcinogenic natural products produced by the moulds *Aspergillus flavus* and *A. parasiticus* on groundnuts (peanuts), chilli peppers, corn, cottonseed, etc. Aflatoxin B_1 is found most frequently and is the most toxic:

Aflatoxin B_1 CAS No. 1162-65-8

$C_{17}H_{12}O_6$ $M = 312.3(R = H)$

M.p. $269°$; λ_{max} 223, 265, 363 nm. Water solubility $10\,g\,l^{-1}$, soluble in ethanol and chloroform. LD_{50} hamster, rat, monkey $2–10\,mg\,kg^{-1}$

Also found with aflatoxin B_1, but rarely alone, are the variants aflatoxin B_2 ($C_{17}H_{14}O_6$) with ring A reduced and aflatoxin G_1($C_{17}H_{12}O_7$), in which ring D is a six-ring lactone. Other minor members of the group have variations of these structures in which $R = OH$ and/or a free OH group occurs on ring C. The reference letters derive from the fluoresence colours, $B =$ blue and $G =$ green.

In warmer climates the fungi producing these toxins are ubiquitous and may contaminate a wide range of foods, but they are not found in the colder climates of Northern Europe. In the USA in 1989 it was estimated that $20 million were lost on the peanut crop alone as a result of mould growth. Effective control requires rapid post-harvest drying of crops and storage with a moisture content below 10%. Poorer communities forced to consume mouldy foods are most at risk; the following table gives a comparison:

Occurrence of aflatoxins in food (Jones, 1992)

Food	Incidence/samples	Average level ($\mu g\,kg^{-1}$)
Peanut butter:		
USA	17/104	14
Phillipines	145/149	213
Corn:		
USA	49/105	30
Philippines	95/98	110
Wheat flour	20/100	up to 150

The toxicity of aflatoxin B_1 was first manifested by the death of turkeys fed on Brazilian peanuts, leading initially to the name 'turkey X disease.' The effects on humans in India and Africa were manifested 3 weeks after the ingestion of mouldy rice or cassava; the symptoms include edema, abdominal pain, liver necrosis and a sensitive liver. In children the lethal dose can be as little as $200\,\mu g\,kg^{-1}$ taken for only 3 weeks. Human fatalities have also been reported in populations in Thailand and Kenya. In animals as little as 15 ppb in diets fed over a few weeks induces liver tumours. In addition to ingestion in food there is a risk of inhalation, as aflatoxin crystals can acquire electrical charge and become attached to dust particles.

In Europe the regulations concentrate on peanut butter, dried figs and animal foods. The legal limit in the UK is $4\,\mu g\,kg^{-1}$.

Aflatoxin B_1 is classed by the IARC as a group 1 carcinogen – a proven carcinogen in humans.

Reference
Jones, J.M. (1992) *Food Safety*, Eagan, St Paul, MN.

Further reading
Eaton, D.L. (ed.) (1994) *The Toxicology of Aflatoxins*, Academic Press, London.
IARC (1976) *Some Naturally Occurring Substances*, IARC, Lyon, Vol. 10, p. 51.

AGENT ORANGE CAS No. 39277-47-9

During the war in Vietnam, the USAF evaluated four pesticide mixtures – Agents Purple, Green, Pink and Orange – for use as defoliants. Agent Orange was selected and some 10^7 gallons were discharged on jungle supply trails, mostly by spraying from aircraft in operation Ranch Hand. Agent Orange was an equimixture of 2,4-D and 2,4,5-T.

$$O—CH_2\overset{\overset{\displaystyle O}{\|}}{C}\text{-OBu}$$

2,4 - D

$$O—CH_2\overset{\overset{\displaystyle O}{\|}}{C}\text{-OBu}$$

2,4,5 - T

These formulations included variable amounts of TCDDs (see **Polychlorodibenzodioxins and Polychlorodibenzofurans PCDD/PCDF),**
but 2,4-D presents no risk as the dioxin-generating side reaction can only produce a compound with a chlorine number less than 4 and these dioxins are not dangerous. On the other hand, overheating of the reaction mixture during the preparation of 2,4,5-trichlorophenol produces the highly toxic 2,3,7,8-TCDD in amounts which vary with the degree of control applied. The levels of this TCDD in Agent Orange varied between 2 and 50 ppm, although this was not appreciated when the operation began in January 1962.

Synthesis of TCDD was described in 1959 and 2,3,7,8-TCDD was identified in 2,4,5-T discharged by an explosion at Coalite and Chemical Products in Derbyshire in 1968. In the USA growing opposition to the massive spraying of this herbicide led to an investigation by the National Institutes of Health which reported in 1969; as a result, the last Ranch Hand mission was flown in May 1970.

Subsequently, the US Veterans Administration conducted an exhaustive survey, in the face of claims from those who had served in Vietnam, of damage to their health from exposure to Agent Orange. Complaints ranged from 'tingling sensations' to skin rashes, rare forms of cancer and birth defects in children. Epidemiological studies were conducted on some 84 000 veterans whose health records were compared with a reference group which included about 10% of the US population. It was concluded that there was no significant difference in 2,3,7,8-TCDD levels in the adipose and blood of the two groups of subjects (Young and Reggiani, 1988). For comment on cancer induction by TCDD, see the entry for **PCDD/PCDF**.

At the end of the war in Vietnam, the USAF was left with a large quantity of Agent Orange largely stored in 208 litre steel drums, which are not secure in the long term; much of this material was shipped to the remote site at Johnstone Island in the Pacific. In the Summer of 1977 8.4×10^6 l of the herbicide were disposed of by incineration at sea from M/T Vulcanus. It was estimated that the amount of TCDD in the whole stock was 23 kg.

Reference

Young, A.L. and Reggiani, G.M. (1988) *Agent Orange and Its Associated Dioxin*, Elsevier, Amsterdam.

Further reading

Tucker, R.T., Young, A.L. and Gray, A.P. (eds) (1983) *Human and Environmental Risks of Chlorinated Dioxins and Related Compounds*, Plenum, New York.

ALACHLOR CAS No. 15972-60-8

(Lasso)

Systemic pre-emergence herbicide. Toxicity class IV. Slightly toxic, $LD_{50} >$ $500 \, mg \, kg^{-1}$; LD_{50} oral rat $930 \, mg \, kg^{-1}$; LC_{50} fish $0.75–2.8 \, mg \, l^{-1}$

$$CH_3-O-CH_2-N-CO-CH_2-Cl$$

$$CH_3-CH_2 \quad CH_2-CH_3$$

$C_{14}H_{20}ClNO_2 \quad M = 269.5$

2-Chloro-2',6'-diethyl-N-(methoxymethyl)acetanilide; Water solubility $240 \, \mu g \, l^{-1}$; $t_{1/2}$ in soil 15 days.

Alachlor was introduced in 1969 by Monsanto and is of major use in the USA on corn, soybeans and sorghum; treatment of these crops requires some $19 \times 10^6 \, kg \, year^{-1}$. The CH_3OCH_2 – group is lost on degradation leading to two principal products of enhanced water solubility: 2-chloro- and 2-hydroxy-2',6'-diethylace-tanilide.

Up to 3% of the herbicide may be lost within 2–6 weeks through run-off and in the cornbelt of the USA levels rose to a maximum of $51 \, \mu g \, l^{-1}$. In early summer of 1991 in Iowa the river at Cedar Rapids held up to $7 \, \mu g \, l^{-1}$ of alachlor but these levels fell rapidly, leaving only traces by autumn. In 1989 the alachlor level in the Mississippi just above Baton Rouge was $170 \, ng \, l^{-1}$; the river carries about 100 tons year^{-1} to the Gulf of Mexico (cf. **Atrazine** and **Metolachlor**).

The WHO recommended level in drinking water is $0.3 \, \mu g \, l^{-1}$ and the EC directs $0.1 \, \mu g \, l^{-1}$; an EC Directive on alachlor is pending for the UK; it is not one of the 18 regulated by the EPA. There are no Codex or EC maximum limits for alachlor in foods.

ALDEHYDES

Contribute to indoor pollution due to combustion of woody fuels and the sandal-wood powder in Chinese incense emits formaldehyde together with acetaldehyde, propionaldehyde and acrolein. In 1992 average levels outdoors in Atlanta, GA, of acetaldehyde and formaldehyde were 2.7–3.0 and 2.6–3.2 ppb, respectively. At Essen in Germany acetaldehyde and formaldehyde occurred in outdoor air at 1.9 and

$2.0\,\mu g\,m^{-3}$ with propionaldehyde and benzaldehyde at 0.35 and $0.61\,\mu g\,m^{-3}$. Average levels of acetaldehyde in air over Long Island are $5.2\,\mu g\,m^{-3}$, with a summer maximum of $15.1\,\mu g\,m^{-3}$; this substance is detectable in Arctic air at levels up to $0.54\,\mu g\,m^{-3}$. The half-life over Los Angeles is about 10 h, owing to its relatively rapid removal by reaction with hydroxyl radicals. In areas where alcohols are used as motor fuels higher levels are found outdoors – in Brazil acetaldehyde and formaldehyde were found at 35 ppb and acetone at 20 ppb.

ALDRIN CAS No. 309-00-2

Non-systemic insecticide. Toxicity class II. Highly toxic LD_{50} 10–50 mg kg^{-1}; LD_{Lo} oral child 1250 $\mu g\,kg^{-1}$; LD_{50} oral rat 33–100 mg kg^{-1}; LC_{50} trout 36 $\mu g\,g^{-1}$; EC_{50} daphnia 28 mg l^{-1}

$C_{12}H_8Cl_6 \quad M = 364.9$

$(1R,4S,5S,8R)$-1,2,3,4,10,10-Hexachloro-1,4,4a,5,8,8a-hexahydro-1,4,5,8-dimethanonaphthalene; v.p. 6.4×10^{-5} mmHg/20°; water solubility 0.01 mg l^{-1} at 20°; bioconcentration factor in molluscs 4571.

This substance was linked to the decline in the peregrine falcon, which fed on dressed seed, in the UK in the years following 1955. In California it was detected in drinking water from wells and there were heavy losses of ducks and other birds in 1972 following treatment of rice seed.

Owing to its toxicity, aldrin was banned in the USA in 1983 and in the UK in 1986.Its use is also restricted throughout Asia. A fall in levels in cattle fat was seen in the USA after 1980; in UK sewage a median value of 0.02 ppm was found in 1981–82. In contrast to the DDT group, levels in sediments at Casco Bay, Maine, had fallen by 1994 to 0.28 ppb, close to the detection limit.

Aldrin is a prohibited pesticide within the EC and is subject to Prior Informed Consent for FAO/WHO countries. The EC maximum residue limits for food are mostly 0.05 mg kg^{-1}. The ADI is 0.0001 mg kg^{-1} body weight. Occupational exposure limits (OEL) require a TLV (TWA) of 0.25 mg m^{-3} in the UK, Germany and France. The same limit is given by NIOSH and the EPA.

ALDICARB CAS No. 116-06-3
(Temic)
Insecticide, acaridide, nematicide. Toxicity class I. Extremely toxic, $LD_{50} <$ 10 mg kg^{-1}; LD_{50} ip rat 280 $\mu g\,g^{-1}$

$C_7H_{13}N_2O_2S \quad M = 190$

2-Methyl-2 (methylthio) propionaldehyde-*O*-methylcarbamoyl oxime; v. p. $9.8\times$ 10^{-6} mmHg (20°); water solubility $6\,g\,l^{-1}$ (25°).

It is used for application to soils at long pre-harvest intervals. Aldicarb is rapidly oxidized to the sulphoxide and sulphone. The ultimate products of its reaction in chlorinated water are dichloromethylamine and chloroform. Its first detection in 1979 in groundwater in Long Island was unexpected and there were then no control guidelines but the level in half the samples exceeded $7\,\mu g\,l^{-1}$. Later sampling of potable wells in 34 US states showed that eight had aldicarb residues.

In 1985, uncontrolled spraying of water melons led to the largest recorded out-break of food-borne pesticide illness in North America. Levels of 0.2 ppm may cause illness and some melons contained as much as 2.7 ppm; 1170 cases were registered from Alaska to California.

In the UK, the drinking water level was restricted to $7\,\mu g\,l^{-1}$ and now meets that of the EC at $0.1\,\mu g\,l^{-1}$; the EPA has set a limit of $10\,\mu g\,l^{-1}$.

ALIPHATIC AMINES

Methylamine CAS No. 74-89-5
LD_{50} sub-cut. rat $200\,mg\,kg^{-1}$

CH_3NH_2 CH_5N $M = 31.1$

Aminomethane, MMA; b.p. $-7°$; v.dens. 1.1; v.p. (20°) 288 kPa. Has an ammoniacal odour, is highly flammable and very soluble in water (108 g/100 g at 25°). Soluble in ethanol and diethyl ether.

Methylamine occurs in fish oils and is widespread in fungi. It is used in tanning and dyeing, as a fuel additive, photographic developer, rocket propellant and in paint removers. Methylamine is an intermediate in the synthesis of caffeine, theophylline and pesticides, e.g. aldicarb and carbaryl.

The industrial production depends upon the vapour phase ammonolysis of metha-nol at 450° catalysed by alumina gel:

$$MeOH + NH_3 \longrightarrow MeNH_2 + Me_2NH + Me_3N$$
$$13.5\% \quad\quad 7.5\% \quad\quad 10.5\%$$

Trimethylamine can be separated from the mixture obtained because it forms an azeotrope with ammonia; the residual methyl and dimethylamines are separable with difficulty by fractional distillation. Depending on the demand of the market, the balance of production can be adjusted by the reforming reactions which, when catalysed, occur at 150–250°:

$$2RNH_2 \longrightarrow R_2NH + NH_3$$
$$RNH_2 + R_2NH \longrightarrow R_3N + NH_3$$

Methylamine is an irritant in air at a level of 25 ppm and can cause chemical bronchitis. It reacts with the OH radical and has a half-life in air of 3–22 h. It has been found in foods, e.g. spinach ($12\,mg\,kg^{-1}$), maize ($37\,mg\,kg^{-1}$) and coffee (27–$80\,mg\,kg^{-1}$). In river water detected levels lie between 1 and 20 ppb.

The ACGIH set the TLV (TWA) at 5 ppm ($6.4\,mg\,m^{-3}$), as there was no risk to the foetus below this level.

Dimethylamine CAS No. 124-40-3
LD_{50} rat 698 mg kg^{-1}

$(CH_3)_2NH$ C_2H_7N $M = 45.1$

DMA; b.p. 7°; v.p. (20°) 170 kPa. Flammable with an ammoniacal odour detectable at 0.5 ppm. Very soluble in water (163 g/100 g at 40°) and soluble in ethanol and diethyl ether.

Dimethylamine occurs in higher plants, fungi and bacteria. It is used as a tanning agent, as an accelerator in vulcanization of rubber and for the synthesis of fungicides and herbicides. Industrial demand is higher than that for the monomethyl and trimethylamines as large amounts are used in the synthesis of the solvents *N,N*-dimethylformamide and *N,N*-dimethylacetamide.

The vapour of dimethylamine is more strongly irritant than that of ammonia and it can cause dermatitis; a high concentration of the vapour can effect the CNS. There is a danger of permanent corneal damage from the liquid in the eye.

Dimethylamine occurs in foods at similar levels to methylamine. In common with all the aliphatic amines it presents a risk of cancer by virtue of the reaction with nitrite *in vivo* and *in vitro* to give the *N*-nitrosoamine. Liver damage and cancers were induced in animals when fed nitrite/dimethylamine.

The TLV has been set at 5 ppm (9.2 mg m^{-3}).

Trimethylamine CAS No. 75-50-3
LD_{50} rat (admin. unreported) 535 mg kg^{-1}

$(CH_3)_3N$ C_3H_9N $M = 59.1$

B.p. 3°; v.p. (20°) 191 kPa; highly flammable; very soluble in water (89 g/100 g at 30°); soluble in diethyl ether, benzene, chloroform.

Widespread in higher plants, fungi and bacteria. The oxide (m.p. 226°) is also found in the tissue of fish, it is unstable and an oxidizing agent.

Trimethylamine is used to produce the basic catalyst Triton B and also in the manufacture of Dowex ion-exchange resins. In common with the other methylamines its aqueous solutions are strongly basic and they all react rapidly and exothermally with acids and oxidizing agents. The production of all methylamines in 1988 world-wide was 395×10^3 t.

The TLV (TWA) of trimethylamine is 5 ppm (12 mg m^{-3}).

Ethylamine CAS No. 75-04-7
LD_{50} rat 400 mg kg^{-1}; LC_{50} (24 h) chub 40 mg l^{-1}

$CH_3CH_2NH_2$ C_2H_7N $M = 45.1$

Aminoethane; b.p. 16.5°; v.p. 400 Torr (116 kPa, 20°); v.dens. 1.56; log P_{ow} − 0.13. Flammable and miscible with water, ethanol and diethyl ether.

Manufactured like the methylamines by ammonolysis of ethanol. It is used as a stabilizer for rubber latex, as a dyestuff intermediate and in resin chemistry. The toxic effects are similar to those of the methylamines.

Ethylamine reacts with OH radicals with a half-life of 8 h. It also reacts in air with ozone, although more slowly. It was found at a level of 16 μg l^{-1} in the river Elbe but does not bioaccumulate significantly in water. It has been detected in spinach ($8\,mg\,kg^{-1}$), cheese ($4\,mg\,kg^{-1}$), hops ($5\,mg\,kg^{-1}$) and as a volatile from boiled beef. The TLV (TWA) is 5 ppm ($9.2\,mg\,m^{-3}$).

Diethylamine CAS No.109-89-7
LD_{50} rat $540\,mg\,kg^{-1}$

$(CH_3CH_2)_2NH$ $C_4H_{11}N$ $M = 73.1$

Ethanamine; b.p. 55°; v.p. 400 Torr (38°); dens. 0.71, log P_{ow} 0.58. Highly flammable. Miscible with water and soluble in most organic solvents.

Used in the rubber and petroleum industries, in flotation agents, resins, pharmaceuticals and in corrosion inhibitors.

The odour threshold is 0.13 ppm and diethylamine is more toxic than ammonia. It is a powerful irritant of the eyes and mucous membrane; repeated exposure may lead to bronchitis and pulmonary edema. There is also a risk of necrosis of the skin and of corneal damage.

The TLV (TWA) is 5 ppm ($15\,mg\,m^{-3}$); the short-term exposure limit is 15 ppm.

Cyclohexylamine CAS No. 108-91-8
LD_{50} rat $400\,mg\,kg^{-1}$

                    $C_6H_{13}N$ $M = 99.2$

B.p. 135°; v.p. 9 Torr (25°); dens. 0.86; log P_{ow} 1.49.

Cyclohexylamine does not occur naturally, it is a strong base, miscible with water and soluble in acetone, diethyl ether and ethanol.

It is used to inhibit corrosion in boilers by attachment via the amino group to form a surface layer; it is also used for the manufacture of insecticides, plasticizers, dry-cleaning soaps and in rubber production. It was found in the effluent of a tyre manufacturer at a level of $0.01\,mg\,l^{-1}$.

The demand in 1984 was 8.5×10^6 lb and NIOSH then estimated that 48 000 workers were exposed to cyclohexylamine.

It is not expected to bioaccumulate and the half-life of the reaction in air with the OH radical is 0.28 days; there is no photolytic degradation as cyclohexylamine does not absorb light of wavelength over 250 nm.

It is rated as a toxic substance; it irritates the skin and shows sympathomimetic activity, although rats withstood 6 h of exposure at 1000 ppm without any sign of toxicity. In mice 0.5% of cyclohexylamine in the diet led to growth retardation and embryotoxicity, but it is not regarded as carcinogenic and this has been extensively tested during screening of cyclamate. The TLV(TWA) is 10 ppm ($41\,mg\,m^{-3}$).

A group of fatty amines derived from fatty acids with 8–18 carbon atoms are also important as surface-active agents. They are made by hydrogenation of the corresponding nitriles, themselves obtained from the acid amides.

Further reading
Kirk–Othmer ECT, 4th edn, Wiley, New York, 1992, Vol. 2, pp. 369–405.

ALLETHRIN CAS No. 584-79-2
LD_{50} mammals 210–4300 mg kg^{-1}; LC_{50} fish 9-90 µg l^{-1}; LD_{50} bees 3-9 µg/bee

$C_{19}H_{26}O_3$ $M = 302.4$

A racemic mixture of equal amounts of eight stereoisomers of the esters of chrysanthemic acid with allethrolone, b.p. ca 140°/0.1 mmHg; insoluble in water, soluble in EtOH, acetone, xylene, CCl_4.

Several hundred tonnes are manufactured annually world wide for the control of household insects. It is used in sprays and aerosols – properly controlled use indoors restricts the level in air to 0.5 mg m^{-3}; it is rapidly destroyed by photolysis.

Allethrin irritates the skin and the eyes and is very toxic to fish and to bees.

Germany has set maximum residue limits in the range 0.5–3 mg kg^{-1} depending on the product.

Further reading
Environmental Health Criteria, World Health Organization, Geneva, 1989, Vol. 87.
The Pesticide Manual, 10th edn, Crop Protection Council, Royal Society of
    Chemistry, Cambridge, 1994, pp. 27–29.

ALLYL ALCOHOL CAS No. 107-18-6
LD_{50} rat, mouse 71, 105 mg kg^{-1}; LC_{50} goldfish (24 h) 1 mg l^{-1}

$$CH_2{=}CH{-}CH_2OH \qquad\qquad C_3H_6O \quad M = 58.1$$

2-propen-1-ol, vinyl carbinol; b.p. 97°; v.p. 10 Torr (10°); v.dens. 2.0; log P_{ow} − 0.25; miscible with water; soluble in EtOH, $CHCl_3$, diethyl ether, petroleum ether. Mustard-like odour.

Obtained by the acetoxylation of propylene using a palladium catalyst:

$$CH_2{=}CH{-}CH_3 + AcOH \xrightarrow[Pd]{[O]} CH_2{=}CH{-}CH_2{-}O{-}\overset{\displaystyle O}{\overset{\|}{C}}{-}CH_3$$

followed by acid-catalysed hydrolysis of this acetate. World production (1990) 70 000 t year^{-1}.

An intermediate for the production of glycerol, epichlorohydrin and acrolein. Used for the production of resins and plasticizers, and as a herbicide in plant nurseries, compost and horticultural crops.

Allyl alcohol is very toxic by inhalation and both the vapour and the liquid irritate the skin and mucous membrane; there is no evidence of chronic or cumulative toxicity. It is readily biodegradable, but is converted *in vivo* to acrolein, which may cause liver damage.

The ACGIH has set a TLV (TWA) of 2 ppm (4.8 mg m^{-3})

Further reading
Environmental Health Criteria, World Health Organization, Geneva, 1980, Vol. 37.

ALLYL CHLORIDE CAS No. 107-05-1
LD$_{50}$ rat 700 mg kg^{-1}; LC$_{50}$ fish 10–20 mg l^{-1}

$$CH_2 = CH-CH_2CCl \qquad\qquad C_3H_5Cl \quad M = 76.5$$

2-Chloropropene; b.p. 44.5°; dens. 0.938; log P_{ow} 1.450. Soluble in organic solvents, EtOH, CHCl$_3$, light petroleum, etc.

Obtained by reaction of the alcohol with hydrogen chloride. Since 1945 the industrial route has been the chlorination of propylene at 500° and 1.5 MPa (218 psi) to yield up to 80%, but with dichloropropane, chloropropenes and 1,5-hexadiene as by-products. World production (1990) 700 000 t year^{-1}.

It is used principally for the production of epichlorohydrin 150 × 10^6 kg in the USA in 1982 and also for epoxy resins, glycerol, allylamines and as the diuretic Ambuside. Diallyldimethylammonium chloride is used with acrylamide in cationic flocculating agents.

Allyl chloride is flammable, damages the lungs and is very toxic when inhaled. The minimum lethal concentration for mice is 72 parts per 1000, comparable to chloroform. Some 5000 workers in the USA are potentially exposed to allyl chloride. Levels at a US manufacturing plant ranged from a minmum of 0.7 mg m^{-3} for administrative staff to a maximum of 19 mg m^{-3} for production workers.

The ACGIH has set a TLV (TWA) of 1 ppm (3 mg m^{-3}). There is limited evidence of its causing cancer in animals and it is rated in group 3 by the IARC.

Further reading
ECETOX (1991) *Technical Report 30*, European Chemical Industry Ecology and Toxicology Centre, Brussels.
International Agency for Research on Cancer (1985) *Allyl Compounds, Aldehydes, Epoxides and Peroxides*, Vol. 36, IARC, Lyon.

ALLYL GLYCIDYL ETHER CAS No. 106-92-3
LD$_{50}$ rat, mouse 390, 920 mg kg^{-1}; LC$_{50}$ (96 h) goldfish 30 mg l^{-1}

$$CH_2{=}CH-CH_2-O-CH_2-CH-CH_2 \qquad C_6H_{10}O_2 \quad M = 114.1$$
$$\diagdown O \diagup$$

Allyl-2,3-epoxypropyl ether; b.p. 154°; dens. 0.97; v.p. 4.7 Torr (25°) v. dens. 3.94. Soluble in most organic solvents.

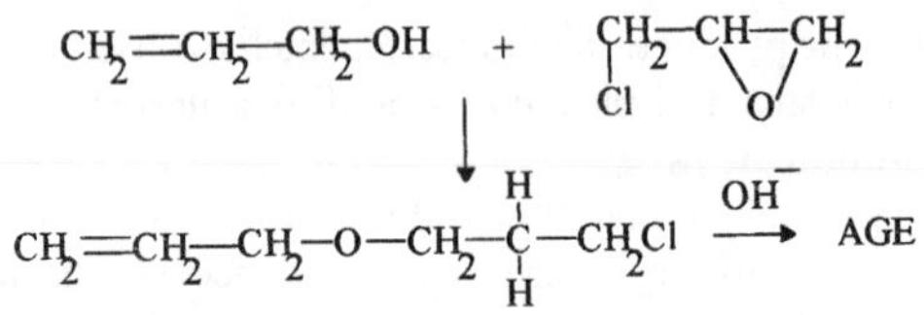

$$CH_2\!=\!CH_2\!-\!CH_2\!-\!OH \quad + \quad \underset{Cl}{CH_2}\!-\!CH\!-\!CH_2 \;\; (O)$$

$$\downarrow$$

$$CH_2\!=\!CH_2\!-\!CH_2\!-\!O\!-\!CH_2\!-\!\underset{H}{\overset{H}{C}}\!-\!CH_2Cl \;\;\xrightarrow{\;OH^-\;}\;\; AGE$$

It is prepared from allyl alcohol and epichlorohydrin (structure **Epichlorohydrin**).

Used as an intermediate for epichlorohydrin rubber when it is copolymerized with the chlorohydrin and ethylene oxide to produce a rubber with high resistance to cold (Saunders, 1988):

Allyl glycidyl ether is clearly carcinogenic in animals and is a potential human carcinogen. Inhalation by rats and mice induced squamous cell carcinomas and carcinomas in the nasal passage.

The ACGIH has assigned a TLV (TWA) of 5 ppm ($23\,mg\,m^{-3}$).

Reference
Saunders, K.J. (1988) *Organic Polymer Chemistry*, 2nd edn, Chapman & Hall, London.

Further reading
ECETOX (1991) *Technical Report 30*, European Chemical Industry Ecology and Toxicology Centre, Brussels.

ALLYL PROPYL DISULPHIDE CAS No. 2179-59-1

$CH_2\!=\!CH\!-\!CH_2\!-\!S\!-\!S\!-\!CH_2CH_2CH_3$ $C_6H_{12}S_2$ $M = 148.3$

Onion oil; b.p. $66°/16\,Torr$.

A constituent of *Allium* species and may be obtained from them by steam distillation. It occurs naturally with methyl propyl and dipropyl disulphides as the odour component of onion; the latter named is also the odour component of garlic.

In garlic-sensitive humans allyl propyl disulphide provokes allergic reactions, including dermatitis. It is used in food and flavourings. Administered after fasting to six normal volunteers it caused a significant change in the level of blood glucose and a rise in serum insulin.

The ACGIH has set the occupational exposure TLV (TWA) at 2 ppm ($12\,mg\,m^{-3}$).

Further reading
Bauer, K., Garbe, D. and Surburg, H. (1990) *Common Fragrance and Flavour Materials*, VCH, Weinheim, p. 133.

ALUMINIUM, CAS No. 7429-90-5

Al

Aluminium, a member of Group 13 of the Periodic Table, has an atomic number of 13, an atomic weight of 26.98, a specific gravity of $2.70\,g\,cm^{-3}$ at 20°, a melting point of 660°, a boiling point of 2519° and only one stable isotope, ^{27}Al. Aluminium is a hard, ductile, silver, white electropositive metal, forming salts in the +3 oxidation state.

Aluminium is the most abundant metal in the earth's crust (8.3% by weight) and is a major constituent of many common igneous rocks including feldspars and micas.

Commercially, the most important mineral is *bauxite*, a mixture of *diaspore*, $HAlO_2$, *bellite*, $AlOOH$, and *hydragyllite*, $Al(OH)_3$. The main anthropogenic sources of aluminium in the environment are mining, solid and liquid emissions from aluminium and some other industries and aluminium components, containers and utensils. Typical concentrations are air $600-1500\,ng\,m^{-3}$ (as aluminosilicates), salt and fresh water $2-300\,mg\,m^{-3}$ (concentrations are up to ten times higher in lakes and rivers than oceans) and soils $150-600\,mg\,kg^{-1}$. Because of the neurotoxicity of aluminium, the recommended level in drinking water is $0.05\,mg\,l^{-1}$, with a maximum admissible concentration of $0.2\,mg\,l^{-1}$. In large industrial centres, raw sludges in the primary settling tanks of biological waste water treatment tanks can contain aluminium concentrations $> 2g\,kg^{-1}$ wet weight. Such high concentrations have implications for disposal and spraying on agricultural land.

There are four categories of exposure to humans: food, air, water or contact with skin. Typically, ingestion and inhalation $\sim 5\,mg\,day^{-1}$; food additives $5-100\,mg\,day^{-1}$, with high amounts due to alum, baking powder and Al^{3+} containing emulsifiers added to processed cheese ; the greatest sources of intake are $Al^{3+}-$ containing antacids, buffered aspirins, antiperspirants and injected vaccines. Citrates facilitate incorporation of Al^{3+} into mammals.

The average daily intake for adult humans is $10-50\,mg$ (1 mg from water), although some aluminium-containing pharmaceuticals may result in daily intakes of 500 mg or higher. Elimination is mainly through the kidneys. The total body content of aluminium in an adult human varies from 50 to 140 mg; in blood, $0.024-0.07\%$ by wt, found in the lung $0.059\,mg\,g^{-1}$, in long bone $0.5\,mg\,g^{-1}$ and in the brain $0.04-0.250\,mg\,g^{-1}$. The two specific sources of aluminium and its compounds in the human body are food and drink. Edible plants containing the following aluminium concentrations are typical ($mg\,kg^{-1}$ dry weight): wheat 42, lettuce leaves 73, carrot roots 78, onion bulbs 6.3, potato 13–76 , apple fruit 7.2 and tea leaves 850–1400 (meat and meat products contain 1.6–20). The aluminium content of some foods was found to double if cooked in aluminium containers. Aluminium is an essential trace element. Humans require 30–50 mg of aluminium daily.

Uses. It is used as the pure metal or in alloys for aircraft and kitchen utensils and as an electrical substitute for copper and a lightweight replacement for steel. Finely ground aluminium is used in photography, explosives and fireworks.

Workers in the aluminium industry are at risk from several diseases. The mining of aluminium exposes miners to the toxic effects of clays. Inhalation of finely divided particles of the oxide can cause lung damage (Shaver's disease). Occupational exposure to bentonite dusts at high concentrations for several years led to chest infections and dysproteinaemia. In ceramics and abrasives manufacture, workers had kaolinosis chest X-rays which showed pathological changes. It is known that chronic exposure of aluminium causes non-nodular pulmonary fibrosis of the lung (aluminosis) which can lead to emphysema. Severe aluminosis has been observed in workers employed in spraying aluminium paint and in the production of pyrotechnic aluminium powders. A study of the exposure of welders to aluminium over a long period revealed a number of neuropsychiatric symptoms, including slower psychomotor reaction and memory loss, and mental and emotional balances were disturbed. Pot room workers in aluminium reduction plants are prone to lung and bladder cancer. This is probably due to fumes of polycyclic aromatic hydrocarbons and dioxins produced during the electrolysis of molten ores with carbon rods, rather than exposure to aluminium.

Most aluminium salts are slightly toxic at low concentrations. High concentrations tend to be poisonous and can cause mutation of cells in laboratory animals. Aluminium and its salts do not constitute a carcinogenic, mutagenic or teratogenic hazard except at extremely high exposure levels. The production and handling of aluminium fluoride and sulphate have been associated with asthma problems for some industrial workers. The toxicity of aluminium in aqueous solution is dependent on a number of factors such as the solubility of aluminium and the effect of pH on the hydrolysis of aluminium salts and whether the water is hard or soft. The chemistry of soluble aluminium is complex and affects bioavailability. Nevertheless, even small concentrations of aluminium salts are generally toxic to aquatic organisms and concentrations of $0.1–0.6 \, \text{mg} \, l^{-1}$ can be fatal to small fish. The aluminium salts concentrate in gill tissue and so interfere with the oxygen uptake. Aluminium is also phytotoxic, e.g. wheat, oats and beans are adversely affected at concentrations $> 1 \, \text{mg} \, l^{-1}$ aqueous solution. Aluminium sulphate is toxic to corn at concentrations $> 10 \, \text{mg} \, l^{-1}$ aqueous solution.

Aluminium produces many of its toxic effects by direct action on the nuclear chromatin (it has a high affinity for plasma proteins), or in an indirect way by replacing other elements or modifying the activity of enzyme systems. The neurotoxicity of the metal may be due to its action on the DNA- dependent RNA synthesis and the inactivation of neurotransmitters. It influences the activity of several enzymes, the reproduction system and both embryonal and post-embryonal development. An important role for aluminium after ingestion is its competitive relationship with calcium ions and phosphorus. When a large excess of aluminium salts is present in the body, calcium retention and phosphorus absorption are both reduced and blood level ATP falls as the phosphorylation processes are depressed. In such cases, aluminium has been found to accumulate to high levels (exceeding the normal value 10–20-fold) in the parathyroid glands, which in severe cases can lead to osteomalacia.

Aluminium salts are used in large amounts in water purification. In July 1988, 20 t of aluminium sulphate was accidently added to the drinking water supply of 20 000 people in the Camelford district of Cornwall, UK, raising the maximum Al level to $620 \, \text{mg} \, l^{-1}$, although most consumers received concentrations between 10 and $50 \, \text{mg} \, l^{-1}$. A wide range of symptoms were reported, including gastrointestinal disturbances, rashes and mouth ulcers. Some of the symptoms persist and are still under study. In addition, thousands of fish were also killed. Similar accidents have also happened at Huntington, Cheshire and Bideford, Devon, UK.

The toxicity of aluminium was first discussed as a complication to dialysis therapy where the gradual build-up of aluminium levels caused renal failure. Aluminium has been shown to impair trans-membrane diffusion and it acts as a membrane toxicant particularly at the blood–brain barrier. It also causes neurofibrillary degeneration in different areas of the brain. Aluminium plays an interfering role in cyclic adenosine-5-triphosphate synthesis and glutamate release. Aluminium in antacids and in waters used for dialysis is known to cause dementia and result in bone deterioration. The high level of aluminium in some waters has been suggested to be the cause of Alzheimer's disease since sufferers tend to have high concentrations of aluminium in their brain. Unfortunately, despite many investigations, there is still much uncertainty about the causal link between the metal and the onset of the disease. As a precaution, it is recommended that the aluminium intake should be decreased with increasing age.

Other sources of aluminium include the leaching of aluminium from cooking pots and drink cans, particularly if they contain acid ingredients, and tea drinking since the tea plant is a natural accumulator of aluminium.

Occupational exposure limit. OHSA PEL (TWA) (soluble salts) $2\,mg\,m^{-3}$ (Al). The finely divided metal is flammable and can cause explosions.

Further reading
ECETOX (1991) *Technical Report 30*, European Chemical Industry Technology and Toxicology Centre, Brussels.
ToxFAQs (1995) *Aluminium*, ATSDR, Atlanta, GA.
Lewis, T. E. (1989) *Environmental Chemistry and Toxicity of Aluminium*, Lewis, Chelsea, MI.
Kirk–Othmer ECT, 4th edn, Wiley, New York, 1992, Vol. 2, pp. 184–345.
King, R. B. (ed.) (1994) *Encyclopedia of Inorganic Chemistry*, Wiley, Chichester, Vol. 1, pp. 103–139.
Thompson, R., (ed.) (1995) *Production and Use of Aluminium Compounds*, Royal Society of Chemistry, Cambridge, Ch. 11, 'Industrial Inorganic Chemicals. Production and Uses'.

Aluminium ammonium sulphate dodecahydrate, CAS No. 7784-26-1

$AlNH_4(SO_4)_2.12H_2O$

Uses. Mordant in dyeing. Water and sewage purification. Food additive and food pH buffer. Bird and animal repellent. Tanning industry. Textile fireproofing.

Toxicity. Low toxicity

Aluminium (III) bromide CAS No. 7727-15-3

$AlBr_3$

Uses. Catalyst in organic chemistry.

Toxicity. Inhalation of powder or vapour causes severe irritation. Mild human poison.

Aluminium (III) chloride CAS No. 7446-70-0

$AlCl_3$

Uses. Catalyst in organic chemistry. Friedel–Crafts reaction. Manufacture of rubbers, lubricants and anti-perspirants.

Toxicity. Corrosive. Causes burns. LD_{50} oral rat $3450\,mg\,kg^{-1}$; LD_{50} oral mouse $390\,mg\,kg^{-1}$.

Aluminium (III) fluoride CAS No. 7784-18-1

AlF_3

Occupational exposure limit: OSHA PEL (TWA) $2.5\,mg\,F^-\,m^{-3}$.

Uses. Manufacture of ceramics, enamels and glazes. Flux.

Toxicity. Can cause fluorosis, fixation of bone calcium by fluoride.

Aluminium(III) hydroxide CAS No. 21645-51-2

$Al(OH)_3$

Uses. Antacid. Source of aluminium metal, alums and aluminium salts. Filler in paper. Manufacture of glass.

Toxicity. Toxic to fish at ~ 1 ppm. LD_{Lo} intraperitoneal rat $150\,mg\,kg^{-1}$. MAK (Germany) fine dust $6\,mg\,m^{-3}$.

Aluminium(III) hydride CAS No. 97-93-8

AlH_3

Uses. Catalyst in organic chemistry. Rocket propellant. Reducing agent in organic chemistry.

Toxicity. Flammable in air. Causes skin burns.

Aluminium (III) nitrate nonahydrate CAS No. 7784-27-2

$Al(NO_3)_3.9H_2O$

Uses. Textile and petroleum industries. Antiperspirant. Corrosion inhibitor.

Toxicity. Irritant. LD_{50} oral rat $3671\,mg\,g^{-1}$; oral mouse $3980\,mg\,kg^{-1}$.

Aluminium(III) oxide CAS No. 1344-28-1

Al_2O_3

Occupational exposure limits: OSHA PEL (TWA) $10\,mg\,m^{-3}$, total dust; respiratory fraction $5\,mg\,m^{-3}$.

Uses. Absorbent in chromatography. Abrasive. Catalyst in organic chemistry.

Toxicity. Corrosive. Irritant when inhaled as dust, can cause lung damage.

Aluminium(III) phosphide CAS No. 20859-73-8

AlP

Uses. Fumigant for killing insects in stored feedstock. Acute rodenticide, moles and rabbits. Source of phosphine.

Toxicity. Inhalation may result in respiratory, cardiac, hepatic and renal changes. LC_{Lo} inhalation rat 1 ppm.

Aluminium(III) selenide CAS No. 12598-14-0

Al_2Sc_3

Uses. Preparation of hydrogen selenide for semiconductors.

Toxicity. Unstable in air. Decomposed by water to give the poison hydrogen selemide.

Aluminium sodium sulphate CAS No. 10102-71-3
$AlNa(SO_4)_2.12H_2O$

Uses. General-purpose food additive. Leavening agent.

Toxicity. Irritant. Can cause dermatitis.

Aluminium(III) sulphate monohydrate CAS No. 17927-65-0

$Al_2(SO_4)_3.H_2O$

Uses. Water treatment. Tanning industry. Paper manufacture. Slug and snail control. Fireproofing and waterproofing fabrics.

Toxicity. Irritant. LD_{50} oral rat 1930 mg kg^{-1}.

4-AMINOBIPHENYL CAS No. 92-67-1;
LD_{50} rat, rabbit, mouse 205–690 mg kg^{-1}

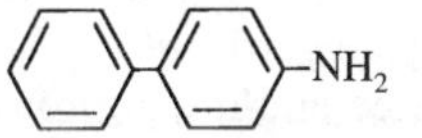

$C_{12}H_{11}N$ $M = 169.2$

p-Aminobiphenyl; m.p. 53°; b.p. 302°; dens. 1.16; log P_{ow} 2.86. Soluble in ethanol, chloroform.

In the 1950s it was used as a rubber antioxidant. It is used in chemical analysis to detect the sulphate ion.

4-Aminobiphenyl occurs in tobacco smoke and was identified as a risk substance in 1972 by the IARC. It is carcinogenic in the mouse, rat, rabbit and dog following oral administration, inducing bladder and liver tumours in mice and bladder papillomas and carcinomas in rabbits and dogs. Examination of the lung tissue of smokers revealed 4-aminobiphenyl DNA adducts; it can cross the human placenta and bind with foetal haemoglobin.

Levels in non-filter cigarettes (ng/cigarette)

	Mainstream	Sidestream
2-Aminobiphenyl	3.0	110
3-Aminobiphenyl	5.0	132
4-Aminobiphenyl	4.6	143

The 2- and 3-aminocompounds present a lower risk and are not certainly carcinogenic in animals. 4-Aminobiphenyl forms DNA adducts *in vivo* in the bladder epithelium of dogs and protein adducts in the serum albumin of rabbits; it induces mutations in human fibroblasts and is mutagenic to bacteria. It is classified by the ACGIH as a group A1 carcinogen and placed in group 1 by the IARC.

Further reading
International Agency for Research on Cancer (1972) *Monograph on the Evaluation of Cancer Risk of Chemicals to Man*, Vol. 1; also (1987) Supplement 7, IARC, Lyon.

2-AMINO-4-NITROTOLUENE CAS No. 99-55-8

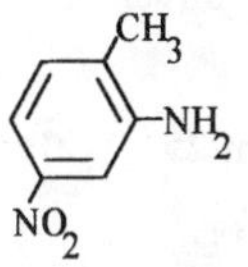

$C_7H_8N_2O_2$ $M = 152.1$

5-Nitro-*o*-toluidine, 2-methyl-5–nitrobenzeneamine, numerous trade names, e.g. Pigment Red 18, Amarthol Fast Scarlet G; m.p. 107° (yellow prisms); slightly soluble in benzene, diethyl ether, soluble in ethanol, acetone, chloroform.

Prepared by the nitration of *o*-toluidine or *o*-benzenesulphontoluidide followed by hydrolysis of the latter product. US production in 1975 was 57 t. It is used to prepare the diazo-coupling component for yellow and red azo dyes.

Formed by the anaerobic reduction of 2,4-dinitrotoluene by activated sludge and detected in the aqueous effluent from the manufacture of TNT, a change which is also affected by human intestinal flora. Reversed-phase HPLC will detect nitrotoluidine at a limit of $1\,ng\,ml^{-1}$ in the wastewater from TNT manufacture; the use of GC–MS permits detection in wastewater at levels of 0.002–$0.10\,ng\,ml^{-1}$ (Spanggord *et al.*, 1982).

2-Amino-4-nitrotoluene emits toxic fumes of NO_x when heated and is dangerous to inhale and on skin contact. It is mutagenic in *Salmonella typhimurium* and causes carcinomas in male and female mice.

Under the MAK regulations it is classed as an A2 carcinogen – clearly carcinogenic in animals under conditions indicative of potential in the workplace.

Reference
Spanggord, R.J., Gibson. B.W., Keck, R.G., Barkley, J.J. and Thomas, D.W. (1982) Effluent analysis of wastewater generated in the manufacture of trinitrotoluene. *Environ. Sci. Technol.*, **16**, 229–232.

Further reading
Chemical Safety Data Sheets (1991) **4b**, Royal Society of Chemistry, Cambridge.
IARC (1990) *Some Flame Retardants and Textile Chemicals and Exposures in the Textile Manufacturing Industry*, IARC, Lyon, Vol. 48, p. 169.

2-AMINOPYRIDINE CAS No. 504-29-0
LD_{50} rat $200\,mg\,kg^{-1}$; LD_{50} mouse $50\,mg\,kg^{-1}$

$C_5H_6N_2$ $M = 94.1$

2-Pyridineamine; m.p. 58°; b.p. 210°; v. dens. 3.25; log $P_{ow} - 0.22$. Soluble in water, ethanol and acetone

Used in the synthesis of pharmaceuticals, especially antihistamines. It may be taken up by inhalation or skin contact leading to headache, nausea and increased blood pressure. A human fatality has been reported (ACGIH).

The occupational limit is 0.5 ppm ($2\,mg\,m^{-3}$, MAK); the same level has been set by the ACGIH.

AMMONIA CAS No. 7664-41-7
LC_{50} rat $350\,mg\,kg^{-1}$; LC_{50} rainbow trout $0.53\,mg\,l^{-1}$; LC_{50} other fish 1.1–$22\,mg\,l^{-1}$

NH_3 $M = 17.0$

M.p. $-78°$; b.p. $-33°$; sp. gr. $0.77\,g\,l^{-1}$ ($760\,mmHg$, 0°), liquifies at 10 atm/25°. Water solubility $530\,g\,l^{-1}$; also soluble in ethanol, methanol, chloroform.

Ammonia is produced by breakdown of nitrogenous animal and vegetable waste and burning of fossil fuels, and is volatilized from the earth's surface at a rate of 10^8 t year^{-1} mostly from natural sources. Typical urban air levels range from 5 to $25\,\mu g\,m^{-3}$, the rural range is from 2 to $6\,\mu g\,m^{-3}$ and over remote oceans it is from 10 to $115\,ng\,m^{-3}$.

Ammonia is manufactured by the Haber–Bosch process for the high pressure combination of nitrogen and hydrogen. Ammonia is stored in refrigerated tanks and piped to reaction sites. In the USA lots of up to 2500 t are transported by barge. It is a toxic and corrosive gas and a common occupational hazard in industry.

Used for pH adjustment in bulk manufacturing, in the production of nitric acid, fertilizers (80%), synthetic fibres (10%) and explosives (5%). The major use in producing ammonium nitrate as a fertilizer derives from the activities of nitrifying microorganisms in soil, which convert NH_3 into NO_3^-. Ammonia is an intermediate in the production of toluene diisocyanate for the production of polyurethane foam and combines with carbon dioxide to give urea for incorporation in urea–formaldehyde resins. It is also formed during the aerobic degradation of organic matter and may be oxidized to nitrogen dioxide which is taken up by water. It is used as a refrigerant and as a chemical intermediate in the production of cyanides, amides and dyestuffs.

Ammonium salts are added to foodstuffs as stabilizers, leavening agents and for flavouring. The daily intake for an adult is about 18 mg. The main source in mammals, including humans, is the biological degradation of protein with the excretion of urea; ammonia is secreted in the body as a means of removing H^+ as NH_4^+.

The odour threshold is about 5 ppm; higher concentrations affect the eyes and mucuous membrane leading to damage to the cornea and the respiratory system. Acute harm will be done to vulnerable subjects at levels of over 2000 ppm threatening death from severe damage to the lungs; the LC_{50} (30 min) for humans is 11 500 ppm.

The EC limit in drinking water (80/778/EEC) for NH_4^+ is 0.05 mg l^{-1} and OSHA gives a TLV (8 h) of 50 ppm (35 mg m^{-3}). The US TLV (TWA) is 25 ppm (17 mg m^{-3}).

Further reading

Chemical Safety Data Sheets, Royal Society of Chemistry, London, 1990, Vol. 3, pp. 33-40.

ECETOX (1991) Technical Report 30(4), European Chemical Industry Technology and Toxicology Centre, Brussels.

Environmental Health Criteria 54. Ammonia. WHO/IPCS, Geneva, 1986.

Kirk–Othmer ECT, 4th edn, Wiley, New York, 1992, Vol. 2, pp. 638–691.

Thompson, R. (ed.) (1995) Industrial Inorganic Chemicals. Production and Uses, Royal Society of Chemistry, Cambridge, Ch. 6.

Ammonium acetate, CAS No. 631-61-8

$CH_3CO_2NH_4$

Uses. Mordant in dyeing wool, reagent in analytical chemistry, preserving meat and in the manufacture of foam rubbers and vinyl plastics.

Toxicity. Irritant to eyes, skin, nose and throat. LD_{50} intraperitoneal rat 632 mg kg^{-1}.

Ammonium chloride CAS No. 12125-02-9

NH_4Cl

Uses. Flux. Batteries. Dyeing, tanning and electroplating industries. Explosives

Occupational exposure limits. ACGIH TLV (TWA) 10 mg m^{-3}

Toxicity. Irritant. LD_{50} oral rat 1650 mg kg^{-1}; intravenous mouse 358 mg kg^{-1}.

Ammonium dichromate CAS No. 7789-09-5

$(NH_4)_2Cr_2O_7$

Uses. Pyrotechnics. Lithography and photoengraving. Mordant. Catalyst. Magnetic recording materials.

Occupational exposure limits. ACGIH TLV (TWA) $0.05 \, mg \, Cr \, m^{-3}$

Toxicity. Harmful, may cause burns. Carcinogen. LD_{50} intravenous rat $30 \, mg \, kg^{-1}$.

Ammonium fluoride CAS No. 12125-01-8

NH_4F

Uses. Chemical reagent. Etching and frosting glass. Printing and dyeing. Wood preservation.

Occupational exposure limits. USA TLV(TWA) $2.5 \, mg \, m^{-3}$.

Toxicity. Toxic by inhalation, contact with skin, and if swallowed. LD_{50} intraperitoneal rat $31 \, mg \, kg^{-1}$.

Ammonium hexachloroplatinate(iv) CAS No. 16919-58-7

$(NH_4)_2PtCl_6$

Uses. Preparation of platinum complexes.

Toxicity. Harmful by inhalation. Prolonged exposure may cause asthma.

Ammonium hydrogencarbonate, CAS No. 1066-33-7

NH_4HCO_3

Uses. Fire extinguishers. Fertilizers. Production of other ammonium salts. Smelling salts, etc.

Toxicity. Low toxicity to humans. Irritant to eyes and skin. LD_{50} intravenous mouse $245 \, mg \, kg^{-1}$.

Ammonium hydrogenfluoride CAS No. 1341-49-7

NH_4HF_2

Uses. Manufacture of magnesium and magnesium alloys. Production of HF. Manufacture of glass and porcelain. Anodizing aluminium. Corrosion inhibitor. Wood preservative.

Toxicity. Severe burns to eyes, skin, and respiratory system. Poison. Damages central nervous system and kidneys, may be fatal. LD_{50} guinea pig $150 \, mg \, kg^{-1}$.

Ammonium hydroxide CAS No. 1336-21-6

NH_4OH

Uses. Preparation of ammonium salts. Detergents. Stain remover.

Toxicity. Highly irritating. LD_{50} oral rat $350 \, mg \, kg^{-1}$.

Ammonium iron(II) sulphate hexahydrate CAS No. 7783-85-9

$(NH_4)_2Fe(SO_4)_2.6H_2O$

Uses. Analytical reagent.

Toxicity. Skin and eye irritant. LD_{50} oral rat $3.25\,g\,kg^{-1}$.

Ammonium metavanadate(V) CAS No. 7803-55-6

NH_4VO_3

Uses. Catalyst. Dyeing and printing. Reagent in analytical chemistry. Glazing ceramics.

Toxicity. Genotoxic destroying enzymes. Possible mutagen. Irritant to eyes and skin. Affects lungs, thorax, gastrointestinal system and blood. LD_{50} oral rat 58–160 $mg\,kg^{-1}$.

Ammonium molybdate(VI) tetrahydrate CAS 13106-76-8

$(NH_4)_6Mo_7O_{24}.4H_2O$

Uses. Catalyst. Dyes. Varnishes. Photography.

Toxicity. Toxic. Irritant. Possible mutagen. LD_{50} oral rat 333 $mg\,kg^{-1}$.

Ammonium nitrate CAS No. 6484-52-2

NH_4NO_3

Uses. Oxidizing agent. Explosives. Fertilizer. Manufacture of nitrous oxide (N_2O).

Toxicity. LD_{50} oral rat 4820 $mg\,kg^{-1}$.

Ammonium perchlorate CAS No. 7790-98-9

NH_4ClO_4

Uses. Primary use is as propellant for solid rocket motors. Pyrotechnics. In solid, slurried and gelled form as blasting formulations.

Toxicity. Hazardous chemical and shock sensitive. Strong oxidizing agent.

Ammonium sulphate CAS No. 7783-20-2

$(NH_4)_2SO_4$

Uses. Manufacture of ammonium alum. Flameproofing fabrics. Food additive.

Toxicity. LD_{50} oral rat 3000 $mg\,kg^{-1}$. TD_{Lo} oral humans 1500 $mg\,kg^{-1}$.

Ammonium thiosulphite CAS No. 7783-18-8

$(NH_4)_2S_2O_3$

Uses. Photographic fixing salt.

Toxicity. LD_{50} oral rat 2890 $mg\,kg^{-1}$.

Further reading
Kirk–Othmer ECT, 4th edn, Wiley, New York, 1992, Vol. 2, pp. 692–708.
Merck Index, 12th edn, Merck, Rahway, NJ, 1996, pp. 87–95.
Thompson, R. (ed) (1995) *Industrial Inorganic Chemicals. Production and Uses*, Royal Society of Chemistry, Cambridge, Ch 6.

ANILINE CAS No. 62-53-3
LD_{50} rat 440 mg kg^{-1}; LC_{50} (96 h) fathead minnow 134 mg l^{-1}

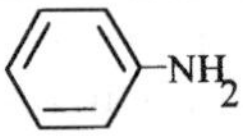

C_6H_7N $M = 93.1$

Aminobenzene, benzeneamine; b.p. 184°; dens. 1.02; v.p. 0.3 Torr (20°); v. dens. 3.2, log P_{ow} 0.90. Colourless when pure. Water solubility 26 g l^{-1} (20°); miscible with most organic solvents.

Aniline occurs in small amounts in coal tar and is mainly manufactured by the catalytic reduction of nitrobenzene. It is widely used industrially in the synthesis of dyestuffs, isocyanates for polyurethane products, antioxidants and for the production of varnishes, resins and perfumes.

Exposure to light and air leads to rapid oxidation and discoloration with the formation of hydrazobenzene, azobenzene, benzidine and aminodiphenylamines. Aniline reacts explosively with strong oxidizing agents, e.g. hydrogen peroxide, nitric acid and tetranitromethane. It is rated as 'extremely hazardous' by the USEPA.

The half-life in air has been estimated at 3.3 h following reaction with OH radicals. In daylight the $t_{1/2}$ in estuarine waters is 27 h. Aniline is degraded by soil bacteria to acetanilide, hydroxyaniline and catechol.

In humans, slight toxic effects, headache and weakness are likely after exposure for several hours at levels in air in the range 7–50 ppm; serious effects are experienced after exposure for 1 h at levels above 100 ppm. In the body aniline causes methaemoglobinaemia due to its oxidation of Fe(II) to Fe(III) and consequent failure of haemoglobin to transport oxygen; this leads to intoxication and possible death by asphyxiation. This condition, unlike carbon monoxide poisoning, is not reversible in fresh air.

Clusters of deaths from bladder cancer in workers in the industry have been attributed to contact with chemicals other than aniline. The IARC classification is group 3 – inadequate evidence of carcinogenicity in animals and humans; however the MAK classification of IIIB is held to require further clarification.

The occupational exposure limit is a TLV (TWA) of 2 ppm (7.6 mg m^{-3}).

Further reading
Chemical Safety Data Sheets (1991) **4a**, *Toxic Chemicals*, Royal Society of Chemistry, Cambridge.

ANISIDINES (methoxyanilines)

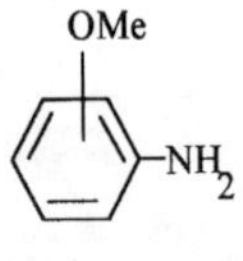

C_7H_9NO $M = 123.1$

Compound	CAS No.	M.p. (°)	B.p. (°)	LD_{50} rat (mg kg^{-1})
2-Anisidine	90-04-0	6	224	1400–2000
3-Anisidine	536-90-3	−10	251	–
4-Anisidine	104-94-9	57	246	2000

All three isomers have low volatility, v.p. < 0.1 Torr $(30°)$ with log P_{ow} 0.95. They are yellow oils which discolour in air/ light and all three are soluble in EtOH, benzene and acetone. 2-Anisidine has a water solubility of $14\,g\,l^{-1}$.

They are prepared by the catalytic hydrogenation of the nitroanisoles and are used as azo dye intermediates, e.g. Pigment Red 190 and Direct Red 24 (Kirk–Othmer, 1992).

In their toxicity and susceptability to oxidation they are similar to aniline and exposure leads to headache, drowsiness and cyanosis; slight symptoms appear at a level of 0.4 ppm for a 3.5 h day over 6 months. Anisidines irritate the respiratory tract and the eyes and may be absorbed through the skin.

The IARC class 2-anisidine as a group 2B carcinogen. The ACGIH set a TLV (TWA) of 0.1 ppm $(0.5\,mg\,m^{-3})$ for the mixed 2- and 4-isomers; a similar limit has been set in the UK and Germany (MAK).

Reference
Kirk–Othmer ECT, 4th edn, Wiley, New York, 1992, Vol. 3, pp. 821–875.

Further reading
Abrahart, E.N. (1977) *Dyes and Their Intermediates*, 2nd edn, Arnold, London.

ANTHRACENE CAS No. 120-12-7
LD_{50} i.p mouse $> 430\,mg\,kg^{-1}$; LC_{50} (96 h) bluegill sunfish $12\,\mu g\,l^{-1}$

$C_{14}H_{10}$ $M = 178.2$

Paranaphthalene, green oil; m.p. $218°$; b.p. $342°$; v.p. 1 Torr $(145°)$; $\lambda_{max} 360\,nm, \epsilon = 6000$. Solubility in water ca $50\,\mu g\,l^{-1}$, slightly soluble in benzene $(16\,mg\,l^{-1})$ and ethanol $(15\,g\ l^{-1})$ log P_{ow} 4.45.

Forms 6–10% of the high-boiling fraction of coal tar, from which it was first isolated by Dumas in 1883. It is oxidized to anthraquinone by a wide range of reagents including air or oxygen/vanadium, chromic acid, selenium dioxide, etc. The preferred industrial route is now the reaction of phthalic anhydride with benzene.

Anthracene is ubiquitous as a product of the incomplete combustion of fossil fuels:

Mainstream cigarette smoke	$2.3\,\mu g/100$ cigarettes.
Gasoline exhaust	$500–600\,\mu g\,l^{-1}$
Charcoal broiled steak	$4.5\,\mu g\,kg^{-1}$
Waste water	$1.6–7.0\,mg\,l^{-1}$
Drinking water	$1–60\,ng\,l^{-1}$

The half-life in soil is 100–175 days and in sunlit water 35 min. Photosensitized oxidation gives the peroxide:

The IARC found no evidence of carcinogenicity in experimental animals. The OSHA PEL (TWA) is $0.2\,\text{mg}\,\text{m}^{-3}$.

Further reading
ECETOX (1991) *Technical Report 30*, European Chemical Industry and Toxicology Centre, Brussels.

ANTIMONY CAS No. 7440-36-0

Sb

Antimony, a member of Group 15 of the Periodic Table, has an atomic number of 51, an atomic weight of 121.76, a specific gravity of 6.69 at 20°, a melting point of 630.7°, a boiling point of 1587° and two natural isotopes with the following percentages abundance: ^{121}Sb 57.21% and ^{123}Sb 42.79%. Over 25 radioactive isotopes have been characterised. Antimony is a brittle, bluish white metalloid (semi-metal) existing in several allotropic forms, the stable form is α-Sb (cf. arsenic.) The three important oxidation states are $-3, +3$ and $+5$.

Antimony is not particularly abundant in the earth's crust, being 62nd in order, with a crustal abundance of 0.2 ppm. Fortunately, *stibnite*, Sb_2S_3, the most important ore of antimony, occurs in large quantities. Complex ores containing lead, copper, silver and mercury sulphides are also important industrial sources of antimony. The metal is obtained commercially by reduction of the oxide or sulphide. Antimony enters the environment through the mining and processing of antimony-containing ores, in the production of antimony metal, alloys and compounds and to a lesser extent by incinerators and coal-burning plants. Typical concentrations are: air 1.4–55 $\text{ng}\,\text{m}^{-3}$, near a smelter 1.0 $\text{mg}\,\text{m}^{-3}$; soil 0.02–12 ppm; fresh water, very low, ~ 5 ppb; seawater $\sim 0.2\,\mu\text{g}\,\text{l}^{-1}$ (mainly in the form of $[Sb(OH)_6]^{3-}$); plants 0.1 ppm. Tobacco contains about 0.1 mg Sb kg^{-1}. Mosses, lichens and fungi accumulate antimony compounds. Some fresh fish can contain $\sim 3.0\,\mu\text{g}\,\text{kg}^{-1}$. Daily intake by adult humans of antimony is usually in the range 10–70 $\mu\text{g}\,\text{day}^{-1}$; total body (70 kg, adult) 70 mg, with average blood levels of $\sim 3\,\mu\text{g}\,\text{l}^{-1}$.

Antimony is used in the manufacture of some special alloys. It is an alloy ingredient in bismuth telluride, Bi_2Te_3, class of alloys which are used in thermoelectric cooling. A lead alloy (4–5% Sb) having properties superior to those of ordinary lead is used for the manufacture of lead plates in storage batteries. Two antimony compounds are formed due to lead-acid manufacture. The casting of battery grids produces antimony(III) oxide. Stibine [antimony(III) hydride] is produced whilst charging, particularly at the end of charging or overcharging. Other uses include antimony-containing bearing alloys and antimony solder. The greatest use of antimony compounds is in flame retardants, plastics, paints, textiles, rubber, ceramics, medicines and semiconductors.

Antimony is non-essential to plants, animals and humans, and many antimony compounds are poisonous. Although metallic antimony may be handled freely, skin contact with the metal and its alloys is not recommended. As for other heavy metals, precautions should be taken to avoid ingestion. Trivalent antimony compounds are invariably more toxic than the corresponding pentavalent compounds. Both antimony and antimony(III) oxide have been reported to cause dermatitis among enamellers and decorators in the ceramics industry. The trioxide also caused eye irregularities and cataracts after a 24 month exposure of $4.5\,\mathrm{mg\,m^{-3}}$ (Sb) to rats. Lung cancer has been found in laboratory animals after prolonged exposure to antimony dust. Antimony is not thought to cause cancer or birth defects or to affect reproduction in humans, but some salts appear to be mutagenic to human and animal cells. Adults who inhaled antimony at a concentration of $3\,\mathrm{mg\,Sb\,m^{-3}}$ over a period from 8 months to 2 years developed heart problems and stomach ulcers. Workers exposed to levels of $9\,\mathrm{mg\,Sb\,m^{-3}}$ developed lung, eye and skin irritations. Prolonged inhalation by antimony workers has been reported to produce a variety of diseases, including pneumoconiosis, pneumonitis, fibrosis, bone marrow damage and carcinomas.

Antimony compounds present in food and drink are only slowly digested in the stomach and most of the antimony is excreted in the faeces. The uptake, retention of antimony in the liver, skeleton and kidneys and the toxicity towards the major organs are highly dependent on the type of antimony compound and its oxidation state. Recently it has been suggested that the presence of antimony compounds in mattresses (as flame retardants) may be a factor in cot deaths, presumably due to the formation of stibine, SbH_3, a highly toxic gas. The latest opinion is that the case is not proven.

Occupational exposure limit. OSHA PEL (TWA) $0.5\,\mathrm{mg\,Sb\,m^{-3}}$.

Futher reading

ECETOX (1991) *Technical Report 30(4)*. European Chemical Industry Technology and Toxicology Centre, Brussels.

Fowler, B.A., Yamauchi, H., Conner, E.A. and Akkerman, M. (1993) *Scand. J. Work Environ. Health*, **19**, 101–103.

Friberg, L. (1986) *Handbook of the Toxicology of Metals*, 2nd edn, Elsevier, Amsterdam, pp. 1–2.

HSE (1993) *Hygienic and Toxicological Properties of Antimony*, Health and Safety Executive London.

King, R.B. (ed.) (1994) *Encyclopedia of Inorganic Chemistry*, Wiley, Chichester, Vol. 1, pp. 170–190.

Kirk-Othmer ECT, 4th edn, Wiley, New York, 1992, Vol. 3, pp. 367–412.

Antimony(III) chloride CAS No. 10025-91-9

$SbCl_3$

Uses. Patent leather mordant. Preparation of other antimony compounds. Catalyst. Bronzing iron.

Toxicity. LD_{50} oral rat $530\,\mathrm{mg\,kg^{-1}}$. LD_{50} intraperitoneal rat 13 mg kg^{-1}. Inhalation causes irritation to nose, throat and upper respiratory tract. 25 ppm of the chloride in water is hazardous to humans.

Antimony(V) chloride CAS No. 7647-18-9

$SbCl_5$

Uses. Chemical reagent. Catalyst.

Toxicity. Mildly toxic to humans. Severe burns to skin and eyes. Evidence of liver and kidney damage after inhalation. LD_{50} oral rat $1120\,\mathrm{mg\,kg^{-1}}$.

Antimony(III) fluoride CAS No. 7783-56-4

SbF_3

Uses. Chemical reagent. Ceramics. Catalyst. Electroplating.

Occupational exposure limits. OSHA PEL (TWA) $0.05 \, mg \, kg^{-1}$

Toxicity. LD_{50} subcutaneous rats $23 \, mg \, kg^{-1}$. LD_{50} oral mouse $804 \, mg \, kg^{-1}$. Toxic by inhalation, affects mouth, nose, stomach and intestines.

Antimony(V) fluoride CAS No. 7783-70-2

SbF_5

Uses. Fluorination of organic compounds

Toxicity. Poison. LC_{50} inhalation mice $270 \, mg \, l^{-1}$.

Antimony(III) hydride (stibine) CAS No. 7803-52-3

SbH_3

Uses. Reducing agent. Source of pure antimony.

Occupational exposure limits. USA PEL (TWA) 0.1 ppm.

Toxicity. Poison.

Antimony(III) oxide CAS No. 1309-64-4

Sb_2O_3

Uses. Flame retardants. Speciality glass and porcelain enamels. Pigment in paints.

Toxicity. LD_{50} oral rat $> 35 \, g \, kg^{-1}$; intraperitoneal mouse $172 \, mg \, kg^{-1}$. Suspected human carcinogen.

Antimony(V) oxide CAS No. 1314-60-9

Sb_2O_5

Uses. Flame retardants. Speciality glass and porcelain enamels. Pigment in paints.

Toxicity. LD_{50} intraperitoneal rat $4 \, g \, kg^{-1}$; intraperitoneal mouse $978 \, mg \, kg^{-1}$.

Antimony(III) sulphide CAS No. 1345-04-6

Sb_2S_3

Uses. Pigment in paints.

Toxicity. LD_{50} intraperitoneal mouse $209 \, mg \, kg^{-1}$.

ARGON CAS No. 7440-37-1

Ar

Argon, a member of Group 18 (the noble gases) of the Periodic Table, has an atomic number of 18, an atomic weight of 39.95, a specific gravity of 1.789 at 0°, a melting point of $-189.3°$ and a boiling point of $-185.9°$. It is a colourless, inert gas and no stable compounds of argon are known. There are three naturally occurring isotopes with the following isotopic percentages abundance: ^{36}Ar 0.337%, ^{38}Ar 0.063% and ^{40}Ar 99.60%. Argon-40 is formed in the decay of the natural radioactive ^{40}K isotope. The noble gases make up about 1% of the earth's atmosphere, the major component being argon. Obtained as the pure element by the liquefaction and fractional distillation of air.

Uses. Most of the world's production of argon is for use mainly as an inert atmosphere in high-temperature metallurgical processes. Smaller amounts are used for filling incandescent lights. Together with neon and krypton it is also used in discharge tubes (neon lights). Further uses are as a cryogenic and as a substitute for nitrogen in synthetic breathing gas for deep-sea diving.

Like the other inert gases, argon acts as a simple asphyxiant, excluding oxygen from the body. Initial symptoms of asphyxiation is reduced mental alertness and slight loss of muscular coordination. A 75:25 mixture of argon and air is fatal in a few minutes.

Toxicity. Asphyxiant.

Further reading
Kirk–Othmer ECT, 4th edn, Wiley, New York, 1993, Vol. 13, pp. 1–53.
King, R.B. (ed.) (1994) *Encyclopedia of Inorganic Chemistry*, Wiley, Chichester, Vol. 5, pp. 2660–2680.
Thompson, R. (ed.) (1995) *Industrial Inorganic Chemicals. Production and Uses*, Royal Society of Chemistry, Cambridge.

ARSENIC CAS No. 7440-38-2

As

Arsenic, a member of Group 15 of the Periodic Table, has an atomic number of 33, an atomic weight of 74.92, a specific gravity 5.73 at 14°, a melting point of 814° (36 atm), a boiling point of 614° (sublimes) and one natural isotope, ^{75}As. The element is a steel-grey, brittle, crystalline metalloid with three allotropes, yellow, black and the stable grey form. Important oxidation states are $-3, 0, +3$ and $+5$.

Arsenic is widely distributed among a great variety of mineral species in the earth's crust, being 51st in order, with a crustal abundance of 2 ppm. It is found in igneous rocks (1–3 ppm), in sedimentary rocks (1–25 ppm, but concentrations as high as 400 ppm have been reported), and in virtually all soils (0.1–40 ppm). The primary sources of arsenic are copper and lead sulphide ores but it is widely found as mixed sulphide–arsenide ores with other metals such as iron, nickel, cobalt, silver and gold. Arsenic is recovered as a by-product during the smelting processes. Arsenic enters the terrestrial environment [mainly as arsenic(III) oxide] through steel and iron production, dissipation to land of steel slag and copper leach liquid, airborne emission from copper smelting, coal burning, etc., and land fill wastes (copper flue dusts, coal fly ash).

Arsenic is distributed widely among plant and animal species. In plant tissues its normal concentration varies from 0.01 to 5 ppm on a dry weight basis. Marine plants have been found to possess much higher concentrations. Seaweed and brown algae concentrate arsenic in their tissues (94 ppm).

Arsenic is always present in all animals: in human body tissues < 0.3 ppm, in blood 0.004 mg kg^{-1} and in urine 0.01 mg l^{-1}. Marine organisms such as cod, shrimps, oysters and lobsters can concentrate arsenic up to 128 ppm on a dry weight basis. More typical values are 3–30 ppm arsenic among marine organisms. Fresh water normally contains 0.15–0.45 µg As kg^{-1}. The maximum allowable As concentration in drinking water is 0.05 ppm. Although common, most foods contain only minute amounts of As; red meat ~ 0.2 ppm dry weight and dairy products 0.0033 ppm. A typical daily diet for an adult is about 0.01–0.3 mg, 12th in abundance in human body. The routes of absorption and exposure include inhalation, ingestion and skin. Concentration in air, 1–20 ng As$_2$O$_3$ m^{-3}.

Uses.   Arsenic is used in metallurgy to harden copper, manufacture of lead alloys. Ceramics and glass. Semiconductors. Electronic applications. Agriculture and medicine. The breakdown of uses is agriculture 80%, ceramics and glass 8%, chemicals 5% and others 5%. Commercial uses include pigments, pesticides and wood preservative.

Arsenic and its compounds must be considered as extremely toxic. Many inorganic and several organic compounds have found uses as poisons, herbicides and pesticides. Organoarsenic compounds are less phytotoxic and are increasingly used in crop protection. Inorganic arsenic compounds are human carcinogens and poisons, with trivalent arsenic more toxic than pentavalent arsenic. In detoxification, trivalent compounds are oxidized to pentavalent compounds and are thereafter transformed by the body to monomethylarsenic acid and subsequently to dimethylarsenic acid and excreted through urine. Biomethylation in mammals takes place in the liver by enzymic transfer of the methyl groups from *S*-adenosylmethionine.

The toxicity of As is due to its chemical similarity with phosphorus, which causes interference in P-containing metabolic pathways by replacing phosphorus in the phosphate groups of DNA or by reacting with thiol enzymes.

In strongly reducing environments, such as soil sediments, methanogenic bacteria reduce As(V) (arsenate) to As(III) (arsenite) and methylate it to give methylarsenic acid, CH$_3$AsO(OH)$_2$ or dimethylarsenic (cacodylic) acid, (CH$_3$)$_2$AsO(OH). These compounds may be further methylated to the very toxic and volatile trimethylarsine and dimethylarsine(III). In contrast, As(V) is stable in an aerobic environment.

An epidemiological study of workers chronically exposed to arsenic/arsenic compounds showed higher death rates than normal due to increased risk of skin and pulmonary cancer and cancer of the bowel. Acute injury invariably involves the blood, brain, heart, liver and kidneys. Symptoms include nervousness, intense thirst, vomiting, diarrhoea, cyanosis, leg cramps, paralysis and coma. Chronic and long-term exposure to arsine affects the bone marrow, skin and peripheral nervous system. Arsine (AsH$_3$), presents a clinical picture which is different from that of other arsenic compounds. Haematuria, abdominal pain and jaundice are the three characteristic symptoms.

Reported cases of arsenic poisoning include the following: Kentucky (USA) arsenic poisoning by ingestion of commercially available crab grass killer (MSMA and DMSA); ingestion of 8–9 g of As resulted in a patient developing asymmetric polyneuropathy, and treatment with BAL (British Anti Lewisite) gave partial recovery; Japanese workers suffering from chronic arsenic poisoning showed increased mortality for all types of cancer; two cases of sub-acute poisoning in glassworkers (arsenic trioxide is used in concentrations of $< 1\%$ in glass as decocolorizer); lung cancer in Ontario uranium miners (arsenic exposure that occurred earlier to miners in gold mines); epidemiology study on death rate in arsenic exposed workers showed an increase in pulmonary cancers and cancers of the bowel.

Occupational Exposure Limits. OSHA TLV(TWA): arsenic and inorganic arsenic compounds $5\mu g\, As\, m^{-3}$. UK long-term limit (MEL): $0.1\, mg\, As\, m^{-3}$

Further reading

Alain, G., Tousignant, J. and Rozenfarb, E. (1993) Chronic arsenic toxicity, *Int. J. Dermatol.* **32**, 899–901.

Chappell, W.R., Abernathy, C.O. and Cothern, C.F. (1994) *Arsenic. Exposure and Health*, Science and Technology Letters, Northwood, UK.

ECETOX (1991) *Technical Report 30(4)*, European Chemical Industry Technology and Toxicology Centre, Brussels.

Fishbein, L. (1984) Overview of analysis of carcinogenic and/or mutagenic metals in biological and environmental samples: arsenic, beryllium, cadmium, chromium and selenium, in Merian, E., Frei, R.W., Härdi, W. and Schlatter, Ch. (eds), *Carcinogenic and Mutagenic Metal Compounds*, Gordon and Breach, New York, pp. 55–112.

Fishbein, L. (1987) Perspective of carcinogenic and mutagenic metals in biological samples, *Int. J. Environ. Anal. Chem.* **28**, 21–69.

Fowler, B.A., Yamauchi, H. and Conner, E.A. (1993) Cancer risks for humans from exposure to the semiconductor metals, *Scand. J. Work Envir. Health*, **19**, 101–103.

IARC (1987) *Monograph*, Suppl. 7, p. 100.

Ishinishi, N., Tsuchiya, K., Vahter, M. and Fowler, B.A. (1986) Arsenic, in Friberg, L., Nordberg, G.F. and Vouk, V.B. (eds), *Handbook on the Toxicology of Metals*, 2nd edn, Elsevier, Amsterdam, Vol. II, pp. 43–83.

Nriagu, J. (ed.) (1994) *Arsenic in the Environment*, Wiley, New York.

ToxFAQS (1993) *Arsenic*. ATSDR, Atlanta, GA.

Arsenic(III) acid CAS No. 1327-52-2
(Arsenious acid)

H_3AsO_3

Uses. Preparation of arsenites.

Toxicity. Poison. Human carcinogen.

Arsenic(V) acid CAS No. 7778-39-4

H_3AsO_4

Uses. Preparation of arsenates and arsenical insecticides. Manufacture of special glass. Soil sterilent.

Toxicity. Confirmed human carcinogen. Poison by injection. Experimental teratogen. Human mutation data. LD_{50} oral mouse $57\, mg\, kg^{-1}$; LD_{50} intravenous rabbit $6\, mg\, kg^{-1}$.

Arsenic(III) chloride CAS No. 7784-34-1

$AsCl_3$

Uses. Preparation of organoarsenics and intermediates for arsenic pharmaceuticals and arsenic insecticides. Chemical agent.

Toxicity. LC_{Lo} (10 min) inhalation mouse 338 ppm.

Arsenic(III) hydride CAS No. 7784-42-1
(Arsine)

AsH_3

Uses. Organic syntheses. Semiconductors.

Occupational exposure limits. OSHA TLV (TWA) $10\,\mu g\,m^{-3}$

Toxicity. LD_{50} oral rat $5\,mg\,kg^{-1}$. LC_{Lo} (15 min) inhalation rat $300\,mg\,m^{-3}$. A level of 500 ppm is lethal after exposure to humans for a few minutes.

Arsenic(III) oxide CAS No. 1327-53-3

As_2O_3

Uses. Glass industry. Manufacture of Paris Green. Rodenticide and insecticide.

Toxicity. Poison. Suspect human carcinogen. LD_{50} oral rat $20\text{--}40\,mg\,kg^{-1}$. Fatal ingested dose for humans $70\text{--}180\,mg\,kg^{-1}$. Target organs are liver, kidneys, skin, lungs and lymphatic system.

Arsenic(V) oxide CAS No. 1303-28-2

As_2O_5

Uses. Preparation of arsenic compounds. Herbicide and fungicide. Manufacture of coloured glass. Dyeing and printing industries.

Occupational exposure limit. OSHA PEL (TWA) $0.01\,mg\,AS\,m^{-3}$

Toxicity. Very poisonous. LD_{50} oral rat $40\,mg\,kg^{-1}$. Select carcinogen. 130 mg of ingested arsenic is usually fatal to humans. 10 ppm of the oxide in water is an acute hazard. Symptoms; nausea, cramps, diarrhoea, liver damage, jaundice and kidney damage.

Arsenic(III) sulfide CAS No. 1303-33-9

As_2S_3

Uses. Specialist glass. Infra-red transmitting glass. Pyrotechnics. Electronics.

Toxicity. LD_{50} oral mouse $185\,mg\,kg^{-1}$.

Dimethylarsenic(V) acid CAS No. 75-60-5
(Dimethylarsinic or cacodylic acid)

$(CH_3)_2AsO(OH)$

Uses. In cotton fields to control weeds/crabgrass in crop and non-crop areas. Herbicide. Medicinal treatment of chronic eczema.

Toxicity. Suspect human carcinogen. LD_{50} oral rat $700\text{--}2000\,mg\,kg^{-1}$.

Phenylarsenic acid CAS No. 98-05-5
(Phenylarsonic acid)

$C_6H_5AsO(OH)_2$

Uses. Growth promoter in pig production.

Toxicity. LD_{50} oral rat $50\,mg\,kg^{-1}$; oral mouse $270\,\mu g\,kg^{-1}$.

ASBESTOS

Asbestos is used as a heat resistant solid to lag hot pipes, for fireproofing, etc. Asbestos is not a single mineral but a collective name that covers several fibrous silicate minerals that are natural such as *actinolite, amosite, anthophylite, chrysotile, crocidolite and tremolite*, as well as man-made minerals. Important asbestos minerals are contained in two mineral groups, amphiboles and serpentines; both groups are hydrated silicates with complex physical structures. *Chrysotile* or white asbestos $[Mg_3(Si_2O_5)(OH)_4]$ is the most important form of asbestos; it is the sole representative of the serpentine layer silicate group. The amphibole group includes the blue asbestos mineral *crocidolite* $[Na_2Fe_3^{II}Fe_2^{III}Si_8O_22(OH)_2]$ and the grey-brown mineral *amosite* $[(Mg, Fe)_7Si_8O_{22}(OH)_2]$.

Man-made asbestos is usually prepared by crystallization of silicates from fluorine-containing melts at 300–350° and very high pressures (up to 1000 atm). They all share specific properties but differ in chemical composition, morphology, durability and, therefore, biological effects. All are carcinogenic to humans. Asbestos fibres can induce neoplastic cell transformations and chromosome changes *in vitro*. Workers exposed occupationally to asbestos have an increased incidence of double-strand breaks in lymphocyte DNA. The human pulmonary system is affected by inhalation. Usually several years of exposure are required before serious lung damage (fibrosis). Airborne dusts, resulting from the mining and milling of asbestos and its production, present the greatest health risk. The inhalation of asbestos dust when deposited in the airways may lead to asbestosis, the severest form of silicatosis, closely associated with lung cancer. For example, the content of mineral fibres in the lungs of asbestos textile workers was reported in the range $1–10\,g\,kg^{-1}$ dry weight, as compared with levels of about $0.3\,g\,kg^{-1}$ dry weight in the general population. The risk of lung cancer is increased by cigarette smoking (additive effect).

Occupational exposure limit. OSHA PEL (TWA) 2×10^6 fibres m^{-3}. ACGIH TLV(TWA) 0.2 fibre cm^{-3} (crocidolite).

Analysis. Electron microscope.

Further reading
Control of Asbestos at Work Regulations 1987, HMSO, London, 1987.
ECETOX (1991) *Technical Report 30(4)*, European Chemical Industry Technology and Toxicology Centre, Brussels.
Environmental Health Criteria 53. Asbestos. WHO/IPCS, Geneva, 1986.
The Environmental Protection(Prescribed Processes and Substances) Regulations 1991, SI 1991 No. 472 HMSO, London, 1991.

ASTATINE CAS No. 7440-68-8

At

Astatine, a member of Group 17 (the halogens) of the Periodic Table, has an atomic number of 85, an atomic mass of ~ 210, a melting point of 302° and a boiling point of 337°. Probable oxidation states of $-1, +1, +3, +5$ and $+7$. Synthesized in 1940 by bombarding bismuth with α particles:

$$^{209}_{83}Bi + ^2_4H \rightarrow ^{211}_{85}At + 2^0_1n$$

Over 20 isotopes are known. The longest lived isotope is ^{210}At with a half-life of only 8.1 h. Minute quantities of ^{215}At, ^{218}At and ^{219}At exist in equilibrium in nature with naturally occurring uranium and thorium isotopes. The total amount of astatine in the earth's crust is probably less than 30 g. Because of this very small concentration, astatine is not a health hazard. Like iodine, the element above in the same Group, it would be probably stored in the thyroid gland of humans.

Further reading
Brown, I. (1987) Astatine: organonuclear chemistry and biomedical applications, *Adv. Inorg. Chem.* **31**, 43–88.
Ullmann's Encyclopedia of Industrial Chemistry, Radionuclides, VCH, Weinheim, 1990, Vol. A15, pp. 5–6; 1993, Vol. A22, p. 524.

ATRAZINE CAS No. 1912-24-9
(Atratol, Gesaprim)
Pre-emergent herbicide. Plant growth regulator. Toxicity class IV, slightly toxic, LD_{50} > 500 $mg\,kg^{-1}$; LD_{50} rat, mouse, rabbit 750–3080 mg kg^{-1}; LC_{50} fish 4.3–8.8 mg l^{-1}

$C_8H_{14}ClN_5$ $M = 215.7$

2-Chloro-4-ethylamino-6-isopropylamino-1,3,5-triazine; water solubility 33 mg l^{-1} at 25°, $t_{1/2}$ in soil 60 days.

Introduced in 1958 by Ciba-Geigy and applied in the range 2–4 lb acre^{-1} principally to corn and sorghum row crops but also to macadamia nuts and pineapples, leading to an estimated global release of 90×10^3 t. Annual usage in the 14 midwestern states of the USA was 21×10^6 kg, consequently, after application in the spring, levels in groundwater peak in June and may rise to 10 times the EPA limit ($3 \mu g\,l^{-1}$), falling by autumn to base levels. It is estimated that in 1989 the river Mississippi transported 429 t to the Gulf of Mexico; a typical level in the river is 400 ng l^{-1}. Eighteen Swiss lakes contained *s*-triazine herbicides and atrazine was the major component occurring in the range 1–460 ng l^{-1}.

Atrazine is not noticeably toxic to fish or invertebrates but it inhibits photosynthesis and so becomes toxic to phytoplankton. Also, by absorbing UV light, it is rapidly degraded to desisopropyl- and desethylatrazine, whose greater solubility in water exacerbates the pollution of ground-water. Levels in chalk aquifers in the UK lie in the range 20–400 ng l^{-1}; hydroxyatrazine, with enhanced water solubility, is the main degradation product environmentally. The EC directive requires a level of 0.1 $\mu g\,l^{-1}$, whilst the WHO recommends 2 $\mu g\,l^{-1}$. Encapsulation in a starch matrix has been shown to reduce triazine run-off and this offers one practical method of control. The UK places some restrictions on its use, e.g. on railways. In the USA the threshold limit value is a TLV (TWA) of 5 mg m^{-3}.

AZIDES CAS No. 14343-69-2

N_3^-

Heavy metal azides such as lead azide, $Pb(N_3)_2$, are used as detonators

Further reading
Lee, J.D. (1991) *Concise Inorganic Chemistry*, 4th edn, Chapman & Hall, London, pp. 487–489.

AURAMINE CAS No. 492-80-8
LD_{50} (i.p.) mouse $103\,mg\,kg^{-1}$

$$Me_2N-\bigcirc-\underset{\underset{H}{\overset{\|}{N}}}{C}-\bigcirc-NMe_2 \qquad C_{17}H_{21}N_3 \quad M = 267.4$$

Bis(*p*-dimethylaminophenyl)methyleneimine, 4-dimethylaminobenzophenone imide, Auramine O (the hydrochloride), CI Solvent Yellow; golden yellow, m.p. 136°. Sparingly soluble in water; soluble in DMF, acetone, EtOH.

Auramine is a non-light-fast dye for solvents, paper, inks, silk, cotton, wool and leather. It is prepared by the condensation of two molecules of *N,N*-dimethylamine with formaldehyde to give the diphenylmethane:

$$2\ \bigcirc + \ H-CHO \longrightarrow \underset{Ar}{\overset{Ar}{\underset{|}{\overset{|}{CH_2}}}}$$

C.I. Basic Yellow 2

This is then heated with sulphur and ammonium chloride in an atmosphere of ammonia, whereby the methylene group is first oxidized to $C=S$ and then converted into the imine $C=NH$ by reaction with ammonia. An alternative synthesis is the reaction of the dimethylamine with phosgene to give the parent Michler's ketone, which is then reacted with $NH_4Cl/ZnCl_2$.

Absorption of auramine through the skin can cause burns, dermatitis, nausea and vomiting. Maintained at pH 5.0 it is hydrolysed to Michler's ketone in a reaction with a half-life of 65 days.

The manufacture of auramine is controlled by the Carcinogenic Substances Regulations (1967) and the process itself has been classed as group 1 by the IARC as proven to cause cancer in humans. The technical grade product is classed in group 2B, proven in animals but with inadequate evidence in humans.

AZIRIDINE CAS No. 151-56-4
LD_{50} rat $15\,mg\,kg^{-1}$; LC_{50} inhalation, rat, rabbit, mouse $100–400\,mg\,kg^{-1}$

$$\underset{\underset{H}{\overset{}{N}}}{\triangle} \qquad C_2H_5N \quad M = 43$$

Ethyleneimine, aminoethylene; colourless liquid with odour of ammonia; b.p. 57°; v.p. 160 Torr (20°); v. dens. 1.48; dens. 0.83. Miscible with water, soluble in EtOH; highly flammable.

Aziridine is used to make polyethyleneimine – a flocculant for water treatment and for cross-linking in coatings and plastics. It is also used to treat textiles to improve their drying and as an anti-static agent.

It is manufactured from (a) ethanolamine or (b) dichloroethane:

$$\underset{\underset{CH_2NH_2}{|}}{CH_2OH} \quad \xrightarrow[2)\,NaOH]{1)\,H_2SO_4} \quad \cdots \longrightarrow \underset{CH_2}{\overset{CH_2}{\diagdown}} NH + \text{sulphate}$$

There is an explosion hazard if aziridine comes into contact with copper or its alloys with silver.

Aziridine is mutagenic in *Salmonella typhimurium* and is classed by the IARC as a group 3 carcinogen – limited evidence of activity in animals.

The ACGIH has set a TLV (TWA) of 0.5 ppm ($0.88\,\text{mg m}^{-3}$). In the UK the same level was set in 1991 but is under review.

Further reading
Chemical Safety Data Sheets (1991) **4a**, Toxic Chemicals, Royal Society of Chemistry, Cambridge.

AZOBENZENE CAS No. 103-33-3
(see also **Azo Dyestuffs**). LD_{50} rat $1.0\,\text{g kg}^{-1}$

$C_{12}H_{10}N_2 \quad M = 182.2$

Benzeneazobenzene, diazobenzene; orange crystals; m.p. 68°; b.p. 299°; v.p. 1 Torr (103°); dens. 1.203 (20°). Insoluble in water, soluble in ethanol, acetic acid.

First obtained by Mitscherlich in 1934 by the action of alcoholic KOH on nitrobenzene – more convenient procedures are the reduction of nitrobenzene by tin or iron in alkaline solution or by electrolysis.

Azobenzene is used as an acaricide and as an intermediate for azo dyestuffs.

The Health and Safety Comission has classed it as harmful by skin contact, by inhalation and if swallowed. Intraperitoneal injection into rats leads to excretion of aniline in the urine.This metabolism may be the cause of methaemoglobinaemia, which is induced by azobenzene and some other azo compounds (see **Aniline**). The IARC classified azobenzene as a group 3 substance – limited evidence of activity in animals, but its use as a pesticide was restricted in the UK in 1975 on the grounds of its carcinogenicity.

AZO DYESTUFFS
These are obtained by coupling a diazonium salt with a wide range of aromatic nuclei activated by electron-donating substituents such as $-NR_2$ or $-OH$:

4-(N,N-diethylamino)-azobenzene

The product in this example is intensely yellow – the electron-donating –NEt_2 group shifts the absorption maximum from that of azobenzene (320 nm, $\epsilon = 21\,000$) to 415 nm ($\epsilon = 29\,500$). A range of colours may be obtained by varying the aromatic nuclei, for example by including variously substituted naphthalenes and quinolines – over 50% of the World's dyestuffs fall within this class. Important colours may also be obtained from doubly diazotized precursors prepared from benzidine, including dimethyl-, dimethoxy- and dichlorobenzidines.

Hazards to dyestuffs workers arise especially from their handling of aromatic amine precursors, since many of these are carcinogenic (see **Naphthylamines, Benzidine**). In addition, toxic amines can be released in the body by the reduction of azo compounds by intestinal bacteria or by reductase in the liver. All azo dyes which can liberate carcinogenic arylamines in this way are themselves classified in the same group (DFG, 1995). Butter Yellow, which was formerly used to colour waxes and polishes, has been banned by NIOSH as a group 2B carcinogen (Sax, 1995) – its reduction *in vivo* gives aniline and N,N-dimethylamino-1,4-phenylenediamine.

Butter Yellow

References
DFG (1995) *List of MAK and BAT Values*, VCH, New York, pp. 97, 102.
Sax (1995) in Lewis, R.J. (ed.), *Dangerous Properties of Industrial Materials*, 9th edn, Van Nostrand Reinhold, New York, p. 1278.

Further reading
Cheremisinoff, N.P., King, J.A. and Boyko, R. (1994) *Dangerous Properties of Industrial and Consumer Products*, Marcel Dekker, New York.

BARIUM CAS Registry No. 7440-39-3

Ba

Barium, a member of Group 2 of the Periodic Table, has an atomic number of 56, an atomic weight of 137.33, a specific gravity of 3.50 at 20°, a melting point of 725° and a boiling point of 1897°. It is a ductile, silver-white metal and forms compounds in the +2 stable oxidation state. Natural isotopes are ^{130}Ba (0.11%), ^{132}Ba (0.1%), ^{134}Ba (2.40%), ^{135}Ba (6.59%), ^{136}Ba (7.81%), ^{137}Ba (11.23%) and ^{138}Ba (71.70%).

Barium is 18th in abundance in the earth's crust with a concentration of ~ 390 ppm. The chief ores of barium are *barite*, $BaSO_4$, and *witherite*, $BaCO_3$. A major source of barium compounds in the environment is effluents from the chemical/

pharmaceutical, petrochemical and paint industries. In large industrial cities, raw sludge in the primary settling tanks of biological waste-water treatment plants can contain as much as ~ 230 mg Ba kg^{-1}. Typical concentrations in the environment are sea water 21 µg l^{-1}, river water 20 µg l^{-1}. Average concentrations in various organisms (mg kg^{-1}, dry weight) are marine algae 3.0, terrestrial plants 1.4, marine animals 0.02–0.3 and terrestrial animals 0.075. Total body weight (70 kg, adult) 20 mg.

Barium and its compounds are used in the ceramics, glass, electronics, paint and paper industries and in X-ray investigations. The isotopes barium-133 and barium-137m are used in gamma spectroscopy.

Water-soluble salts of barium such as Cl$^-$, NO$_3^-$ and S^{2-} are highly poisonous. Water and stomach acids solubilize barium salts such as the sulphate and can also cause poisoning. Symptoms are stimulation of all forms of muscle, vomiting, colic, diarrhoea, convulsions, tremors and muscular paralysis. In severe cases death can occur within a period of a few hours to days. In general, the Ba^{2+} ion is toxic or inhibitory to cellular processes in bacteria, fungi, mosses and lichens. For aquatic organisms BaCl$_2$ is lethal at 150 mg l^{-1} (72 h). For humans, BaCl$_2$ is toxic in doses of 0.2–0.9 g when ingested, with a lethal dose of 0.8–0.9 g. In acute oral poisoning by barium compounds the myocardia, nervous system and vessels suffer most, giving rise to a variety of clinical symptoms, nausea, vomiting, colic, blood pressure elevation, vision and speech disorders, and paralysis of the trunk and lower limbs may develop. In severe cases, death occurred up to 24 h after oral intake, while chronic poisoning may affect the bones, bone marrow, gonads and liver. The action of barium on the myocardia is similar to that of digitalis. It blocks a group of enzyme systems and causes degeneration changes, thereby impairing cardiac conductivity and rhythm. It can also produce a cholinolytic effect and induce hypokalemia giving rise to high concentrations in human organs (0.004–0.03% , dry ash weight). Barium is able to cross the placental and blood–brain barriers, and there is some evidence it may be teratogenic and mutagenic. Barium can displace phosphorus and calcium from bones, leading to osteoporosis. The half-life for barium in bones is ~ 50 days. Its chemical and physiological properties enable it to compete with and replace calcium in processes mediated by the latter. Excess of barium in soil, water and animal foods may result in impaired calcium metabolism and severe diseases of the peripheral joints and spine. Epidemiological studies indicate that in communities having high barium levels in drinking water (2–10 mg l^{-1}), death rates from all cardiovascular diseases were higher than in those consuming water with low barium levels (< 0.2 mg l^{-1}). Barium has not been shown to be an essential trace element but it is found in many tissues of a healthy body (human blood concentration 0.006–009 mg%). An adult man of 70 kg weight contains about 22 mg of barium with $> 90\%$ of this amount in the bones. Daily intake is 0.6–1.7 mg day^{-1}. Elimination of barium occurs principally in the faeces.

There are now several cases of occupational hazards from working with barium compounds: (i) exposure to soluble barium compounds occurred in workers exposed to welding fumes during arc welding and that contained 20–40% of such compounds, and urine sample concentrations reached 31–234 µg l^{-1} after 3 h of exposure compared with 1.8–4.7 µg l^{-1} in control samples; (ii) lethal poisoning from mixtures of barium carbonate/barium sulphate used in radiographic examinations; (iii) accidental ingestion of barium sulphide led to cardiac arrest; (iv) a patient suffered acute renal failure associated with barium chloride poisoning; (v) barium cyanide is used in metallurgy and electroplating processes, and workers used as electroplaters and picklers can develop cyanide rash; (vi) baritosis (a benign pneumoconiosis) was reported among workers exposed to finely ground barium sulphate in the USA, Germany and Czechoslovakia; (vii) exposure to barium sulphate dust (17–313 mg m^{-3}) [UK TLV (TWA) 10 mg m^{-3}] was severe enough that workers developed laryngitis, bronchitis, emphysema and breathing abnormalities; (viii) it was found that over 90% of the barium additives in diesel fuels is emitted in vehicle exhaust as barium sulphate; (ix) some of the workers engaged in the crushing, grinding, mixing or synthesis of materials based on barium titanates and exposed to their dusts with mean concentrations of 8.4 mg m^{-3} had functional changes in the CNS and/or neuralgia in the upper extremities; (x) in one reported case of where more than 100 people had consumed sausages made with barium carbonate instead of potassium carbonate meal, 19 people were hospitalized with symptoms ranging from mild vomiting and diarrhoea to partial paralysis, and one patient died.

Occupational exposure limits. US TLV (TWA) 0.5 mg m^{-3}. Barium powder is flammable.

Further reading
Environmental Health Criteria 107, Barium, WHO/IPCS, Geneva.
Kirk–Othmer ECT, 4th edn, Wiley, New York, 1992, Vol. 3, pp. 902–931.
ToxFAQS (1995) *Barium*. ATSDR, Atlanta, GA.

Barium(II) acetate CAS No. 543-80-6

$Ba(O_2CCH_3)_2$

Uses. Paint and varnish drier. Catalyst in organic chemistry.

Toxicity. Poison; target organs are eyes, skin, central nervous system and heart. LD$_{50}$ oral rat 240 mg kg^{-1}. LD$_{50}$ intravenous mouse 23 mg kg^{-1}.

Barium(II) bromate hydrate CAS No. 13967-90-3

$Ba(BrO_3)_2.H_2O$

Uses. Analytical agent. Oxidizing agent. Corrosion inhibitor in steel.

Toxicity. Toxic. Strongly oxidizing.

Barium(II) carbonate CAS No. 513-77-9

$BaCO_3$

Uses. Ceramics. Optical glass. Chemical industry. Brick and clay products. Manufacture of ferrites.

Toxicity. Poison. LD$_{50}$ oral mouse 200 mg kg^{-1}. LD$_{Lo}$ oral human 57 mg kg^{-1}.

Barium(II) chloride dihydrate CAS No. 10326-27-9

$BaCl_2.2H_2O$

Uses. Ceramics. Pest control. Textile industries. Manufacture of glass and white leather. Removal of sulphate ions in some electrolytic plant productions of chemicals such as chlorine, sodium hydroxide, magnesium metal.

Toxicity.   Poison. Irritant to eyes and skin. LD_{50} oral rat $188\,mg\,kg^{-1}$. LD ingestion human $80–100\,mg\,kg^{-1}$.

Barium(II) chromate CAS No. 10294-40-3

$BaCrO_4$

Uses.   Pigment

Toxicity.   Poison. Human carcinogen

Barium(II) cyanide CAS No. 542-62-1

$Ba(CN)_2$

Uses.   Metallurgy. Electroplating processes

Toxicity.   Hazardous compound and poison.

Barium(II) fluoride CAS No. 7787-32-8

BaF_2

Uses.   Insecticide

Toxicity.   Toxic by inhalation, or if swallowed. Target organs are heart and CNS. LD_{50} oral rat $250\,mg\,kg^{-1}$.

Barium(II) hydroxide octahydrate CAS No. 12230-71-6

$Ba(OH)_2.8H_2O$

Uses.   Lubricating greases. Manufacture of glass and barium salts. Ba–Cd stabilizers for PVC formulations.

Toxicity.   Poison. LD_{50} intravenous chicken $613\,mg\,kg^{-1}$.

Barium(II) nitrate CAS No. 10022-31-8

$Ba(NO_3)_2$

Uses.   Pyrotechnics and incendaries (green colour). Ammunition. Ceramic glazes. Rodenticide (former use).

Toxicity.   Poison. LD_{50} oral rat $355\,mg\,kg^{-1}$. LD_{Lo} oral rabbit $150\,mg\,kg^{-1}$.

Barium(II) oxide CAS No. 1304-28-5

BaO

Uses.   Drying solvents and gases. Manufacture of lubricating oil detergents.

Toxicity.   Poison. LD_{50} subcutaneous mouse $50\,mg\,m^{-3}$

Barium polysulphide CAS No. 50864-67-0

Ba_4S_7

Uses.   Insecticide and fungicide.

Toxicity. LD_{50} oral rat $375\,mg\,kg^{-1}$. TD_{Lo} oral human $226\,mg\,kg^{-1}$.

Barium(II) sulphate CAS No. 7727-43-7

$BaSO_4$

Uses. Filler in paints, rubber, paper and textiles, in ceramics and as a radiopaque medium. Petroleum drilling fluid, added to increase density of mud.

Toxicity. LC_{50} (96 h) salmon $78\,mg\,l^{-1}$.

Barium sulphide CAS No. 21109-95-5

BaS

Uses. Manufacture of lithopone and cadmium lithopones (pigments). Dehairing hides. Flame retardant. Luminous paints.

Toxicity. N.a. (low except in acidic media which produce H_2S)

Barium(II) titanate CAS No. 12047-27-7

$BaTiO_3$

Uses. The ferroelectric and piezoelectric behaviour arising from the perovskite structure gives rise to important applications in the production of compact capacitors, and as a ceramic transducer in devices such as microphones and gramophone pickups.

Toxicity. N.a.

BENOMYL CAS No. 17804-35-2
(Benlate, BBC)
Systemic fungicide, acaricide. Toxicity class III, slightly toxic, $LD_{50} > 500\,mg\,kg^{-1}$; LD_{50} oral rat $10\,g\,kg^{-1}$; TD_{Lo} oral rat $3500\,\mu g\,kg^{-1}$; LC_{50} rainbow trout $170\,\mu g\,l^{-1}$

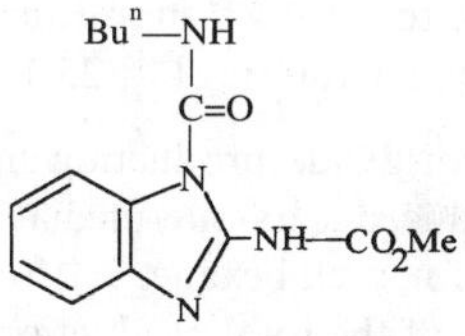

$C_{14}H_{18}N_4O_3 \quad M = 290.6$

Methyl 1-(butylcarbamoyl)benzimidazol-2-ylcarbamate. Water solubility $2\,mg\,l^{-1}$ (25°).

Used on fruit,vines,vegetables, soybeans and wheat; the maximum daily intake is 0.02 ppm. It is immobile in soils but is toxic to earthworms and has a half-life in soil of 6–12 months; the metabolites in plants, soil and water include 2-benzimidazole carbamate and 2-aminobenzimidazole.

A law suit has been brought against the manufacturer Du Pont claiming that exposure of a mother to Benlate during pregnancy led to anopthalmia(eyeless birth) in her baby. Benlate has caused anopthalmia in rats and at high doses it is also a possible human carcinogen.

The maximum residue limits in food are as for carbendazim. The ACGIH threshold limit value (TWA) is $10\,mg\,m^{-3}$ (0.84 ppm, 25°/760 mmHg).

BENZAL CHLORIDE CAS No. 98-87-3
LD_{50} rat $3.2\,g\,kg^{-1}$

$C_7H_6Cl_2$ $M = 161.0$

Benzylidene chloride, α,α-dichlorotoluene; b.p. 214°; v.p. $0.13\,kPa$ (35°). Miscible with most organic solvents.

A lachrymatory liquid prepared together with benzotrichloride by the chlorination of boiling toluene, also by the reaction of benzaldehyde with PCl_5 or $SOCl_2$. The reaction between toluene (1 mol) and chlorine (2.18 mol) gave the following product balance: benzyl chloride 0.016 mol, benzal chloride 0.77 mol and benzotrichloride 0.21 mol. It is hydrolysed by aqueous acids and alkalis to benzaldehyde, a reaction which is used for commercial production as an alternative to the direct chlorination of toluene. The production in the EC in 1979 was estimated at $5 \times 10^6\,kg$.

Benzal chloride vapour is a powerful irritant. It has been given a MAK rating as a carcinogen of group III A2, clearly carcinogenic in animals under conditions indicative of potential in the workplace.

Further reading
Kirk–Othmer ECT, 4th edn, Wiley, New York, 1993, Vol. 3, pp. 113–122

BENZENE CAS No 71-43-2
LD_{50} oral rat $3\,g\,kg^{-1}$. Human carcinogen

C_6H_6 $M = 78.1$

Annulene, benzine, benzole; b.p. 80.1°; m.p. 5.5°; v.p. $13\,kPa$ (25°); flash point -11°. Detectable by GC–MS down to $0.01\,\mu g\,m^{-3}$ (30 ppt) in air and to $1\,ng\,kg^{-1}$ in soil and water. Miscible with common organic solvents, solubility in water $1800\,mg\,l^{-1}$ (25°).

Occurs naturally in crude petroleum up to $4\,g\,l^{-1}$ and worldwide production in 1983 was $15 \times 10^6\,t$. In Western Europe 50% of benzene produced is an intermediate in styrene production, 20% is used for cumene/phenol, 13% for cyclohexane, 7% for nitrobenzene/aniline and 2% for chlorobenzenes. About 10% of the total production is used in motor fuel; benzene is present at 1% by weight in the USA and 2.5–3.0% v/v in Europe. Releases to air arise from petroleum processing, from use as an intermediate, e.g. the manufacture of toluene, and from vehicle exhausts. An inventory taken in California indicated that 70% of the benzene released to the environment came from this source – in 1982 the EC estimate for exhausted benzene was as high as $171 \times 10^3\,t$ or 86% of the total.

Levels of up to 3 ppm ($10\,mg\,m^{-3}$) have been recorded at filling stations and may reach 160 ppb ($500\,\mu g\,m^{-3}$) in residences, although the median is 4.2 ppm compared with a median outdoors of 2.7 ppm. Tobacco smoking contributes to pollution indoors – active smokers (30 cigarettes) produce $1800\,\mu g\,day^{-1}$. Occupational exposures may rise to a TWA of $15\,mg\,m^{-3}$. Rural outdoor levels are about $0.2\,\mu g\,m^{-3}$ but may rise to $349\,\mu g\,m^{-3}$ in heavily trafficked areas; 155 ppb was recorded recently on the M1 motorway near Luton, UK.

The detection limit for benzene in these studies was 30 ppb ($0.09\ \mu g\,m^{-3}$) and the residence time in air varies with climate, ranging from a few hours to a few days depending on the concentrations of ozone and hydroxyl radicals. In an urban area when the concentration of hydroxyl radicals was 10^8 molecules cm^{-3} the half-life of benzene was 3–10 days. NO_x participated in the system and the products included phenol, nitrophenol and nitrobenzenes. Benzene absorbs ultraviolet light at 256 nm and is photolysed in sunlight.

Human exposure was once high from use of benzene as a solvent in paint stripping, levels in a paintworks reaching 20 ppm; it was also used in cleaning, rubber cements and artists materials. With the realization of its toxicity, its use as a solvent now accounts for < 2% of production, other solvents such as toluene taking its place.

Exposure at levels of $80\ mg\,m^{-3}$ (25 ppm) is without clinical effect, 160–$480\ mg\,m^{-3}$ for 5 h causes headache. Illness follows exposure for 60 min at $1600\ mg\,m^{-3}$; such acute toxicity leads to CNS depression and eventually to respiratory failure. The single lethal dose in humans is estimated at 10 ml (8.8 g).

Benzene is readily absorbed by animals and humans through the skin and it can cross the placental barrier. After a death from 'glue sniffing' it was found in blood ($20\ mg\,l^{-1}$), the brain ($390\ mg\,kg^{-1}$), the liver ($16\ mg\,kg^{-1}$) and abdominal fat ($22\ mg\,kg^{-1}$). Metabolism occurs primarily in the liver through the cytochrome P-450 enzymes by a complex pathway initiated as follows:

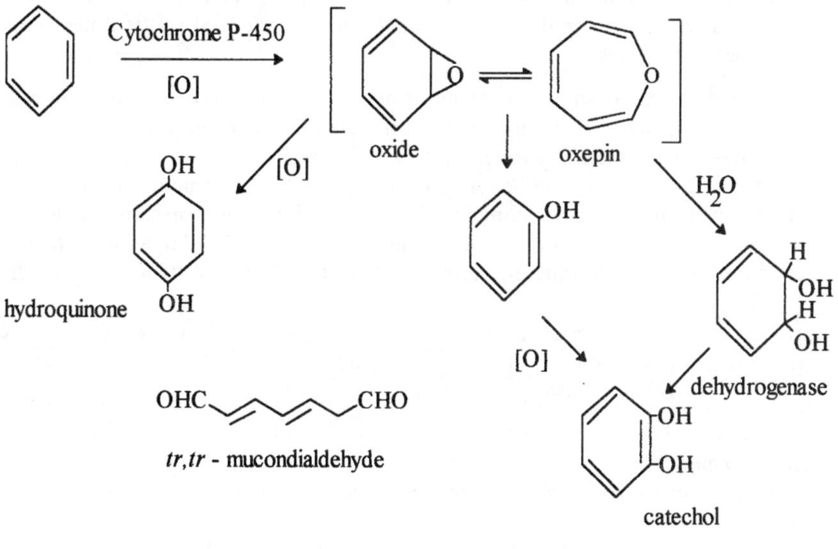

Further products include conjugates formed with sulphuric acid and GSH, which are excreted in the urine; *trans,trans*-mucondialdehyde and muconic acid are also formed by ring cleavage. MAK give a correlation between $6.5\ mg\,m^{-3}$ in workplace air, a blood level of $14\ \mu g\,l^{-1}$ and excreted *trans,trans*-muconic acid at $3\ mg\,l^{-1}$.

The carcinogenic effects are well established in humans through a number of case studies based on exposure in chemical plants, coke ovens and those handling adhesives (shoe workers). In a cohort study from China, which included 28 460 workers in 233 factories, 30 cases of leukaemia were found compared with four cases in a similar cohort from textile and machine production with no exposure to benzene (Yin *et al.*, 1989).

Benzene is rated by the ACGIH as an A2 carcinogen with a TLV (TWA) of 10 ppm $(33\,mg\,m^{-3})$.

Reference
Yin, S.-N. *et al.* (1989) *Environ. Health Perspect.*, **82**, 207–213.

Further reading
Berlin, A. Draper, M., Krug, E. and Van der Venne, M. Th. (eds) (1989) *The Toxicology of Chemicals*, Commission of the EC, Brussels.
Nielsen, I.R., Rea, J.D. and Howe, P.D. (eds) (1991) *Environmental Hazard Assessment: Benzene*, DOE, Watford.

BENZENETHIOL CAS No. 109-98-5
LD_{50} rat $46\,mg\,kg^{-1}$

                                    C_6H_6S $M = 110.2$

Thiophenol, phenyl mercaptan; b.p. 169.5°; v.p. 1 Torr (19°); dens 1.973; odour threshold 0.2 ppb. Solubility in water $470\,mg\,l^{-1}$ (15°); soluble in most organic solvents. Has a repulsive putrid odour which may be removed by treatment with aqueous permanganates.

The first stage oxidation product in air or oxygen is the disulphide Ph-S-S-Ph; with stronger agents, such as permanganates, nitric acid or peroxide, benzenesulphonic acid, $PhSO_3H$, is formed. Treatment with concentrated sulphuric acid gives thianthrene. Originally made from phenol by Beckmann in 1878 by reaction with P_2S_5. A more recent route is the treatment of phenol with a thiocarbamyl chloride, $R_2NC(S)Cl$. Thiophenol is more acidic than phenol with pK_a 7.76 in 48% MeOH–H_2O. It forms a range of salts, insoluble in water, with heavy metal ions of Pb, Hg, Zn and Cu, etc.

It is poisonous by ingestion, inhalation and skin contact. Thiophenol is an eye irritant and exposure causes headache, dizziness and vomiting. It is a dermatitic agent listed by the USEPA as 'extremely hazardous'.

The ACGIH has set the TLV (TWA) at 0.5 ppm.

Further reading
Croner's Substances Hazardous to Health, Croner Publications, Kingston, Surrey, 1996.

BENZIDINE CAS No. 92-87-5
LD_{50} rat $309\,mg\,kg^{-1}$; LD_{Lo} dog $200\,mg\,kg^{-1}$. Inhalation man TC_{Lo} $17.6\,mg\,m^{-3}$

$H_2N-\langle\ \rangle-\langle\ \rangle-NH_2$ $C_{12}H_{12}N_2$ $M = 184.2$

(1,1'-Biphenyl)-4,4-diamine, 4,4'-diaminobiphenyl; A crystalline solid, often discoloured brown, with isotropic forms m.p. 116° and 129°; b.p. 402°; dens. 1.25;

$\log P_{ow}$ 2.007. Solubility in water $400 \, \text{mg} \, \text{l}^{-1}$ ($12°$), $9.4 \, \text{g} \, \text{l}^{-1}$ ($100°$); soluble in diethyl ether, dimethyl sulphoxide, slightly soluble in ethanol. The dihydrochloride is soluble in water and ethanol.

Benzidines are formed as the principal products of the acid-catalysed rearrangement of hydrazobenzenes, but four others are known (Coffey, 1973); their proportions depend upon the conditions and substituents on the starting material. When hydrazobenzene is itself treated in aqueous hydrogen chloride the product is 92% benzidine, as the dihydrochloride, and 8% 2,4'-diaminobiphenyl:

Benzidine is used as a precursor for tetraazo dyes and in laboratories for the detection of blood and of hydrogen peroxide in milk, and also as a spray reagent for sugars.

The first benzidene diazo dye was Congo Red, produced in 1884 by coupling tetraazobenzidine with two molecules of naphthionic acid:

Congo Red is still used in large amounts for the dyeing of cotton and wool in India. Subsequently over 250 related dyestuffs have been synthesized.

Any degree of exposure is considered dangerous; benzidine is poisonous by inhalation and through the skin. Ingestion leads to nausea, vomiting and damage to the liver and kidneys; it can cause damage to the blood–haemolysis and depression of the bone marrow. In 1989 a report described the medical records of 2926 leather tannery workers from Tuscany who had been exposed to benzidine during 1950–83. They experienced significant increased mortality from cancers of the lung, kidney and pancreas and leukaemia. An increase in bladder cancer was also established for workers in the USSR.

Investigations in China showed that Congo Red (CAS No. 573-58-0) and 11 other benzidine dyes were mutagenic and required regulation. Wastewater from a plant using 3,3-dichlorobenzidine dyes was discharging benzidines at levels of $12 \, \text{ng} \, \text{l}^{-1}$ when GC with EC detection was sensitive to $3 \, \text{ng} \, \text{ml}^{-1}$ (Whang et al., 1988). A study of dyestuffs workers in the USA and Japan revealed a significant increase in bladder cancer compared with that in a control group.

Benzidine is no longer manufactured in the USA or Japan, but production continues in France, Brazil, India, China and Russia. Owing to the toxic risk, production may be conducted without isolation of intermediate stages. The production of the direct colours continues; Direct Black 38 (CAS No. 1937-37-7) is an example:

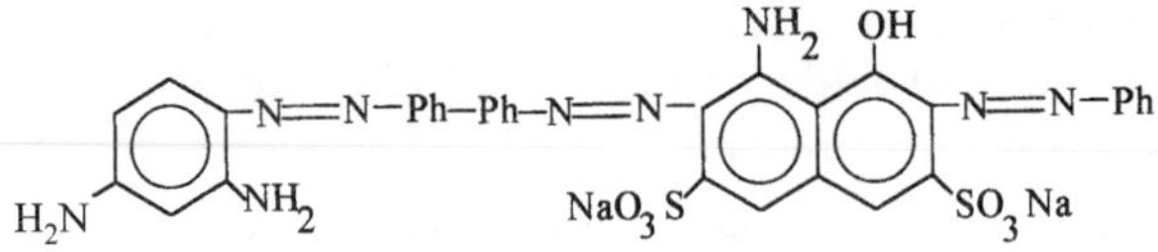

152 t of Direct Blue 80, worth $2200 t^{-1} and 10 t of Direct Blue 98, worth $1800 t^{-1}, were produced under control in the USA in 1988 (Kirk–Othmer, 1993).

The ACGIH and MAK rate benzidine as a group 1 carcinogen, proven in man. The use of benzidine is prohibited in the UK.

References
Kirk–Othmer ECT, 4th edn, Wiley, New York, 1993, Vol. 3, p. 832.
Coffey, S. (ed.) (1973) *Rodd's Chemistry of Carbon Compounds*, 2nd edn, Elsevier, Amsterdam, Vol. IIIC, pp. 195, 228.
Whang, C.W. and Yang, L.L. (1988) *J. Chin. Chem. Soc. (Taipei)*, **35**, 109–118.

Further reading
IARC (1982) *Some Industrial Chemicals and Dyestuffs*, IARC, Lyon, Vol. 29, pp 149–184.

BENZOIC ACID CAS No. 65-85-0
LD_{50} rat 1700 mg kg^{-1}; LD_{Lo} oral human 500 mg kg^{-1}

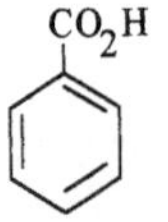

$C_7H_6O_2$ $M = 122.1$

Phenylformic acid, benzenecarboxylic acid; m.p. 122°; b.p. 249°; dens 1.27; v.p. 1 Torr (96°, sublimes); log P_{ow} 1.87. Soluble in water (0.29 g/100 g at 20°, 6.8 g/100 g at 95°), soluble in ethanol (58 g/100 g at 20°), diethyl ether, benzene, CCl$_4$ (94 g/100 g at 20°).

Benzoic acid is obtained industrially by the cobalt catalysed oxidation of liquid toluene and it is the product of autoxidation of benzaldehyde. A commercial route to phenol is the decomposition of benzoic acid in air/steam in the presence of copper and magnesium benzoates:

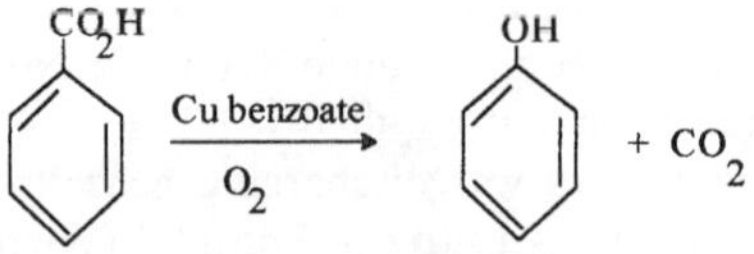

Benzoic acid is added to alkyd resin coatings to improve their properties and it is an intermediate in the synthesis of caprolactam and in the production of terephthalic acid. The benzoyl ester of ethylene glycol is used as a plasticiser for PVC flooring and is an alternative to phthalate esters. Sodium benzoate is water soluble (61 g/100 g), as is potassium benzoate (73 g/100 g at 20°). They are inactive but in solutions with pH below 4.5 they are largely dissociated to the free acid, which acts as an antimicrobial agent, e.g. in soft drinks.

Benzoic acid is removed from wastewater by activated sludge within 24 h; it is also removed ($t_{1/2}$ = 2 days) from air by reaction with OH radicals. It is a severe eye irritant and is moderately toxic by ingestion. A value for occupational exposure is pending (MAK).

Further reading
Kirk–Othmer ECT, 4th edn, 1992, Wiley, New York, Vol. **4**, p. 103–115.

BENZOYL CHLORIDE CAS No. 98-88-4
LD_{50} rat 2460 mg kg^{-1}; TC_{Lo} human 1 min 2 ppm; LC_{50} fathead minnow (96 h) 35 mg l^{-1}

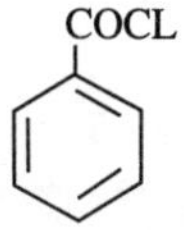

C_7H_5ClO $M = 140.6$

Benzene carbonyl chloride; m.p. − 1.0°, b.p. 197°, dens.1.207, v.p. 1 Torr (32°); v. dens. 4.9. 1 ppm = 0.174 mg m^{-3}. Hydrolysed by water, soluble in benzene, diethyl ether, CS_2.

Benzoyl chloride was first prepared by the chlorination of benzaldehyde (1832) but the most used method is the reaction between benzoic acid and benzotrichloride:

$$PhCO_2H + PhCCl_3 \longrightarrow 2PhCOCl + HCl$$

USA production in 1982 was 18×10^6 kg and 80% of this was treated with sodium peroxide to produce benzoyl peroxide, an initiator for vinyl and styrene-based polymers and an additive for acrylic dentures. Smaller amounts of benzoyl chloride are used in the pesticide and dyestuffs industry and for the synthesis of benzophenone.

Benzoyl chloride is strongly lachrymatory and burns the skin. Respiratory cancer was reported amongst Japanese workers in contact with it during 1952–74, but other compounds such as toluene and α-chlorotoluenes could have been responsible.

The IARC classification is in group 3, but in Germany it is classified as a possible human carcinogen in a mixture with α-chlorotoluenes (MAK).

Further reading
IARC (1982) *Some Industrial Chemicals and Dyestuffs*, IARC, Lyons, Vol. 29, pp. 83–92.

BENZYL ALCOHOL CAS No. 100-51-6
LD_{50} rat 1230 mg kg^{-1}; LC_{50} rat inhalation (2 h) 200 ppm; LC_{50} fathead minnow (96 h) 460 mg l^{-1}

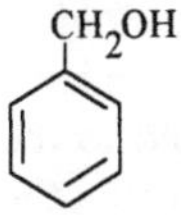

C_7H_8O $M = 108.1$

α-Hydroxytoluene, phenylmethanol, benzenemethanol; b.p. 205°; dens. 1.045; v.p. 1 Torr (58°); v. dens. 3.72; log P_{ow} 1.10. Water solubility 35 g l^{-1} (20°), soluble in ethanol, diethyl ether, acetone, benzene.

Occurs naturally, free and as esters, in numerous essential oils, including jasmine. Obtained industrially by the hydrolysis of benzyl chloride with aqueous sodium carbonate; this amounted to 21 000 t in 1988 for the USA, Japan and the EC and accounted for 25% of their production of the chloride (91 000 t).

Benzyl alcohol is mostly used in photography in developing baths and as a dispersing agent. It has antiseptic properties and is added to soaps, pharmaceuticals and insect repellants with additional use as a degreasing agent in the polymer industry and during tyre manufacture.

It is rapidly metabolized by animals and humans to benzoic acid. Benzyl alcohol is an eye irritant and is moderately toxic by inhalation and by skin contact. It is listed on the inventory of the US Toxic Substances Control Act.

Further reading
Kirk–Othmer ECT, 4th edn, Wiley, New York, 1992, Vol. 4, pp. 18–125.

BERYLLIUM CAS No. 7440-41-7

Be

Beryllium, a member of Group 2 of the Periodic Table, has an atomic number of 4, an atomic weight of 9.013, a specific gravity of $1.85 \, \text{g cm}^{-3}$ at 20°, a melting point of 1287°, a boiling point of 2471°, and one stable isotope, ^9Be. Beryllium is a light, silver-grey, metal with principal oxidation state +2.

The most important source of the element and its compounds is *beryl* (beryllium silicate, $3\text{BeO.Al}_2\text{O}_3.6\text{SiO}_2$). Other primary minerals include *chrysoberyl* (BeAl_2O_4) and *phenacite* (Be_2SiO_4). The average crustal abundance of beryllium (45th among elements) is $\sim 3 \, \text{ppm}$ but amounts in soil can vary from 1 to 15 ppm depending on locality. Metallic beryllium can be obtained by reduction of BeF_2 with magnesium at about 1300°. The primary environmental source of beryllium is coal combustion. It has been estimated that combustion of 500×10^6 t of coal containing an average concentration of 2.5 ppm would yield about 1250 t of beryllium. Typical values include the following: air (urban) $1\text{–}10 \, \text{ng m}^{-3}$; daily dietary intake (70 kg adult) 0.01 mg Be ; mass in humans 0.036 mg Be.

Uses.   Hardening agent in alloys, e.g. Be_{12}Nb and Be_{12}Ti. Neutron moderator in nuclear reactors. Aerospace and space industries. Window material in X-ray tubes.

Beryllium and its compounds are toxic and carcinogenic to all species, including humans. Effects have been reported in persons living near processing plants and in families of beryllium workers. Exposure to beryllium compounds can lead to dermatitis, skin ulcers and, by inhalation, lung fibrosis, dyspnea, weight loss, rhinitis and in severe cases acute pneumonitis. Acute pulmonary beryllium disease and chronic beryllium disease (berylliosis) can be caused by airborne particulates of soluble beryllium salts (e.g. sulphate and fluoride). Symptoms can appear many years after exposure and lung disease has been reported in workers associated with the manufacture and use of beryllium and its compounds. Disease is generally limited to the upper respiratory tract. There have been less than 10 total confirmed cases within the past 40 years. Since beryllium forms stable complexes with organic compounds, complexes are readily formed with plasma proteins. The intake, transport and accumulation in the body can affect cell division, DNA synthesis and gene expression and change the activity of phosphatases and kinases. Beryllium salts are phytotoxic but may be beneficial at low concentrations to certain plants, e.g timothy and tobacco plant. Hickory is a good accumulator of the metal. Typical concentrations, in plants are > 0.1 ppm (dry weight). It accumulates primarily in the roots and upon translocation accumulates more in the leaves rather than fruit or stems. The metal apparently interferes in plant nutrition, not only in the inhibition of certain enzymes, but also through antagonism of calcium and magnesium and manganese nutrition. There are a number of reports on the effects of over-exposure to beryllium in the workplace, with workers in beryllium ceramics suffering from berylliosis of the lungs as well as damage to lymphocyte cells. For workers handling or machining beryllium metal there is the possibility of dermatitis and increased risk of lung cancer. Also, one case has been reported of chronic beryllium pneumonitis resulting from spot welding of beryllium alloys in Scotland.

In 1948, after a major fire in a beryllium manufacturing facility at Lorain, Ohio, which included the ignition of 2.5 t of magnesium, no firefighters, police or spectators reported any evidence of the symptoms of beryllium poisoning. However, on 12 September 1990 an explosion and fire occurred at the nuclear fuels plant at Ust Kamenogorsk, Kazakhstan, causing the discharge of large amounts of beryllium over the town. Many firemen who attended the scene required treatment for mild lung disorders. Of 1760 people examined up to end of September, 110 had beryllium concentrations of 0.3–$0.8\,\mathrm{mg\,l^{-1}}$ in their urine (normal levels are in the $\mathrm{ng\,l^{-1}}$ range). Beryllium takes several years to be cleared from the body, and can cause muscle wasting, liver and kidney failure and cancer. In addition, there are neoplastigenic and tumorigenic data for some beryllium compounds.

Occupational exposure limits. OSHA PEL (TWA) $0.002\,\mathrm{mg\,Be\,m^{-3}}$.

Toxicity. Beryllium and its compounds are in IARC Group 2A (limited carcinogenic data)

Further reading

Aller, J.A. (1990) The clinical significance of beryllium, *J. Trace Elem. Electrolytes Health Dis.* **4**, 1–6.

ECETOX (1991) *Technical Report 30(4)*, European Chemical Industry Ecology and Toxicology Centre, Brussels.

Environmental Health Criteria 106. Beryllium, WHO/IPCS, Geneva, 1990.

Kirk–Othmer ECT, 4th edn, Wiley, New York, 1992, Vol. 4, pp. 126–153.

Léonard, A. and Lauwerys, R. (1987) Mutagenicity, carcinogenicity and teratogenicity of beryllium, *Mutat. Res.*, **186**, 35–42.

Reeves, A.L. (1986) Beryllium, in Friberg, L., Nordberg, G.F. and Voux, V.B. (eds), *Handbook on the Toxicology of Metals*, 2nd edn., Elsevier, Amsterdam, Vol II, pp. 95–116.

ToxFAQs (1993) Beryllium, ATSDR, Atlanta, GA.

Beryllium chloride CAS No. 7787-47-5

$BeCl_2$

Uses. Manufacture of beryllium catalysts and beryllium metal.

Occupational exposure limit. USA TLV (TWA) $0.002\,\mathrm{mg\,Be\,m^{-3}}$.

Toxicity.   Very toxic (skin and lung damage). Carcinogenic agent. LC_{50} (96 h) fathead minnow $150 \mu g \, l^{-1}$ (soft water); LD_{50} oral rat $86 \, mg \, kg^{-1}$.

Beryllium fluoride CAS No. 7787-49-7

BeF_2

Uses.   Commercial production of the pure element. Manufacture of glass.

Toxicity.   Poison and carcinogen. Irritant, causes eczema LD_{50} oral rat $98 \, mg \, kg^{-1}$; LD_{50} subcutaneous mouse $20 \, mg \, kg^{-1}$.

Beryllium nitrate, CAS No. 13597-99-4

$Be(NO_3)_2$

Uses.   As a stiffener for mantles in gas and acetylene lamps.

Toxicity.   Irritant. Chemical agent. LD_{50} intraperitoneal guinea pig $100 \, mg \, kg^{-1}$.

Beryllium oxide CAS No. 1304-56-9

BeO

Uses.   Electrical insulator.

Toxicity.   Suspected human carcinogen (lungs). LD_{50} intraperitoneal rat $5.32 \, mg \, kg^{-1}$.

Beryllium sulphate tetrahydrate CAS No. 7787-56-6

$BeSO_4.4H_2O$

Uses.   In X-ray media.

Toxicity.   Highly toxic. Carcinogen, IARC Groups 2A and 2B. LD_{50} oral mouse, rat $80 \, mg \, kg^{-1}$; LD_{50} subcutaneous rat $1.5 \, mg \, kg^{-1}$.

BIOCONCENTRATION FACTOR (BCF)

In water this is (concentration of compound in fish lipid) / (concentration of compound in water) when equilibrium is established between intake and elimination. The bioconcentration factor for PCBs in trout $= 1.0 \times 10^4$ and for hexachlorobenzene in rainbow trout $= 5.5 \times 10^3$. For plants the BCF = (concentration in plant tissue) / (concentration in the soil). In plants the BCF is inversely proportional to the square root of the octanol/water partition coefficient of the pollutant – one expects an inverse relationship because transport from soil to the plant depends on the substances solubility in water.

Pollutant	Log K_{ow}	Log BCF (plants)
Aldrin	5.52	−1.67
1254-Aroclor	6.47	−1.77
DDT	5.76	−1.80
Lindane	3.66	−0.41
Simazine	2.22	0.22
Cyanazine	2.02	−0.06

In foodstuffs the BCF = (concentration in the food) / (daily intake); here the BCF is directly proportional to K_{ow}.

Pollutant	Log K_{ow}	Log BCF (beef)
1254-Aroclor	6.47	−1.28
DDT	5.76	−1.55
Aldrin	5.52	−1.07
Hexachlorobenzene	5.45	−1.35
Malathion	2.89	−4.74
2,4-D	2.81	−5.32

For examples of bioconcentration through a food chain, see Alloway and Ayres (1997) and **Polychlorobiphenyls**.

Reference
Alloway, B.J. and Ayres, D.C. (1997) *Chemical Principles of Environmental Pollution*, 2nd edn, Blackie, London, pp. 289–290.

Further reading
Travis, C.C. and Arms, A.D. (1988) Bioconcentration of organics in beef, milk and vegetation, *Environ. Sci. Technol.* **22**, 271–274.

BIOLOGICAL OXYGEN DEMAND (BOD)

This gives an estimate of the degree of contamination of sewage and industrial waste; it is defined as the amount of dissolved oxygen ($mg\,l^{-1}$) required to stabilize decomposable organic matter by aerobic chemical action. It is determined by diluting the sample with oxygen-saturated water and measuring the oxygen level initially and after incubation for 5 days.

Substances which are readily biodegradable include ethanol, *n*-propanol, benzaldehyde and benzoic acid. Less affected are ethylene glycol, 2-propanol, cresols and pyridine, while methanol, aniline, methyl ethyl ketone and acetone are resistant. Most non-ionic detergents are biodegradable. Organophosphate pesticides are more susceptible than organochlorines such as DDT and its derivatives.

The BOD value gives guidance about the risk to the aqueous environment from the release of organic compounds, but it provides no understanding of the ultimate products of biodegradation. Wastes with high BOD include sulphite waste liquors from delignification, bleaching plant effluent, crop residues, e.g. rice straw, corn crops and other starch-based natural products.

At Renwick, near Penrith, UK, on 16 June 1995, a pump failure lead to cattle slurry entering a tributary of the river Eden. About 4000 trout died as a result of the solution of ammonia combined with lack of oxygen. This incident resulted in prosecution, a fine of £2200, costs of £2400 and restocking at £3400. In the same month a milk tanker suffered a tyre blow-out, which caused a rupture in the outlet pipe and discharge of $22 \times 10^3\,l$ of milk into a tributary of the river Tyne. Milk is highly polluting because of the high oxygen demand of the bacteria acting upon it.

See also **Chemical oxygen demand (COD)**.

Further reading

Lewis, R.J. (ed.) (1993) *Hawley's 'Condensed Chemical Dictionary'*, 12th edn Van Nostrand, New York.

Environment Agency (1996) *Water pollution Incidents in England and Wales*, HMSO, London, pp. 7–8.

BIPHENYL CAS No. 92-52-4

LD$_{50}$ rat, rabbit 2.4–3.2 g kg^{-1}

                      $C_{12}H_{10}$ $M = 154$

Diphenyl, phenylbenzene; m.p. 71°; b.p. 255°; dens. 1.041; v.p. 25 kPa (200°); log P_{ow} 3.98. Slightly soluble in water (17 mg l^{-1}), soluble in ethanol, benzene, diethyl ether.

First reported by Fittig in 1862. It is the major product of the pyrolysis of benzene at 700° and this is the basis of the industrial process run at 800° with benzene diluted in steam: higher temperatures lead to the formation of terphenyls ($C_{18}H_{14}$). Biphenyl is also obtained commercially from the product of hydrodealkylation of toluene, which affords 1 kg of biphenyl for 100 kg of benzene. US production (1990), 16×10^6 kg year^{-1}, having declined owing to the cessation of polychlorobiphenyl production.

Residues may be detected by HPLC, GC or by UV spectroscopy (also terphenyls) by virtue of the strong absorption, $\epsilon = 19\,300$ at 246 nm.

Biphenyl was once used to fumigate fruit cargoes during shipment; the current major use is as a heat-transfer medium mixed with diphenyl ether (Dowtherm A, 26% biphenyl, 74% diphenyl ether), also as a dye carrier.

Biphenyl irritates the eyes but is not toxic by ingestion. It is metabolized by mammals by hydroxylation to 4-hydroxy- and dihydroxybiphenyls. Humans exposed to levels in air above 35 mg m^{-3} suffered damage to the CNS and liver.

The ACGIH sets a TLV (TWA) of 0.2 ppm (1.3 mg m^{-3}). The STEL for the UK is 0.6 ppm.

BIS-(2-CHLOROETHYL) ETHER CAS No. 111-44-4

LD$_{50}$ rat, mouse 75–112 mg kg^{-1}; LC$_{50}$ (96 h) bluegill sunfish 600 mg l^{-1}

ClCH$_2$CH$_2$—O—CH$_2$CH$_2$Cl $C_4H_8Cl_2O$ $M = 143.0$

2,2'-Dichloroethyl ether, *sym*-dichloroethyl ether; b.p. 178°; dens.1.222; v.p. 0.7 Torr (20°); v. dens. 4.93; log P_{ow} 1.29. Slightly soluble in water and miscible with common organic solvents.

Manufactured by the chlorination of ethylene oxide/ethylene or by the action of thionyl chloride on diethylene glycol. It is used as a soil fumigant, in paints and varnishes, as a solvent and for the manufacture of 1,3-butadiene.

Bis(2-chloroethyl) ether is flammable and forms explosive peroxides on exposure to air/light. It is lachrymatory, may be absorbed through the skin and can cause liver and kidney damage. The human response to a level of 550 ppm for brief periods was intolerable irritation of the eyes and nose with nausea and retching; a severe response followed exposure at 100 ppm.

The ACGIH has set a TLV (TWA) of 5 ppm (29 mg m^{-3}) with a STEL of 10 ppm. Dichloroethyl ether is classed by the IARC as a group 3 substance – limited evidence of carcinogenicity in animals.

Further reading
Chemical Safety Data Sheets (1991), **4a**, Royal Society of Chemistry, Cambridge, pp. 60–63.

Bis (CHLOROMETHYL) ETHER CAS No. 542-88-1
LD$_{50}$ rat 210 mg kg^{-1}; LC$_{50}$ inhalation rat (7 h) 7 ppm

ClCH$_2$—O—CH$_2$Cl   C$_2$H$_4$Cl$_2$O $M = 115.0$

Chloromethyl ether monochloromethyl ether, α, α'-dichlorodimethyl ether; b.p. 106°; dens. 1.315 (20°); v.p. 30 Torr (22°); v. dens. 4.0. It reacts rapidly with water to give hydrogen chloride and formaldehyde. It is soluble in benzene, ethanol and diethyl ether.

It was formerly used as an alkylating agent but is now of principal concern as a contaminant (1–8%) in chloromethyl methyl ether. It strongly irritates the eyes and skin.

The carcinogenic properties of the haloethers as capable of alkylating DNA and so inducing mutagenesis was first mentioned by Van Duuren *et al.* (1968). Bis-CME is a highly potent inducer of respiratory tract cancer in mice and rats. In the early 1960s six relatively young industrial workers in a group of 100 died from lung cancer and further experiments on mice and rats showed that cancers were induced at LC$_{50}$ inhalation 5–7 ppm (6 h).

The ACGIH has set a TLV (TWA) for occupational exposure at 0.001 ppm (0.0047 mg m^{-3}). *Bis*-(chloromethyl) ether has been classed as an A1 carcinogen in the MAK list.

Reference
Van Duuren, B.L., Goldschmidt, B.M., Katz, C., Langseth, G.M. and Sivak, A. (1968) α-Haloethers: a new type of atkylating carcinogen, *Arch. Environ. Health*, **16**, 472–476.

Further reading
Berlin, A., Draper, M., Krug, E., Roi, R. and Van der Venne, M. Th. (eds) (1989) *The Toxicology of Chemicals*, Commission of the EC, Brussels, Vol. 1, pp. 33–35.

BISMUTH, CAS No. 7440-69-9
Bi
Bismuth, a member of Group 15 of the Periodic Table, has an atomic number of 83, an atomic weight of 208.98, a specific gravity of 9.80 at 20°, a melting point of 271.3°, and a boiling point of 1560°. In nature Bi exists as one stable radioactive nuclide, ^{20}Bi, half-life 2×10^{17} years. Bismuth is a silver-white crystalline metal, with stable oxidation states of +3 and +5.

Although bismuth occurs as the minerals *bismite* (Bi$_2$O$_3$) and *bismuthinite* (Bi$_2$S$_3$), it is obtained as a byproduct from lead and copper ores. It is 69th in order of abundance in the earth's crust, occurring at 0.008 ppm on average. The pure metal is produced by the reduction of the oxide by iron or charcoal.

Uses. Bismuth compounds are used in pharmaceuticals and cosmetics, as industrial catalysts and industrial pigments and in the electronics industry. The metal is used as a metallurgical additive to steel, aluminium and cast iron, particularly in the manufacture of free machining steel and aluminium. It is also used in the manufacture of fusible alloys. A mixed metal alloy of Bi–Pb–Sn–Cd–In–Sb has a melting point of 47°. The radio nuclide ^{212}Bi($t_{1/2} = 1$h) is used in radio(immuno)therapy.

Several bismuth compounds are used in medicine-see below. Small quantities of bismuth interfere with growth but intake of large quantities is highly toxic, resulting in a variety of symptoms. It is recommended that the administration of bismuth compounds should be stopped when inflammation of the gums (gingivitis) appears. Overdosage of bismuth may result in symptoms such as malaise, albuminuria, diarrhoea and skin reaction. It is generally considered to be one of the least toxic metals. Large doses of salts can cause kidney damage and are poisonous to humans. It has similar pharmacology and toxicology to lead. Thus, absorption of bismuth by industrial workers results in a dark line in gums similar to lead poisoning. No other industrial poisoning by bismuth or its compounds has been reported. However, precautions should be taken regarding the handling of bismuth and its compounds, inhalation and ingestion of dusts and fumes should be avoided.

Occupational exposure limits.   UK TLV (TWA) N.a. Bismuth powder is flammable.

Toxicity.   UNR humans LD_{Lo} 221mg kg^{-1}. LD_{50} oral rat 5–7 g kg^{-1}.

Further reading
King, R. B. (ed.) (1994) *Encyclopedia of Inorganic Chemistry*, Wiley, Chichester, Vol. 1, pp. 280–292.
Kirk–Othmer ECT, 4th edn, Wiley, New York, 1992, Vol. 4, pp. 237–270.

Basic bismuth carbonate CAS No. 5892-10-4.

$(BiO)_2CO_3$

Used as an antiacid. Irritant.

Bismuth(III) oxide CAS No. 1304-76-3

Bi_2O_3

Uses.   Manufacture of glasses, ceramics and porcelain. Catalyst for acrylonitrile production.

Toxicity.   Harmful if inhaled, on contact with skin or if swallowed. Target organs, liver and kidneys.

Bismuth nitrate CAS No. 1304-85-4
$BiONO_3$

Cosmetics ingredient, and used in medicine.

Bismuth oxychloride CAS No. 7787-59-9

$BiOCl$

Uses.   Cosmetic preparations.

Toxicity. LD_{50} oral rat 22 g kg^{-1}.

Bismuth subgallate CAS No. 99-26-3.

$C_7H_7O_7Bi$

Used for skin disorders and treatment of haemorrhoids.

Bismuth telluride CAS No. 1304-82-1

Bi_2Te_3

Uses.   In electronics as a thermoelectric cooling agent.

Occupational exposure limits.   OSHA PEL (TWA) total dust $15\,mg\,m^{-3}$; respirable fraction $5\,mg\,m^{-3}$; ACGIH (TLV) TWA $10\,mg\,m^{-3}$.

Toxicity.   N.a.

Further reading
Kirk–Othmer ECT, 4th edn, Wiley, New York, Vol. 8, pp. 223–237.
The Pesticide Manual, 10th edn, British Crop Protection Council/Royal Society of Chemistry, Cambridge, 1994, pp. 105– 106.

BISPHENOL A CAS No. 80-05-7
LD_{50} rat $3.2\,g\,kg^{-1}$

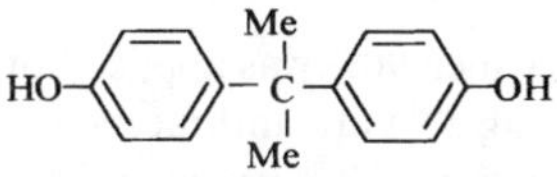

$C_{15}H_{16}O_2$ $M = 228.3$

2,2-bis-(4'-hydroxyphenyl)propane, 4,4'-dihydroxyphenylpropane, DIAN; m.p. 157°, b.p. 252°/13 Torr. Insoluble in water, soluble in alkalis and EtOH, slightly soluble in CCl_4

It is manufactured by the acid-catalysed condensation of phenol with acetone. Material of 95% purity is acceptable for the production of epoxy resins by combination with epichlorohydrin. For the production of polycarbonate the contaminating *o, p*-isomer is removed by distillation/crystallization and material of > 99% purity is combined with phosgene.

These epoxy resins are used for the protection of metals (food cans), in circuit boards, adhesives and sealants. Applications of the polycarbonates include compact discs, business machine cases, telephones and window glazing.

Bisphenol A is a skin and eye irritant; it is moderately toxic by ingestion, inhalation and by skin contact.

The glycidyl ether:

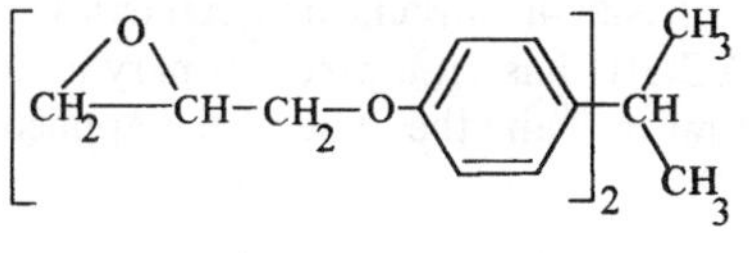

m.p. 43°, CAS No. 1675-54-3, is an intermediate in the production of epoxy resins and may occur in drinking water conveyed in pipes sealed with epoxy resin. It is known to cause rashes on the eyelids, face and hands of human subjects.

Further reading
Wittcoff, H.A. and Reuben, B.G. (1996) *Industrial Organic Chemicals*, Wiley, New York, p. 241.

BITUMENS CAS No. 8052-42-4

Asphalt, petroleum asphalt, asphalt oxidized (CAS No. 64742-93-4). Typical composition: C 79–88%, H 7 13%, S trace–8%, O 2–8%, N trace–3%.

The term 'asphalt' is sometimes reserved for manufactured materials from petroleum processing while bitumens are classified as native or natural asphalt (SPEIGHT), the two are closely comparable, but they should not be confused with 'petroleum pitches,' which are the hard, ductile and highly aromatic residues from the thermal cracking of petroleum. Natural asphalts are also obtained from the refining of oil shales (CAS No. 68308-34-9), of which the largest is the Green River deposit in Colorado, Utah and Wyoming. The 'tar sands' of Athabasca and Nigeria are also major sources.

The product which makes the greatest environmental impact is asphalt from petroleum processing. Types are graded according to ASTM (American Society for Testing and Materials) and vary in hardness and viscosity. The fractions which are soluble in low boiling petrol fractions are known as asphaltenes. Straight-run asphalts are the involatile runoff from the refining of crude oil of b.p. 300–400° and are used for roadmaking. Air-blown material is used for roofing materials and is obtained from asphalt heated in air at 200–275°.

The world usage in 1985 was over 60×10^6 t year^{-1}, of which 90% was applied hot, so leading to the emission of PAH. Some typical levels (μg g^{-1}) are anthracene 0.3, phenanthrene 7.0, pyrene 1.57, fluoranthene 0.7, benzo[*a*]pyrene 1.8, chrysene 1.5 and benz[*a*]anthracene 1.1. During periods of heavy rainfall runoff introduces these compounds into streams. Wear on roadways will also introduce substances added to asphalt to improve service life; these include sulphur, polymers and metal complexes. Quaternary derivatives of imidazoline are also liable to abrasion, a general structure is

where R is a fatty acid residue. This associates strongly with the asphalt layer, whilst the quaternized base residue becomes strongly attracted to the metalling material of the road.

The US Occupational Hazard Survey (1974) estimated that 2 million workers were exposed to contact with bitumen and 33 000 were exposed to its fumes. However, fumes from hot roofing asphalt did not induce cancer in animals, nor was it carcinogenic to the skin of mice and rabbits – this is in contrast to the fumes from coal tar. There was no significant difference in the health of asphalt workers in 25 refineries as compared with those in a control group. The ACGIH has set a precautionary level for occupational exposure at a TLV (TWA) of 5 mg m^{-3} and the same level is applied in the UK as a long-term exposure limit.

BORANES CAS No. 13283-31-3

Boranes are used in welding and brazing metals, as rocket fuels and as reducing agents in organic syntheses. Boranes are many times more toxic than other boron compounds and even more toxic than phosgene or hydrogen cyanide. Pentaborane-9 (CAS No. 19624-22-7) [OSHA (PEL) TWA 0.0005 ppm, STEL 0.015 ppm; ACGIH TLV (TWA) 0.005 ppm] is particularly toxic and causes severe damage to the nervous system, kidneys and liver. For example, hamsters, rabbits, rats and dogs died within 4 weeks on repeated exposure to pentaborane-9 at a level of $0.0026 \, mg \, B \, l^{-1}$. Ingestion of pentaborane-9 and decaborane-14 **(CAS No. 17702-41-9)** led to vomiting, paresthesia, meningism, tremors and, in more severe cases, to convulsions and coma. In some cases death occurred. Pentaborane-9 and decaborane-14 are very irritating to the skin and mucous membranes. In the acute phase of poisoning, boron levels in the blood rose to $0.24 \, \mu g \, ml^{-1}$ (normal 0.04–$0.05 \, \mu g \, ml^{-1}$). Examination of 14 individuals 4–12 weeks after a brief exposure to pentaborane-9 showed some workers to have mild brain dysfunction.

Further reading
King, R.B. (ed.) (1994) *Encyclopedia of Inorganic Chemistry*, Wiley, Chichester, Vol. 1, pp. 338–357.

BORIC(III) ACID CAS No. 10043-35-3

H_3BO_3 (orthoboric acid) $M = 61.8$

Uses. Preparation of borates. Weatherproofing and flameproofing. Electrolytic capacitors. Cements. Manufacture of porcelain and glass. Additive to cooling water for nuclear reactors. Medical and pharmaceutical uses. Photography.

Toxicity. Human poison; 250–$330 \, mg \, kg^{-1}$ can be lethal to humans.

The metaborates of calcium, lead and barium and the perborates of potassium and sodium are used in the manufacture of borosilicate glass, textile-grade glass fibres, glazes, enamels, rubber, plastics and in the textile industry. Borates and perborates are used in the production of bleaching agents and as additives to washing powders. Orthoboric acid and sodium tetraborate (anhydrous and decahydrate), TLV (TWA) $1 \, mg \, m^{-3}$, are important commercial compounds and used in the glass and ceramics industry, as preservatives for food and wood and in medicine. They are also used in the manufacture of biological growth agents, water treatment, algicides, herbicides and insecticides. Orthoboric acid is a mild antiseptic used as an eyewash.

Further reading
King, R.B. (ed.) (1994) *Encyclopedia of Inorganic Chemistry*, Wiley, Chichester, Vol. 1, pp. 318–327.

BORIDES CAS No. 24389-64-8
Borides are used as alloys with certain transition metals and for coating steel and other metals to increase their hardness and resistance to wear and corrosion, as catalysts and as semiconductor material. Workers exposed to aeresols of metal borides had reduced blood levels of albumin.

Further reading
King, R.B. (ed.) (1994) *Encyclopedia of Inorganic Chemistry*, Wiley, Chichester Vol. 1, pp. 327–338.

BORON CAS No. 7440-42-8

B

Boron exists as two allotropes: crystalline boron is extremely hard, being inferior only to diamond in hardness ; it can also exist as a yellow or brown amorphous powder. Boron, a member of Group 13 of the Periodic Table, has an atomic number of 5, an atomic weight of 10.81, a specific gravity 2.34 at 20°, a melting point of 2075° and a boiling point of ~ 4000°. The oxidation state of B in most compounds is ︱3. It exists in nature as two isotopes, ^{10}B(19.9%) and ^{11}B(80.1%).

Boron is 51st in order of relative abundance in the earth's crust, ~ 11 ppm. The major boron minerals are *borax*, $Na_2B_4O_7.10H_2O$, and *kernite*, $Na_2B_4O_7.4H_2O$. Important sources of emission are wastewaters from the metallurgical, textile, glass, ceramics and leather goods industries as well as household sewage saturated with dissolved washing powders. In sewage effluent the level is ~ 1 ppm mainly from the use of borates and perborates in detergents. In sludge this can rise to > 200 ppm. Coal can contain up to 400 ppm boron but the average amount is ~ 60 ppm. Coal-fired power stations are boron emitters.

The following are typical concentrations in the environment: sea water 4450 µg l^{-1} (4.6 ppm); river water ~ 100 ppb–0.5 ppm; soils $1 \times 10^{-3}\%$ (10 ppm), of which about 10% is usually water-soluble boron; terrestrial plants $25 \times 10^{-4}\%$ dry weight. Reported average boron concentrations in organisms (mg kg^{-1} dry weight): marine algae 12; terrestrial plants 5; marine animals 2.5; terrestrial animals 0.05. Boron is an essential element for higher plants but questionable for humans and mammals. Plants can contain 2–3875 ppm, human blood 0.003–0.04 ppm and urine 0.04–6.6 ppm. The average daily intake by humans is ~ 3 mg, but over 98% of ingested boron is excreted in the urine. Normal levels are 0.8 mg l^{-1} in the blood and 0.7 mg l^{-1} in the urine; total body content (adult, 70 kg), 10 mg.

Uses. Boron is added in small amounts to steels to increase their resistance to heat and corrosion. It is used in the nuclear power industry, e.g. in control rods and as a constituent of nuclear shielding material for nuclear reactors, in uranium and solar batteries, as semiconductor material and in amorphous magnetic alloys that are based on B, Fe, Ni and Co and used in power transformers as a soft magnet to convert from high to low voltage.

Although boron compounds are fairly toxic, boron appears to be an essential trace element since it is believed to facilitate the transport of sugars through membranes and to participate in the synthesis of nucleic acids. Boron is the only non-metal among the plant micronutrients. For plants, boron concentrations > 700 mg kg^{-1} dry tissue weight are toxic to most crops, while those ~ 100 mg kg^{-1} are toxic to the more sensitive plants, including legumes, cabbage and some fruit trees. Boron concentrations in water > 4 mg l^{-1} cause injury to most crops. Plants which accumulate boron without harm are beet and cotton plants (500–3300 mg kg^{-1} dry weight). Boron-deficient soils (< 1 ppm) occur widely (normal concentration 30 ppm). The toxicity of boron and its compounds to fish is low, e.g. borax ($Na_4B_4O_7$) ~ 3000–5000 mg l^{-1}.

Boron and its compounds can enter the human body by inhalation, ingestion and absorption through mucous membranes or skin burns. Absorption of boron and soluble borates from the gastrointestinal tract and skin burns is almost complete.

Boron is a component of all tissues but its role in the body is not clear. About 86% of body boron is stored in the skeleton. The remainder is concentrated in the soft tissues, predominantly the brain, liver and fatty tissues. Nearly all boron found in tissues may be bound to carbohydrates as found in the carbohydrate fraction of the liver. Boron entering the body in excessive amounts has been shown to lower the activities of various oxidative and digestive enzymes and to inhibit nucleic acid and carbohydrate metabolism. Boron(III) oxide and orthoboric acid are highly toxic substances with multiple sites of reaction. In particular, they display hepatotoxic and gonadotropic activity and can be embryotoxic by virtue of being able to pass through the placental barrier. There is evidence that orthoboric acid is carcinogenic. Orthoboric acid and borates of metals are potent cholinolytic agents.

The lethal oral dose of orthoboric acid for adult humans is 15–20 g and for an infant 5–6 g. The early symptoms of acute poisoning are nausea, gastric pains and severe vomiting. The blood boron levels rise to $40\,\text{mg}\,l^{-1}$ or more. In fatal cases, severe gastroenteritis, hepatitis and congestion and edema in the brain and myocardium are symptomatic. Autopsy showed increased boron concentrations in the brain, kidneys, liver and urine. Fatal poisoning of children has been caused by the accidental substitution of boric acid for powdered milk.

After prolonged exposure to boron, signs of reduced sexual activity, neurasthenia and neuralgia were symptoms frequently observed in workers exposed to orthoboric acid dust at $20–50\,\text{mg}\,m^{-3}$ (which is 2–5 times higher than the present MAC). Manifest in female workers were inflammatory diseases of the genital organs. Male workers exposed to $10–35\,\text{mg}\,B\,m^{-3}$ for 10 years showed reduced ejaculate volumes and low sperm counts. Likewise, long term consumption of water containing $1\,\text{mg}\,B\,l^{-1}$ over 5 years was consistently associated with reduced sexual potency. Workers producing borates and perborates are at risk from dust bronchitis and pneumoconiosis. Orthoboric acid and sodium tetraborate can be absorbed through the skin. Cited are 109 cases of poisoning, with a fatality rate of 70% occurring in infants treated with powders, emulsions or solutions containing these substances.

Occupational exposure limit. Germany (DFG) MAK. $0.5\,\text{mg}\,m^{-3}$. Boron dust is a severe fire and explosion hazard.

Toxicity. LD_{50} oral mouse $2000\,\text{mg}\,kg^{-1}$.

Further reading

King, R.B. (ed.) (1994) *Encyclopedia of Inorganic Chemistry*, Wiley, Chichester, Vol. 1, pp. 357–374.

Kirk–Othmer ECT, 4th edn, Wiley, New York, 1992, Vol. 4, pp. 360–437.

Loomis, W.D. and Durst, R.W. (1992) Chemistry and biology of boron, *Biofactors*, **3**, 229–239.

ToxFAQs (1995) *Boron*, ATSDR, Atlanta, GA.

Boron(III) bromide CAS No. 10294-33-4

BBr$_3$

Uses. Manufacture of diborane and pure boron. Preparation of metal bromides. Catalyst. Semiconductor manufacture.

Occupational exposure level.   OSHA PEL(CL).   1 ppm.

Toxicity.   Toxic by inhalation, contact with skin, or if swallowed. Destructive to mucous membranes and internal organs.

Boron carbide CAS No. 60063-34-5

B_4C

Acute and chronic inflammatory diseases of the upper respiratory tract were detected in workers employed in the carbide producing department of a factory over a period of 12–14 years.

Boron(III) chloride CAS No. 10294-34-5

BCl_3

Uses.   Manufacture of pure boron. Catalyst. Semiconductors.

Toxicity.   Toxic by inhalation, contact with skin, or if swallowed. LC_{Lo} inhalation mouse 20 ppm.

Boron(III) fluoride CAS No. 7637-07-2

BF_3

Uses.   Catalyst in organic chemistry.

Toxicity.   Severe irritant to mucous membranes. TLV (TWA) $2.8 \, mg \, m^{-3}$. 25 ppm danger to life and health. LC_{50} (4 h) inhalation rat $1180 \, mg \, m^{-3}$.

Boron(III) iodide CAS No. 13517-10-7

BI_3

Uses.   Catalysts. Preparation of boron compounds.

Toxicity.   A poison.

Boron nitride CAS No. 10043-11-5

BN

Used to make fillings for furnaces and in rocket and nuclear engineering.

Boron(III) oxide CAS No. 1303-86-2

B_2O_3

Uses.   Soldering flux. Manufacture of glass, enamels and glazes.

Occupational exposure limits.   OSHA TLV (TWA). $10 \, mg \, m^{-3}$.

Toxicity. LD_{50} oral mouse $63 \, mg \, kg^{-1}$.

BROMATES CAS No. 15541-45-4

BrO_3^-

Strong oxidizing agents. Toxic, suspect carcinogens and can cause CNS paralysis.

BROMIDES CAS No. 24959-67-9

Inorganic bromides (Na, K, NH_4, Ca and Mg) are depressants and in severe cases cause psychosis and mental deterioration. Bromides can also cause dermatitis.

Organic bromides such as bromomethane (methyl bromide) and bromoethane (ethyl bromide) are toxic. Methyl bromide is a CFC- like halon and can potentially interfere with the ozone layer. The action of UV radiation on this compound produces the reactive species, the bromine atom.

BROMINE CAS No. 7726-95-6

Br stable as Br_2

Bromine, a member of Group 17 (the halogens) of the Periodic Table, has an atomic number of 35, an atomic weight of 79.90, a specific gravity of 3.13 at 20°, a freezing point of −7.25°, and a boiling point of 58.8°. It is the only non-metallic element that is liquid at standard conditions. The most common oxidation states are −1 and +5, but +1, +3, +7 are also observed. Bromine has two stable isotopes, ^{79}Br (50.69%) and ^{81}Br (49.31%). Isotopes usable as radioactive tracers are ^{77}Br, ^{80}Br and ^{82}Br.

Bromine is widely distributed in nature but in relatively small amounts, 40th in order in the earth's crust at an average concentration of 2.5 ppm. Bromine occurs in the form of bromide in seawater (0.0085% by weight) and in natural brine deposits. In all current methods of bromine production, chlorine, which has a higher reduction potential than bromine, is used to oxidize bromide to bromine.

The following concentrations are typical: 7 ppm (Br) in terrestrial plants; edible foods up to 20 ppm; animal tissues 1–9 ppm; blood 5–15 ppm. In adult human males the bromine content in serum varies from 3.2 to 5.6 μg ml^{-1}, in urine from 0.3 to 7.0 mg ml^{-1} and in hair from 1 to 49 μg ml^{-1}. Total body content (70 kg, adult), 200 mg. Bromine may be an essential element. The recommended WHO daily intake of bromine is 1 mg kg^{-1} of body weight per day.

Bromo compounds are used in fire extinguishers, in inorganic and organic synthesis, for bleaching silk and other fibres and in the manufacture of ethylene dibromide, $BrCH_2CH_2Br$, an antiknock agent in petrol. Photography.

Bromine has a sharp pungent odour and is poisonous by inhalation. Air concentrations of ~ 1 ppm are irritating and cause eye watering; ~ 10 ppm levels are intolerable. Inhalation of 20 ppm bromine and higher concentrations causes severe burns to the respiratory tract and is highly toxic. Symptoms include coughing, nose bleed, dizziness, headaches and abdominal pains. Pneumonia may also follow. Mail workers in the Royal Mail Sorting Office, Bradford, UK, were in a scare after bromine fumes were noted escaping from a parcel. A number of bromine compounds other than bromides are used commercially.

Occupational exposure limits. OSHA PEL (TWA) 0.1 ppm (0.66 mg m^{-3}); ACGIH (TLV) TWA 0.1 ppm.

Toxicity LD_{Lo} oral human 14 mg kg^{-1}. LC_{50} (9 min) inhalation mouse 750 ppm.

Further reading

Kirk–Othmer ECT, 4th edn, Wiley, New York, 1992, Vol. 4, pp. 560–589.

Thompson, R. (ed) (1995) *Industrial Inorganic Chemicals. Production and Uses*, Royal Society of Chemistry, Cambridge, Ch. 1.

Bromine chloride, CAS No. 13863-41-7

BrCl

Uses.   Used in organic addition and substitution reactions.

Toxicity. Poisonous by inhalation. LC_{50} (96 h) shrimp $0.70\,\mathrm{mg\,l^{-1}}$.

Bromine(V) fluoride, CAS No. 7789-30-2

BrF_5

Uses.   As a fluorinating agent in organic synthesis and in the preparation of inorganic fluorides such as **uranium hexafluoride, UF_6**, and **uranium tetrafluoride, UF_4**.

Occupational exposure limits.   OSHA PEL (TWA) 0.1 ppm $(0.72\,\mathrm{mg\,m^{-3}})$; TWA $2.5\,\mathrm{mg\,F\,m^{-3}}$. ACGIH TLV (TWA) 0.1 ppm.

Toxicity.   A poisonous, corrosive and extremely reactive gas. Causes burns.

BROMOFORM CAS No, 75-25-2
LD_{50} rat $1150\,\mathrm{mg\,kg^{-1}}$; LC_{50} bluegill sunfish (96 h) $29\,\mathrm{mg\,l^{-1}}$

CHBr₃ $M = 252.7$

Tribromomethane; m.p. 8°; b.p. 151°; v.p. 5.6 Torr (25°); v. dens. 8.7; log P_{ow} 2.38. Chloroform-like odour. Solubility in water $2\,\mathrm{g\,l^{-1}}$; soluble in most organic solvents. Decomposes in air/light and acquires a yellow colour due to the release of bromine, but may be stabilized by the addition of EtOH (1%) or diphenylamine.

Prepared by the action of $AlBr_3$ on chloroform below 60° or by the action of hypobromite on ethanol or methyl ketones. Bromoform occurs naturally in seawater due to release from red algae *Asparagopsis* spp. and *Falkenbergia rubolanosa*.

Bromoform is used for coal flotation and for mineral separation, also in fire-resistant formulations and as a solvent for greases and waxes.

It is formed with chloroform on the chlorination of groundwater when levels may reach $240\,\mathrm{\mu g\,l^{-1}}$; the median level in surface water is $4\,\mathrm{\mu g\,l^{-1}}$. The range in drinking water is $1–10\,\mathrm{\mu g\,l^{-1}}$ although it has reached $92\,\mathrm{\mu g\,l^{-1}}$ in some US cities. A study has been made of the levels of halomethanes in Los Angeles drinking water (Hartwell *et al.*, 1987), where spring levels $(\mathrm{\mu g\,l^{-1}})$ were chloroform 33, bromodichloromethane 24, dibromochloromethane 32 and bromoform 3.0; winter levels lay between 20% and 50% of spring levels. In the UK levels in drinking water are generally below $2.5\,\mathrm{\mu g\,l^{-1}}$. Above beaches with high algal growth air levels may reach $68\,\mathrm{ng\,m^{-3}}$; in air over swimming pools levels range from 0.1 to $20\,\mathrm{\mu g\,m^{-3}}$. Pulses of bromoform enter the air during the 3 months of spring in the Arctic. Air levels at Point Barrow, Alaska, are typically $6.3\,\mathrm{ng\,l^{-1}}$ and at Cape Kumukahi, Hawaii, $3.1\,\mathrm{ng\,l^{-1}}$.

Analysis of air samples is by adsorption on charcoal or Tenax; GC–MS gives a sample limit of 20 ng. Bromoform is ubiquitous in water, which is analysed by purging with an inert gas adsorption on Chromosorb; GC–MS gives a limit of $0.12\,\mathrm{\mu g\,l^{-1}}$.

Exposure to the vapour irritates the respiratory tract, pharynx and larynx; after inhalation the highest concentration is in the brain. Daily treatment of mice (14 days) at $50\,\mathrm{mg\,kg^{-1}}$ caused liver damage. Samples of human milk contained up to $3\,\mathrm{mg\,l^{-1}}$

of $CHCl_3$, up to $0.3\,mg\,l^{-1}$ of $CHBr_2Cl$ but only traces of $CHBr_3$. A study of Japanese housewives revealed a daily intake from 90 meals of $20\,ppb\,day^{-1}$ of $CHCl_3$ and $0.5\,ppb\,day^{-1}$ of $CHBr_3$. There is inadequate evidence of carcinogenicity in humans and limited evidence in animals – bromoform is a group 3 carcinogen.

Reference

Hartwell, T.D., Pellizzari, E.D., Perritt, R.L., Whitmore, R.W., Zelon, H.S., Sheldon, L.S., Sparacino, C.M. and Wallace, L. (1987) Results from the total exposure assessment study in selected communities in Northern and Southern California, *Atmos. Environ.*, **21**, 1995–2004.

Further reading

IARC (1991), *Chlorinated Drinking Water; Chlorination Byproducts*, IARC, Lyon, Vol. 52, pp. 213–242.

1,3-BUTADIENE CAS No. 106-99-0

LD_{50} rat $3.2\,g\,kg^{-1}$; LC_{50} (2 h) inhalation mouse $270\,g\,m^{-3}$

$$CH_2{=}CH{-}CH{=}CH_2 \qquad\qquad C_4H_6 \quad M = 54$$

Biethylene, divinyl; b.p. $-4.5°$, v.p. $760\,Torr$ at b.p., $1840\,Torr$ at $21°$. Highly flammable, flash point $-70°$, bioconcentration factor 1.9, accumulation unlikely.

Commercial production began in 1930; it is now obtained as a fraction from the steam cracking of hydrocarbon feedstocks, e.g. propane yields ethylene (42%), propylene (17%) and butadiene (3%). It is also obtained by the dehydrogenation of *n*-butane at $550°$ over an Al_2O_3/Cr_2O_3 catalyst with 30–40% conversion.

Butadiene tends to dimerize by Diels–Alder addition and it must be refrigerated in storage. It reacts with oxygen to form insoluble explosive polymeric peroxides – *t*-butylcatechol (100 ppm) is added to remove free radicals and so prevent exothermic polymerization.

In 1993 in the USA, 540 000 kg, 32% of the total production, was used for the production of styrene–butadiene rubber; synthesis of polybutadiene rubber accounted for a further 23%. Other products are acrylonitrile–butadiene–styrene (ABS) and styrene–butadiene latex. 13% of production goes to making adiponitrile and thence hexamethylenediamine.

$$CH_2{=}CH{-}CH{=}CH_2 \xrightarrow{Cl_2} \underset{Cl}{CH_2}{-}CH{=}CH{-}\underset{Cl}{CH_2} \xrightarrow{NaCN} \underset{CN}{CH_2}{-}CH{=}CH{-}\underset{CN}{CH_2} \xrightarrow{H_2} \underset{NH_2}{CH_2}{-}(CH_2)_4{-}\underset{NH_2}{CH_2}$$

In 1993 3.7% of butadiene produced was converted into dichlorobutenes by direct chlorination at $250°$; isomerization and dehydrochlorination then gave 2-chloro-1,3-butadiene or chloroprene, $CH_2{=}CH{-}CHCl{=}CH_2$, the precursor for polychloroprene rubber.

In 1984 it was estimated that 65 000 workers were exposed to butadiene. Adverse effects from inhalation include cough, hallucinations and irritation of the eyes and mucous membrane.

In the early 1980s the permissible exposure limit at work was 1000 ppm (OSHA), but this was revised following a finding in 1981 that rats and mice developed multiple organ cancer after exposure at the lower level of 625 ppm. At present both the MAK and the ACGIH classify butadiene as a proven carcinogen in animals and suspect in humans (group A2). The latter have set an occupational TLV (TWA) at 2 ppm $(4.4 \, mg \, m^{-3})$, far more stringent than the previous OSHA limit.

In the UK the AEA Technology group now monitors butadiene continuously by capilliary GLC, taking hourly samples at seven sites. In winter, in London and Cardiff, levels of 10 ppb were recorded.

Further reading

Bower, J.S., Broughton, G.F.J., Willis, P.J. and Clark, H. (1995) *Air Pollution in the UK: 1993/4*, AEA Technology, Streets, Baldock.

Chemical Safety Data Sheet (1991), **4a**, Royal Society of Chemistry, Cambridge, pp. 85–87.

BUTANES
See **Liquified Petroleum Gas**.

BUTANETHIOL CAS No. 109-79-5
LD_{50} rat $1.5 \, g \, kg^{-1}$; LC_{50} bluegill sunfish $7.4 \, mg \, l^{-1}$ (24 h)

$$CH_3(CH_2)_2CH_2SH \qquad\qquad C_4H_{10}S \quad M = 90.1$$

1-Butanethiol, thiobutyl alcohol; b.p. 100°; dens. 0.837; v.dens. 3.1; water solubility $590 \, mg \, l^{-1}$ (22°), soluble in ethanol, diethyl ether; log P_{ow} 2.28.

A foul-smelling liquid which is the defence secretion of the skunk. It is highly flammable and reacts violently with strong oxidizing agents, e.g. nitric acid. Butanethiol irritates the eyes, skin and respiratory tract.

The C_8–C_{18} analogues are used to modify synthetic rubbers: a styrene–butadiene rubber was obtained from the following emulsified components: styrene 25 parts, butadiene 75 parts, water 180 parts; lauryl mercaptan $(C_{12}H_{25}SH)$ 0.5 parts and potassium persulphate initiator 0.3 parts.

ACGIH has set the TLV (TWA) for butanethiol at 0.5 ppm $(1.8 \, mg \, m^{-3})$.

BUTANOLS
$$CH_3CH_2CH_2CH_2OH; \; CH_3CH_2CH(OH)CH_3; \; (CH_3)_3COH; \; (CH_3)_2CHCH_2OH$$

$$C_4H_{10}O \; M = 74.1$$

Butanol	CAS No.	B.p. (°)	v.p. (20°) (kPa)	Water solubility (%)	Other
1-Butanol	71-36-3	118	0.56	7.7	Misc. EtOH, Et$_2$O
2-Butanol	78-92-2	98.5	1.66	12.5	As above
t-Butanol	75-65-0	81.5 m.p. 25°	0.4	Soluble	As above
Isobutanol	78-83-1	108	1.17	8.7	As above

The butanols are determined by the NIOSH method by trapping from an air stream on charcoal, desorption with CS^2 followed by GLC with FI detection with a sensitivity of $1 \, mg \, m^{-3}$.

1-Butanol is used in perfumes and flavourings and as a solvent for paints, coatings and resins. It is an intermediate in the synthesis of dibutyl phthalate, butyl acetate and herbicide esters, also in ore-flotation agents.

As a flavouring agent in beverages (rum, whisky) it is limited to $12\,mg\,l^{-1}$, in cordials to $1\,mg\,l^{-1}$ and in ice-cream to $7\,mg\,l^{-1}$. It strongly flavours drinking water at a concentration of $20\,mg\,l^{-1}$ and has an odour threshold of $1\,mg\,l^{-1}$.

It is rapidly degraded and makes a considerable oxygen demand. 1-Butanol was found within half of a group of mobile homes at a concentration in air of 5 ppb. In 1983 total emissions over The Netherlands were estimated at 616 t.

In humans it causes symptoms of alcohol intoxication, irritation of the nose and throat, headache and vertigo. The LD_{50} oral rat is in the range 0.8–$2.0\,g\,kg^{-1}$. Case studies of industrial workers led the ACGIH to set a TLV of 50 ppm ($154\,mg\,m^{-3}$).

2-Butanol is principally used as an intermediate for methyl ethyl ketone. It is low in toxicity towards fish and in the rat the oral LD_{50} is $6.4\,g\,kg^{-1}$. It has a similar effect on humans as 1-butanol and has a low health risk.

t-Butanol is used primarily as a dehydrating agent and solvent and in the manufacture of *t*-butyl compounds such as *t*-butyl chloride. The 1983 estimate of release over The Netherlands was 207 t; it has low human toxicity.

Isobutanol is used as a solvent, plasticizer, in perfumes and flavourings and for the synthesis of isobutyl esters. It is readily degraded via isobutyraldehyde and isobutyric acid and has a high oxygen demand. Severe exposure produces similar effects in humans to those arising from 1-butanol although in general the risk is low.

It is formed naturally by the fermentation of carbohydrates and is found in fruit (cherry, raspberry, grapes) and beverages (brandy, gin, cider) and also in coffee. In the USA maximum residue levels have been set: beverages 17, ice-cream 7 and sweets $30\,mg\,kg^{-1}$.

The LC_{50} (9 h) is in the range 1–$4\,g\,l^{-1}$; in the rat the oral LD_{50} is $2.5\,g\,kg^{-1}$ with LC_{50} on 4 h inhalation $8\,g\,m^{-3}$.

Further reading
Environ. Health Criteria, World Health Organization, Geneva, 1987, Vol. 65.

BUTYL ACETATES

$AcO(CH_2)_3CH_3$	$AcOCHEt$ $|$ CH_3	$AcOC(CH_3)_3$
n-butyl	*sec*-butyl	*tert*-butyl
CAS No. 123-86-4	105-46-4	540-88-5
B.p.(°) 126	112	96
Flash point (°) 33	31	16–22
Dens. 0.88 (20°)	0.86 (25°)	0.87 (20°)
V.p. (Torr) 10 (20°)	16 (20°)	–
Water solubility (g 100 g) 1.6	–	–
US TLV (ppm) 20	200	200

All three are soluble in common organic solvents and are themselves used as solvents for nitrocellulose and cellulose acetate and for acrylic, urethane and polyester coatings. They may be mixed with ketones, glycol ethers, alcohols or hydrocarbons to ensure the correct flow and viscosity on application. tert-Butyl acetate is a gasoline additive.

These acetates are released to air when applied in paints and varnishes. They are readily biodegraded and do not persist in the environment. TLV's are largely based on their irritation of the eyes and nasal passages; however, n-butyl acetate is an experimental teratogen and has narcotic properties; its TLV has recently been reduced by the ACGIH.

Further reading
Archer, W.L. (1996) *Industrial Solvents Handbook*, Marcel Dekker, New York.

BUTYLATED HYDROXYANISOLE CAS No. 25013-16-5
LD_{50} rat $2\,g\,kg^{-1}$; TD_{Lo} rat $30\,g\,kg^{-1}$

$C_{11}H_{16}O_2$ $M = 180.27$

BHA, BFW 750, BQ 1000, butylhydroxyanisole, 3-*tert*-butyl-4- hydroxyanisole; m.p. 105°; b.p. 270°/733 mmHg.

The commercial product is a mixture of m.p. 48–55° which typically contains 85% of the 3-*tert*-butyl and 15% of the 2-*tert*-butyl isomer, which can be separated by chromatography on Sephadex LH20.

The preparation is by partial methylation of hydroquinone followed by alkylation with *tert*-butanol –H_3PO_4. Commercial production is by the action of isobutylene on hydroquinone methyl ether and was 300×10^3 t year^{-1} in 1982.

Samples may be obtained from human plasma and urine by petrol ether extraction followed by GC–FID and from foodstuffs by steam distillation, extraction (CH_2Cl_2) and GC.

BHA has been used since 1947 as an antioxidant in fatty foodstuffs (butter, lard, etc.) up to a present concentration limit of 0.02% and is also used to protect cereals, sweets and beer. It is much used in cosmetics including over 1000 different types of lipstick and eye shadow.

The levels in vegetable fats and oils range between 19 and 91 mg kg^{-1} (0.009%) and in flour and cereals between 7 and 22 mg kg^{-1} (0.002%). The average estimated intake of an individual in the USA is 4.3 mg or 0.1 mg kg^{-1} body weight. BHA and **BHT** (see following entry) show weak activity as 'endocrine disruptors' – chemicals which when ingested have effects related to the female sex hormones and which contribute to a decline in male sperm counts in the technologically advanced countries (Environmental Data Services, 1995). BHA is moderately toxic by ingestion and has been classified as group C by the IARC with evidence of carcinogenicity in animals but with no human data. The EC limit (1978) for an adult is a daily intake of 30 mg (combined with BHT); the WHO(1983) retains a limit of 0.5 mg kg^{-1} (with BHT).

Reference
Environmental Data Services (1995) *The ENDS Report*, No. 243, pp. 15–20.

Further reading
IARC (1986) *Some Naturally Occurring and Synthetic Food Compounds*, IARC, Lyon, Vol. 40, pp. 123–159.
Ash, M. and Ash, I. (1996) *Handbook of Paint and Coating Raw Materials*, Gower, Aldershot, Vol. 2, pp. 36–39.

BUTYLATED HYDROXYTOLUENE CAS No. 138-37-0
TD_{Lo} rat $6 g kg^{-1}$; LD_{50} rat $890 mg kg^{-1}$

$C_{15}H_{24}O$ $M = 220.4$

BHT, butylhydroxytoluene, 2,6-di-*tert*-butyl-*p*-cresol, 2,6-bis-(1,1-dimethylethyl)-4-methylphenol; m.p. 71°; b.p. 265°,, 136°/10 mmHg; dens. 1.048; v. dens 7.6; v.p. 0.01 mmHg (20°); log P_{ow} 5.80. Solubility in water $0.4 mg l^{-1}$ (20°), soluble in most hydrocarbon solvents, ketones and alcohols. $1 mg m^{-3} = 9.01$ ppm.

Produced commercially by the alkylation of *p*-cresol with isobutylene. US production in 1984 was 12×10^6 kg; Europe and Japan are also major sources.

Food grade BHT is 99% (w/w) pure with a maximum level of heavy metals of $10 mg kg^{-1}$. Like butylated hydroxyanisole, it is effective as an antioxidant in food, but the main use now is in plastic food packaging and in rubber and gasoline (up to 0.008%).

Analytical procedures are similar to those for BHA and the BHT limit in food is set by the WHO at $0.5 mg kg^{-1}$ (including any BHA). Students at the University of California took large doses of BHT (Shlian and Goldstone, 1986) as a treatment for genital herpes simplex virus infections (HSV) at doses of $2–6 g day^{-1}$; it was available as 250 mg capsules in health food shops. A 22-year-old woman patient ingested 4 g of BHT on an empty stomach and experienced vomiting and dizziness and lost consciousness. The drug is not approved by the FDA as a treatment for herpes and the source was a bestselling book by Pearson and Shaw (1984). This use was inadvisable because in addition to irritating the skin and eyes, BHT may cause skin rashes and is moderately toxic by ingestion – the estimated lethal dose in humans has been estimated as $20 g day^{-1}$, with a TD_{Lo} oral human of 80 mg. Further, it induces lung tumours in mice and liver and bladder cancer in rats. Butylated hydroxytoluene accumulates in human adipose tissue and is rated a group 3 substance by the IARC – limited evidence for carcinogenicity in animals.

Compare with the preceding entry on **BHA**.

References
Pearson, D. and Shaw, S. (1984) *The Life Extension Companion*, Warner, New York.
Shlian, D.M. and Goldstone, J.(1986) Toxicity of butylated hydroxytoluene, *N. Eng. J. Med.* **314**, 648–649.

Further reading
IARC (1986) *Some Naturally Occurring and Synthetic Food Components*, IARC, Lyon, Vol. 40, pp. 161–206.

tert-BUTYL HYDROPEROXIDE CAS No. 75-91-2
LD_{50} rat 406 mg kg^{-1}; LC_{50} inhalation (4 h) rat 500 ppm

$$CH_3-\overset{\overset{\displaystyle CH_3}{|}}{\underset{\underset{\displaystyle CH_3}{|}}{C}}-O-OH$$

$C_4H_{10}O_2$ $M = 90.1$

1-Dimethylethyl hydroperoxide, perbutyl, TBHP-70; m.p. $-8°$; b.p. $40°/23$ mmHg; v.p. 174 mmHg at $60°$. Slightly soluble in water, soluble in ethanol, diethyl ether, chloroform.

TBHP is supplied as a 70% or 90% solution and is prepared by the alkylation of hydrogen peroxide with t-butanol or isobutylene using sulphuric acid as a catalyst. Over 100 organic peroxides are formulated as solids, liquids, pastes, powders, solutions and emulsions; t-butyl derivatives are a significant fraction of the whole, for example:

t-butyl peroxypivalate (A) t-butyl peroxyisobutyrate (B) 1,1- di(t-butylperoxy)cyclohexane

$$CH_3-\overset{\overset{\displaystyle CH_3}{|}}{\underset{\underset{\displaystyle CH_3}{|}}{C}}-O-O-\overset{\overset{\displaystyle O}{\|}}{C}-\overset{\overset{\displaystyle CH_3}{|}}{\underset{\underset{\displaystyle CH_3}{|}}{C}}-CH_3 \quad (A)$$

$$CH_3-\overset{\overset{\displaystyle CH_3}{|}}{\underset{\underset{\displaystyle CH_3}{|}}{C}}-O-O-\overset{\overset{\displaystyle O}{\|}}{C}-\overset{\overset{\displaystyle CH_3}{|}}{\underset{\underset{\displaystyle H}{|}}{C}}-CH_3 \quad (B)$$

CAS No. 927-07-1   CAS No. 109-07-1

Worldwide production of all organic peroxides in 1993 was 103 600 t including US production of 33 700 t; about 5% of these were alkyl hydroperoxides.

When stored in a volume of 5 gallons or more TBHP will undergo self-accelerating decomposition at $88°$ with the explosive release of CO and CO_2. It is a powerful oxidizing agent which must not contact powdered metals, strong acid or alkali or combustible materials.

It is used to peroxidize organic molecules and with its derivatives to initiate low-temperature polymerization in industry, including ethylene, styrene, vinyl chloride, acrylonitrile and butadiene, and for cross-linking polyethylene and copolymers.

Further reading
Kirk–Othmer ECT, 4th edn. Wiley, New York, 1994, Vol. 18, p. 230.

p-tert-**BUTYLPHENOL** CAS No. 98-54-4
LD_{50} rat 3250 µg l kg^{-1}

$C_{10}H_{14}O$ $M = 150.2$

4-t-butylphenol, PTBP, 4-(1,1-dimethylethyl) phenol; m.p. $99°$; b.p. $238°$; v.p. 1 mmHg ($70°$); v. dens. 5.1.

Used as a plasticizer for cellulose acetate and as an intermediate for antioxidants, phenolic resins and paints and as a motor oil additive. It is readily oxidized.

PTBP is moderately toxic, irritates the eyes and skin and carries the risk of cutaneous absorption, which can lead to a far higher uptake of toxic substances than inhalation (cf. aniline, nitrobenzene, ethylene glycol, phenols). The MAK value is set at a TLV (TWA) of 0.08 ppm ($0.5\,mg\,m^{-3}$)

CADMIUM CAS No. 7440-43-9

Cd

Cadmium, a member of Group 12 of the Periodic Table, has an atomic number of 48, an atomic weight of 112.40, a specific gravity of 8.65 at 20°, a melting point of 320.9° and, a boiling point of 767°. It is a soft, ductile, silver-white metal, with eight naturally occurring isotopes, ^{106}Cd (1.25%), ^{108}Cd (0.89%), ^{110}Cd (12.49%), ^{111}Cd (12.80%), ^{112}Cd (24.13%), ^{113}Cd (radioactive, 12.22%), ^{114}Cd (28.73%) and ^{116}Cd (7.49%). Like the other Group 12 members (Zn, Hg), cadmium exists in a stable oxidation state of $+2$.

In the earth's crust the cadmium level is ~ 0.20 ppm ($0.1\,mg\,kg^{-1}$) but varies widely according to type of rock. Cadmium, 67th in relative abundance, is obtained as a by-product in the extraction of zinc from *zinc blende*, ZnS. Metallic cadmium is obtained by reducing the oxide with carbon. Sulphide ore roasting, coal burning, phosphate fertilizers (2–20 ppm) and sewage sludge (up to 22 ppm Cd) are the main sources of cadmium in the environment, leading to local and regional pollution of soils and river sediments. Average soil concentration ~ 2 ppm. High levels of cadmium were found in spoil from former zinc mines at Shipham, Somerset, UK. In solution, cadmium is highly toxic to plants and animals. Despite being phytotoxic, some plant species accumulate cadmium, particularly leafy plants – lettuce, spinach, turnips greens – the average cadmium content of cigarettes is reported to be $\sim 2\,\mu g\,g^{-1}$ Normal cadmium levels in plants are below $0.5\,mg\,kg^{-1}$ (dry weight). Cd is present in drinking water at ppb levels (0.01–$0.15\,\mu g\,l^{-1}$). Further, since the concentration of cadmium in air is usually low ($ng\,m^{-3}$), ingestion is the main way cadmium enters the human food chain. Only about 6% of the estimated 40–50 $\mu g\,day^{-1}$ of ingested cadmium is absorbed by the body, whereas 25–50% of the 2–10 $\mu g\,day^{-1}$ of cadmium in inhaled dust is absorbed. The average daily diet is 25–60 μg (WHO recommends that the daily intake of cadmium should not exceed $1\,\mu g\,kg^{-1}$ of body weight). The average body burden is 15–30 mg cadmium, stored in the liver and kidneys, where it is predominantly bound to metallothioneins, which are cysteine-rich proteins. In whole blood, cadmium is mainly bound to erythrocytes. Typical concentrations are blood serum 0.012 μg % and urine 0.06 μg %. Elimination from the body is slow and the biological half life is of the order of 10–20 years.

Uses. Cadmium and its compounds are used in the manufacture of batteries, pigments/paints, polymer stabilizers, alloys, printing and graphics, electroplated goods and safety rods in the nuclear industry. Cadmium is used in the production of special alloys melting at low temperatures. Wood's alloy (50% Bi, 25% Pb, 12.5% Cd, 12.5% Sn) melts at 70° and is used in fuses for telephone installations and fire prevention systems. Babbit metals (Cd–Ag or Cd–Ag–Cu) are used in producing special antifriction bearings. CdS is an important yellow pigment and CdS and CdSe are semiconductors. Cadmium tungstate, X-ray screen production. CdO, ceramic glaze. $CdBr_2$ and $CdCl_2$ are used in photocopying.

In the EC, USA and other countries a number of directives have been issued to reduce or eliminate the use of cadmium, e.g. replacement of cadmium in electroplating by zinc, cadmium pigments by other metals/organics and cadmium-containing stabilizers in plastics by Pb, Ba, Zn or Ca/Zn. Recycling programmes for spent Ni–Cd batteries have been started by several battery manufacturers in some countries, e.g. France and Japan. Recent UK directives include the following: the cadmium content in manufactured material not to exceed 0.01% by weight; for incineration of waste and a plant capacity of $1–3\,t\,h^{-1}$, the cadmium concentration should be less than $0.2\,mg\,m^{-3}$.

Cadmium and its compounds are highly toxic and usually carcinogenic and have been designated as one of the 100 most hazardous substances. Very toxic by ingestion and inhalation. Acute ingestion of cadmium concentrations $\gtrsim 0.1–1.0\,mg\,kg^{-1}day^{-1}$ produces symptoms of nausea, vomiting, abdominal cramps and headache. Inhalation may cause acute or chronic lung/renal diseases. Chemical pneumonitis or pulmonary oedema may result from acute over-exposure to cadmium fumes, as oxide and chloride aerosols, at a dose of $5\,mg\,m^{-3}$ over an 8 h period. $1\,mg\,m^{-3}$ inhaled over the same time period gives rise to clinically evident symptoms in sensitive individuals. Deaths from acute cadmium poisoning have resulted from inhalation of cadmiun oxide smokes and fumes, usually from welding operations on cadmium-plated steels in poorly ventilated areas. Lethal doses of cadmium have been calculated to be 2600–$2900\,mg\,m^{-3}min^{-1}$. Chronic poisoning: long-term exposure to cadmium may affect body organs and systems, especially the kidneys. Long-term inhalation of cadmium fumes resulted in chronic rhinitis and pharyngitis and nasal bleeding. Increased risk of lung and prostate cancer. Toxic effects are largely due to its ability to inhibit various enzymes systems by blocking the carboxyl, amine and especially sulphydryl groups in protein molecules. Cadmium reduces the activity of the digestive enzymes trypsin and pepsin. Affects carbohydrate metabolism and inhibits glycogen synthesis in the liver. Competition between cadmium and its group congener zinc can be either antagonistic or synergistic. Acute poisoning can result if cadmium displaces zinc in a number of metalloenzymes. A serious case of chronic cadmium poisoning occurred in the Zinzu river basin, Toyama, Japan, where Pb–Zn mining has caused widespread Zn–Cd contamination of paddy fields and drinking water sources and led to a high intake of $\sim 600\,\mu g$ Cd day^{-1}. Structural changes in bones (osteoporosis and osteomalacia) were observed in post-menopausal women. This was named the Itai-Itai ('Ouch! Ouch!') disease. Symptoms of the disease are aching joints and kidney tubule dysfunction.

Examples of cadmium poisoning include the following. Food contamination by cadmium may be due to pots and pans with Cd glazing. Three men died after drinking half a glass of a liquid containing 17.5% of $CdCl_2$. One death from 20 ml of a 0.9% $CdSO_4$ solution. When food or drink with a concentration of $15\,mg\,l^{-1}$ is ingested, vomiting and other gastrointestinal symptoms occur. Cadmium-induced pneumonitis: in this case was described one elderly brazer who developed chemical pneumonitis and respiratory insufficiencies after brazing with a Cd-containing lead solder (Cu,Zn + 15% Cd). In 1963, a welder developed pulmonary oedema after welding a Cd–Ag alloy; 17 years later this man developed severe progressive pulmonary fibrosis. Another man spent 2 weeks brazing the propeller of a ship with a Cd-containing solder in a confined area with poor ventilation; he developed fibrosing alveolitis.

Occupational exposure limits. US TLV (TWA) $0.05\,mg\,m^{-3}$. UK long-term limit MEL $0.05\,mg\,m^{-3}$. ACGIH PEL (TWA) total dust $0.01\,mg\,m^{-3}$; respirable fraction $0.002\,mg\,m^{-3}$. Cadmium powder is flammable.

Toxicity. LD_{50} oral rat 225 mg kg^{-1}
LC_{50} (30 min) inhalation rat 25 mg m^{-3}.

Further reading

ECETOX (1991) *Technical Report 30(4)*, European Chemical Industry Ecology and Toxicology Centre, Brussels.

Environmental Health Criteria 134. Cadmium, WHO/IPCS, Geneva, 1992.

Environmental Health Criteria 135. Cadmium – Environmental Aspects, WHO/IPCS, Geneva, 1992.

Friberg, L. (1986) *Handbook of the Toxicology of Metals*, 2nd edn, Elsevier, Amsterdam, pp. 1–2.

Norberg, G.F., Herber, R.F.M. and Alessio, L. (1992) *Cadmium in the Human Environment: Toxicity and Carcinogenicity*, IARC Lyon, vol. 118.

Cadmium acetate CAS No. 543-90-8

$(CH_3CO_2)_2Cd$

Uses. Electroplating. Colouring agent in glass and ceramics

Toxicity. Harmful chemical. Toxic to small fish at conc. 10 mg l^{-1} (3–24 h). LD_{50} ip mouse 14 mg kg^{-1}.

Cadmium bromide tetrahydrate, CAS No. 7789-42-6

$CdBr_2.4H_2O$

Uses. Lithography and photography.

Toxicity. Harmful chemical. Carcinogenic when inhaled. Toxic to small fish at 10 mg l^{-1} (3–24 h).

Cadmium carbonate CAS No. 513-78-0

$CdCO_3$

Uses. Preparation of cadmium red/yellow, cadmium sulphide.

Toxicity. Poison and carcinogenic. Target organs: respiratory system, kidneys, prostate and blood.

Cadmium chloride CAS No. 10108-64-2

$CdCl_2$

Uses. Fungicide. Preparation of cadmium salts and cadmium yellow pigment.

Toxicity. Poison and carcinogenic. LD_{50} oral rat, mouse 60, 88 mg kg^{-1} respectively.

Cadmium fluoride CAS No. 7790-79-6

CdF_2

Uses. Glass manufacture. Catalyst. Nuclear reactor controlling agent.

Toxicity. Toxic by all routes. Carcinogenic.

Cadmium iodide, CAS No. 7790-80-9

CdI_2

Uses.   Lithography and photography. Manufacture of phosphors.

Toxicity.   Toxic by all routes. Carcinogenic. LD_{Lo} oral humans $81\,mg\,kg^{-1}$. LD_{50} oral mouse $166\,mg\,kg^{-1}$.

Cadmium nitrate, CAS No. 10325-94-7

$Cd(NO_3)_2$

Uses.   Preparation of cadmium salts. Photographic emulsions. Nuclear reactors. Colouring glass.

Toxicity.   Highly toxic to fish and plants; $\sim$9 g is fatal to humans.

Cadmium oxide, CAS No. 1306-19-0

CdO

Uses.   Batteries. Glass. Semiconductors. Ceramic glazes. Preparation of cadmium salts. Catalyst. Manufacture of Teflon and PVC plastics.

Occupational exposure limit.   US TLV (TWA) $0.01\,mg\,m^{-3}$ total dust as Cd.

Toxicity.   Toxic by all routes. Carcinogenic. LD_{50} oral rat, mouse $72\,mg\,kg^{-1}$; LC_{50} (15 min) inhalation rabbit $3000\,mg\,m^{-3}$.

Cadmium(II) selenide CAS No. 1306-24-7

CdSe

Uses.   Photoelectric cells, photoconductors, rectifiers.

Occupational exposure limits.   ACGIH TLV (TWA) $0.002\,mg\,Cd\,m^{-3}$ (respiratory dust).

Toxicity.   Carcinogen. Acute effects, produces nausea and headaches. Changes to the kidneys, lungs and liver.

Cadmium sulphate CAS No. 10124-36-4

$CdSO_4$

Uses.   In the manufacture of long chain fatty acids used as stabilizers for plastics. Electroplating. In phosphors. Catalyst.

Toxicity.   Poison and suspect carcinogen. LD_{50} oral mouse $88\,mg\,m^{-3}$. LD_{Lo} oral dog $105\,mg\,kg^{-1}$.

Cadmium(II) sulphide CAS No. 1306-23-6

CdS

Uses.   Pigment in plastics, paints, enamels, rubber, ceramics, etc.

Toxicity.   Possible selective carcinogen. LD_{50} oral rat $7080\,mg\,kg^{-1}$.

CAESIUM CAS No. 7440-46-2

Cs

Caesium, a member of Group 1 in the Periodic Table, is a soft, ductile, silver-white electropositive metal. Like sodium and potassium, caesium is an alkali metal and forms only one stable cation, Cs^+, in solution. It has an atomic number of 55, an atomic weight of 132.91, a specific gravity of 1.87 at 20°, a melting point of 28.5° and boiling point of 670°; it has only one stable isotope in nature, ^{133}Cs. Radioisotopes $^{134}Cs(t_{1/2} = 2.1$ years) and $^{137}Cs(t_{1/2} = 30$ years) are produced in nuclear reactors. $^{129}Cs(t_{1/2} = 32$ h) is used in heart diagnosis.

A rare element, being around 3 ppm in the earth's crust (similar to bromine and uranium.). Two rare minerals are *pollucite*, $(Cs, Na)[AlSi_2O_6].nH_2O$, and *avogadrite*, $(K, Cs)BF_4$. Also found in *beryl* and *carnallite* but the main source of caesium is as a byproduct in lithium production. Caesium is emitted from coal burning power stations. Measureable amounts are found in plants and animals, water, soils and emissions from coal-fired power stations. Typical concentrations are seawater $1 \mu g l^{-1}$, river water $0.05 \mu g l^{-1}$, tea leaves $0.5-1 mg kg^{-1}$ dry ash; daily intake $0.004-0.03$ mg; total body concentrations in adults ~ 1.25 mg with a blood level content of $\sim 2.8 \mu g l^{-1}$.

Uses. Caesium and its compounds are used in photo devices, scintillation counters, storage batteries, semiconductors, glass and ceramics, and atomic clocks, which are accurate to 5 s in 300 years. The radioisotope ^{132}Cs is used in the treatment of cancer.

The caesium ion, Cs^+, is more toxic to plants and animals than sodium ions but less toxic than potassium, rubidium and lithium ions. The solubility of caesium salts is much greater in marine waters than fresh waters and the uptake by fish is controlled by the K^+ ion content in water. Caesium enters the body via ingestion where it is readily absorbed through subcutaneous tissue, lungs, muscle, neural tissues, placental and biological barriers. Excessive intakes of caesium are largely stored and accumulated in the muscle, leading to systemic poisoning which is detected by changes in blood serum and blood sugar levels. Chronic effects on workers in caesium production show abnormalities of cardiac activity and changes in the upper respiratory tract. The clinical picture of acute poisoning was marked by excitations, convulsions and breathing difficulties, followed by death. It has been found that completely replacing potassium ion in the diet of rats with Cs^+ ion caused death after 10–17 days. The majority of caesium salts are moderately toxic by ingestion or via the intraperitoneal route.

Caesium-137 (β emitter, $t_{1/2} = 30.2$ years), used in the treatment of cancer, is a product of atomic fusion of uranium. In 1986, a nuclear accident occurred at the Chernobyl power station near Kiev in the Ukraine which released large amounts of ^{137}Cs (and ^{134}Cs, β emitter, $t_{1/2} = 2.06$ years) into the environment. More than two orders of magnitude greater than pre-accident levels were recorded in foliage and soils in many areas of Europe including the UK. Once deposited on land, caesium becomes tightly bound by the clay minerals in the soil, and root uptake is slight. However, it has been reported that the absorption of caesium ions is inversely proportional to potassium ion concentrations in the soil so uptake increases if the soil is potassium deficient and vice versa. Because Cs is only slowly released from clay minerals, foliar absorption by grasses, cereals, lichens, etc., and the feeding on these plants by livestock, providing meat and milk, is the most important way ^{137}Cs enters the human food chain. Apropos Chernobyl, even 2 months after the disaster, grasslands in the Lake District, North Wales and Scotland were so polluted that one million sheep and lambs were too radioactive to be slaughtered for food.

Occupational exposure limit. N.a. Metallic caesium is extremely flammable.

Further reading

SI 1991 No. 472 The Environmental Protection (Prescribed Processes and Substances) Regulations 1991, HMSO, London, 1991.

Alloway, B.J. and Ayres, D.C. (1997) *Chemical Principles of Environmental Polution*, 2nd edn, Blackie, London.

Kirk–Othmer ECT, 4th edn, Wiley, New York, 1993, Vol. 5, pp. 749–764.

Caesium(I) bromide CAS No. 7787-69-1

CsBr

Uses. Lenses and prisms.

Toxicity. LD_{50} oral rat 600–800 mg kg^{-1}; intraperitoneal rat 1400 mg kg^{-1}.

Caesium(I) chloride CAS No. 7647-17-8

CsCl

Uses. Scintillation counters. Purification of nucleic acids.

Toxicity. LD_{50} oral rat 2600 mg kg^{-1}; intraperitoneal rat 1500 mg kg^{-1}.

Caesium(I) hydroxide monohydrate CAS No. 35103-79-8

CsOH.H$_2$O

Uses. Battery electrolyte. Catalyst in the polymerization of siloxanes.

Occupational exposure limit. USA TLV(TWA) 2 mg m^{-3}

Toxicity. Irritant. Caustic agent. Above 5 mg m^{-3} skin scarring and eye irritant. LD_{50} oral rat 600 mg m^{-3}. LD_{50} intravenous mouse 910 mg kg^{-1}.

Caesium(I) iodide CAS No. 7789-17-5

CsI

Uses. Scintillation counters. Lenses and prisms for infrared spectrometers.

Toxicity. Moderately toxic. LD_{50} oral rat 2386 mg kg^{-1}.

Caesium(I) nitrate CAS No. 7789-18-6

CsNO$_3$

Uses. Changing refractive index of silicate/borosilicate glass.

Toxicity. Moderately toxic. LD_{50} oral rat 2350 mg kg^{-1}; intraperitoneal rat 1200 mg kg^{-1}.

CALCIUM CAS No. 7440-70-2

Ca

Calcium, a member of Group 2 (the alkaline earths) of the Periodic Table, has an atomic number of 20, an atomic weight of 40.08, a specific gravity of 1.55 at 20°, a melting point of 842° and a boiling point of 1484°. It is a bright silvery, soft and ductile metal, forming salts with a +2 oxidation state. In nature there are six isotopes:

^{40}Ca (96.94%), ^{42}Ca (0.65%), ^{43}Ca (0.14%), ^{44}Ca (2.09%), ^{46}Ca (0.004%) and ^{48}Ca (0.19%).

It occurs abundantly (fifth in order), constituting 3.63% of the earth's crust. Found as *calcite*, $CaCO_3$ (limestone, marble, chalk), *gypsum*, $CaSO_4.2H_2O$, *dolomite*, $CaCO_3.MgCO_3$, and many other minerals. Oceans contain a vast amount of calcium, $\sim 400\,mg\,l^{-1}$.

Uses. Reducing agent for the production of less common metals (e.g. Hf, Th, W, Zr). Alloys with iron, lead and aluminium. Manufacture of electronic vacuum tubes. Treatment of calcium deficiency.

It is an essential element for plant and animal life. It is present in living organisms, as a constituent of bones, teeth, shells and coral. In humans, calcium is an essential body electrolyte, regulated by the parathyroid hormone calcitonin and vitamin D. Calcium compounds are mostly non-toxic, except where they contain a toxic element as part of the anion. The recommended daily intake for adult humans is $\sim 800\,mg$. The mass of calcium in the human body is between 800 and 1400 g. Cow's milk contains on average about $1.30\,g\,l^{-1}$. The natural uptake of calcium from the gastrointestinal tract can vary from normal to two extremes. More calcium than the body normally absorbs can result in hypercalcaemia, which is commonly caused by neoplastic disease or hyperparathyroidism. Conversely, reduced absorption from the gastrointestinal tract may result in hypocalcaemia, which can be caused by reduced parathyroid hormone activity and vitamin D deficiency.

Further reading

ECETOX (1991) *Technical Report 30(4)*, European Chemical Industry Technology and Toxicology Centre, Brussels.

King, R.B. (ed.) (1994) *Encyclopedia of Inorganic Chemistry*, John Wiley, Chichester.

Kirk–Othmer ECT, 4th edn, Wiley, New York, 1992, Vol. 4, pp. 777–826.

SI 1991 No. 472 The Environmental Protection (Prescribed Processes and Substances) Regulations 1991, HMSO, London, 1991.

Thompson, R. (ed.) (1995) *Industrial Inorganic Chemicals. Production and Uses*, Royal Society of Chemistry, Cambridge, Ch 1.

Calcium arsenate CAS No. 7778-44-1

$Ca_3As_2O_8$

Uses. Cotton fields insecticide. Fruit and vegetable pesticide. Its use is widely prohibited.

Toxicity. Poison. Human carcinogen. Toxic via skin absorption.

Calcium carbide CAS No. 75-20-7

CaC_2

Uses. Acetylene generation on a small scale. Preparation of calcium cyanamide. Desulphurizing and deoxidizing agent in metallurgy.

Toxicity. Reacts with water to give flammable gas. Irritant to eyes and skin.

Calcium chloride CAS No. 10043-52-4

$CaCl_2$

Uses. Pavement de-icer. Fire extinguishers. Drying agent.

Toxicity. LD_{50} oral rat $1\,g\,kg^{-1}$; oral mouse $1940\,mg\,kg^{-1}$.

Calcium cyanamide CAS No. 7440-70-2

CaNCN

Uses. Fertilizer. Herbicide. Pesticide. Manufacture of melamine. Manufacture and refining of iron.

Toxicity. LD_{50} oral rat $158\,mg\,kg^{-1}$. LC_{Lo} (4 h) inhalation rat $86\,mg\,m^{-3}$. Toxic to humans.

Calcium fluoride CAS No.7789-75-5

CaF_2

Uses. Main source of fluorine and its compounds since it occurs as a mineral. Oral hygiene products.

Toxicity. Non-carcinogenic to humans or animals. Excess fluoride produces changes in teeth and skeleton. LD_{30} oral rat $4250\,mg\,kg^{-1}$

Calcium hypochlorite CAS No. 7778-54-3

$Ca(OCl)_2$

Uses. Fungicide. Algicide. Bactericide. Disinfectant. Oxidizing agent. Bleaching agent.

Toxicity. Moderately toxic. Causes severe irritation of skin, eyes and mucous membranes. Emits fumes capable of causing pulmonary oedema. Mutation data reported for laboratory animals. LD_{50} oral rat $850\,mg\,kg^{-1}$.

Calcium nitrate CAS No. 13477-34-4

$Ca(NO_3)_2.4H_2O$

Uses. Manufacture of explosives, fertilizers, matches and pyrotechnics.

Toxicity. Toxic to humans at high concentration. LD_{50} oral rat $3900\,mg\,kg^{-1}$.

Calcium oxide CAS No. 1305-78-8

CaO

Uses. Building materials. Fungicide and insecticide. Desiccant.

Toxicity. Strongly caustic. Severe exposure can cause pulmonary oedema.

Calcium polysulphides CAS No. 1344-81-6

CaS_x

Uses. Fungicide. Insecticide.

Toxicity. Irrritant.

Calcium tungstate(VI) CAS No. 7790-75-2

$CaWO_4$

Uses.   Fluorescent lighting. X-ray screens. Luminous paints. Scintillation counters.

Toxicity.   May be harmful by all routes. Eye and skin irritant.

CAMPHOR synthetic CAS No. 76-22-2
             dl-form CAS No. 21368-68-3
             (−)-form CAS No. 464-48-2
             (+)-form as shown CAS No. 464-49-3
LD$_{50}$ oral mouse 1.31 g kg^{-1}; LC$_{50}$ fathead minnow (96 h) 110 mg l^{-1}

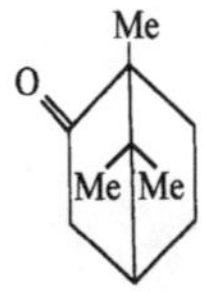

$C_{10}H_{16}O$ $M = 152.2$

1,7,7-Trimethylbicyclo[2.2.1]-2-heptanone, 2-oxobornane; m.p. 179°; b.p. 204°; dens.
0.992 (25°); v.p. 0.18 Torr (20°). Slightly soluble in water, very soluble in ethanol,
diethyl ether. Note the limited range as a liquid.

The optically active (+)-form occurs in the gum of the camphor tree, *Cinnamomum
camphora*, and also in members of the Lauracaea, Labaiateae and Compositae. The
(−)-form is less common; it is found in *Matricaria parthenium*.

The starting materials for the preparation of camphor are otained from turpentine
which contains 60–70% of α-pinene and 20–25% of β-pinene. Synthetic camphor is
obtained by the rearrangement of α- or β- pinene to give camphene in the absence of
water with a surface- active catalyst such as TiO$_2$. A further acid-catalysed change in
acetic acid affords bornyl acetate, which gives camphor following hydrolysis and
oxidation:

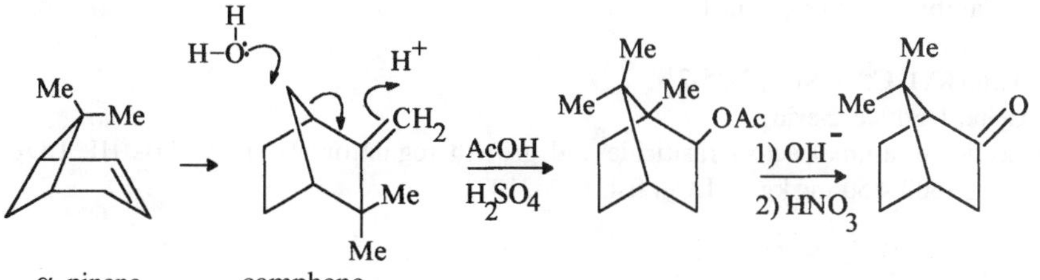

Camphor is used as a plasticizer for cellulose nitrate and other cellulose esters and
for explosives and lacquers, also in moth and mildew proofing, insecticides, flavouring, toothpowders and pyrotechnics.

The fatal dose of the (+)-form in a 1 year old child was ca 1.0 g; in adults 2 g produce
toxic symptoms and 4 g may be fatal. A seriously poisoned subject had a plasma level
of 1.7 μg ml^{-1}. Symptoms in humans are nausea, vomiting, vertigo, mental confusion
and convulsions arising from the stimulation of the cerebral cortex. The cause of death
is respiratory failure, chronic exposure may cause hepatic and renal damage.

Camphor is metabolized in the liver to hydroxy derivatives which are excreted as
glucuronic acid conjugates. It has been banned from internal use in the USA, where

the TLV (TWA) is set at 2 ppm ($12\,\mathrm{mg\,m^{-3}}$), low enough to avoid irritation of the eyes, nose and throat.

Further reading
Dictionary of Natural Products, Chapman & Hall, London, 1994, entry C-00194.
Kirk–Othmer ECT, 3rd edn, Wiley, New York, 1983, Vol. 22, pp. 710, 746.

CAPTAN CAS No. 133-06-2
(Merpan, hexacap)
Fungicide, orthocide, bacteriostat in soap. Toxicity class EPA IV. LD_{50} rat, mouse 7–$9\,\mathrm{g\,kg^{-1}}$; LC_{50} trout $0.034\,\mathrm{mg\,l^{-1}}$. Toxic to bees

$C_9H_8Cl_3NO_2S$ $M = 300.6$

N-(Trichloromethylthio)cyclohex-4-ene-1,2-dicarboximide; m.p. 178°; water solubility $3.3\,\mathrm{mg\,l^{-1}}$; $t_{1/2}$ on hydrolysis at pH $7.0 = 32\,\mathrm{h}$.

Captan is used on fruits and as a seed treatment for cereals and for the control of scab and mildew on vines.It irritates the eyes and skin.

Codex Alimentarius gives acceptable daily intake $0.1\,\mathrm{mg\,kg^{-1}}$ body weight. Maximum residue levels for apples 25, dried grapes 5 and tomatoes $15\,\mathrm{mg\,kg^{-1}}$. The OSHA standard (California) is a PEL of $5\,\mathrm{mg\,m^{-3}}$. Limits in workplace air are UK and USA (ACGIH): $5\,\mathrm{mg\,m^{-3}}$.

Further reading
European Directory of Agricultural Products, Royal Society of Chemistry, Cambridge, 1990, Vol. 1.

CARBARYL CAS No. 63-25-2
(Alpha, Dethlac, Sevin)
Insecticide, animal ectoparasiticide and growth regulator. Toxicity class III. LD_{50} oral rat 500–850 $\mathrm{mg\,kg^{-1}}$; LC_{50} fish 690–7100 $\mathrm{\mu g\,l^{-1}}$

$C_{12}H_{11}NO_2$ $M = 201.2$

1-Naphthyl-*N*-methylcarbamate; m.p. 142°; water solubility 120 mg $\mathrm{l^{-1}}$ (30°); v.p. $< 4 \times 10^{-5}\mathrm{mmHg}(25°)$. Toxic to bees, bioconcentration factor in algae 4000; cholinesterase inhibitor.

Carbaryl is used to limit fruiting of apple trees and against insects in cotton, soft fruit and vegetables and to control Colorado beetle on potatoes. It is rapidly metabolized by mammals to 1-naphthol but has adverse effects on wildlife in forest and grassland communities and, although not banned in Asia, its use is restricted in

Indonesia and the Philippines. It is in the EC 'Red List.' There was a disastrous release of a synthetic intermediate at Bhopal (see **methyl isocyanate**).

TWA (8 h) in the UK and USA (ACGIH) $5 \, mg \, m^{-3}$. *Codex Alimentarius* gives acceptable daily intake 0.0005 $mg \, kg^{-1}$ (1986). Maximum residue levels: 0.002 mg kg^{-1} for milk and 0.02 $mg \, kg^{-1}$ for many other products.

CARBENDAZIM CAS No. 10605-21-7
(Carbendazol, Bavistin)
Fungicide with preventative and curative action. Toxicity class EPA IV. LD_{50} rat $15 \, g \, kg^{-1}$; LC_{50} (96 h) trout $2-3 \, mg \, l^{-1}$

$C_9H_9N_3O_2 \quad M = 191.2$

Methyl benzimidazol-2-ylcarbamate; m.p. 302–307°; water solubility 28 mg l^{-1} (20°).

Persists in soil for 4 months and may be formed from Benomyl. Used to control mildew on cereals, scab on fruit and mould on vines. Its degradation products include 2-aminobenzimidazole and 5-hydroxy-2-aminobenzimidazole.

Codex Alimentarius acceptable daily intake 0.01 $mg \, kg^{-1}$. Maximum residue levels: taro(edible root) 0.1, wheat straw 5.0 and potato 0.1 $mg \, kg^{-1}$. EC limit in drinking water 0.1 $\mu g \, l^{-1}$.

Further reading
Thomson, W. T. (1976–7) *Agricultural Chemicals*, Thomson, London, Vols **1–4**.

CARBIDES CAS No. 12385-15-8
Carbon forms binary compounds with most elements. The so-called interstitial carbides are important since they are very hard and are extensively used as abrasives. Examples include TiC, SiC, Fe_3C, WC and TaC.

Further reading
King, R. B. (ed.) (1994) *Encyclopedia of Inorganic Chemistry*, Wiley, Chichester, Vol. 2, pp. 519–530.
Kirk–Othmer ECT, 4th edn, Wiley, New York, 1992, Vol . 4, pp. 841–911.

CARBON CAS No. 7440-44-0
C
Carbon, a member of Group 14 of the Periodic Table, has an atomic number of 6, an atomic weight of 12.01 and a melting point of ca 3550°. Graphite sublimes at 3825°. The principal oxidation state is +4. Natural carbon comprises of two stable isotopes; ^{12}C (98.89%) and ^{13}C (1.11%). Several radioactive isotopes are known; the longest lived, ^{14}C ($t_{1/2} = 5730$ years), is produced by cosmic ray bombardment of the upper atmosphere and by nuclear explosions.

Carbon occurs mainly as carbon dioxide, as the hydrogen carbonate ion HCO_3^- in sea and fresh water and as carbonate minerals. Carbon in the form of fossil fuels is found in large concentrated areas as coal, oil and natural gas. Carbon is found in nature in three allotropic forms: amorphous, graphite and diamond. Coal is an impure form of carbon. Total body content (adult, 70 kg) 12 600 g.

Uses.   As a fossil fuel. In the manufacture of carbon black.

The main danger of working in the manufacture or extraction of carbon in its various forms is the inhalation of very fine particles of carbon. The combustion of diesel oil also produces minute particles of carbon. Experiments have shown that laboratory animals are more sensitive than humans to carbon particles. However, prolonged exposure of carbon dusts (from coal, carbon black, etc.) to humans showed respiratory symptoms including impaired lung function from dust deposition in the lungs and in severe cases emphysema. German DFG-MAK value fine dust 6 mg m^{-3}.

Further reading
IARC Monogr. (1987), **Suppl. 7**, 59.
IARC Monogr. (1984), **33**, 1–232.
SI 1991 No. 472 The Environmental Protection (Prescribed Processes and Substances) Regulations 1991, HMSO, London, 1991.
Kirk–Othmer ECT, 4th edn, Wiley, New York, 1992, Vol. 4, pp. 949–1117.

CARBON BLACK CAS No. 1333-86-4
Amorphous graphite, $M = 12.01$, dens. 1.8–2.1, b.p. 4200°, insoluble all solvents.
Carbon black was known to ancient civilizations in Egypt and China. Its formation in candle flames was discovered by Michael Faraday in the 1860s. Most of today's production is by the oil furnace process, where an aromatic petroleum distillate is injected into a hot flame with a limited oxygen supply (Kirk–Othmer, 1992). The characteristics of the product can be varied by choice of conditions, e.g. the maximum surface area of 145 m^2 g^{-1} is obtained at a temperature of 1800° with a brief residence time (0.008 s). At 1400° and a residence time of 1.5 s the product has a surface area of 25 m^{2}g^{-1}.

Carbon black is formed by the aggregation of polycyclic hydrocarbons and its structure is related to that of graphite, although it does not have the same high degree of three-dimensional symmetry. In 1988 the total world production capacity was 7000 × 10^3 t, of which 1210 × 10^3 t was in the USA, 1230 × 10^3 t in Western Europe and 1800 × 10^3 t in Asia. In 1994 carbon black was 37th in the list of top chemicals in the USA when production was 1504 × 10^3 t. Some 90% of US production goes into the rubber industry and most of this is used in car tyres to provide mechanical strength and wear resistance. Carbon black is also used to colour PVC and to make it electrically conducting, a property which is enhanced by small particle size.

The dust is flammable and can cause dust explosions; precautions must also be taken to ensure that it is not inhaled. In view of the method of production, carbon black is expected to include carcinogenic PAHs, including the dangerous benzo[a]-pyrene. However, there is no evidence of cancer induction in the workplace. Extraction by solvents such as benzene can remove 300–2000 ppm (0.03–0.26%) of material, including benzo[a]pyrene. The PAH's are not extracted by hot or cold water, gastric juices or blood plasma and they are apparently so strongly absorbed on the carbon black that it is inactive in animal tests.

Occupational exposure level.   ACGIH PEL (TWA) 3.5 mg m^{-3} was set on the basis that the level of extractable compounds could not exceed that (0.2 mg m^{-3}) set for PAHs.

Reference
Kirk–Othmer ECT, 4th edn, Wiley, New York, 1992, Vol. 4, pp. 1037–1074.

CARBON DIOXIDE, CAS No. 124-38-9

CO_2 $\qquad\qquad\qquad\qquad\qquad M = 44.01$

Carbon dioxide is a gas, melting point $-56.5°$ at 3952 mmHg, boiling point $-78.5°$. It is present in the atmosphere, 0.03% of the mass of the atmosphere, 0.01–0.9 mg m^{-3}. CO_2 is by far the most abundant carbon compound in the atmosphere. The gas enters the atmosphere through a variety of processes: degassing of the gas from the oceans; weathering of carbonate minerals; oxidation or organic carbon (this includes decay and burning the biomass, fossil fuel burning and respiration, industrial fermentation and mining processes, etc.). CO_2 is removed from atmosphere by dissolution in the oceans and by photosynthesis. The increasing rate of burning of fossil fuels (about 0.4% p.a.) produces increasing quantities of carbon dioxide which contribute to the 'greenhouse effect' (CO_2 absorbs infrared radiation). Atmospheric levels show a steady increase from around 1870. The average atmospheric concentration rose from 290 ppm in 1950 and will rise to 390 ppm by volume by 2000. A rise of CO_2 to 600 ppm would entail an increase of 4.5° in the mean temperature of the earth, with a consequent rise in the levels of the world's oceans by 5–7 m.

Uses. Carbonation of soft drinks. Preparation of carbonates. Dry ice as a coolant.

Carbon dioxide irritates skin and mucous membranes with increasing concentration. Breathing air containing around 1% CO_2 causes changes in respiration and blood circulation. At concentrations around 7%, symptoms include dizziness and disorientation. Examples of death have occurred in a variety of indoor situations, e.g. cleaning of fermentation vats, tobacco storages sites, in silo towers containing wheat, etc., rotten potatoes stored in cellars.

Occupational exposure limits. OSHA (TLV) TWA 9000 mg m^{-3}. UK short-term limit 27 000 mg m^{-3}.

Toxicity. LC_{Lo} (5 min) inhalation unspecified mammal 90 000 ppm. An asphyxiant.

Further reading
B.J. Alloway and D.C. Ayres, (1997) *Chemical Principles of Environmental Pollution*, 2nd edn, Blackie, London.
ECETOX (1991) *Technical Report 30(4)*, European Chemical Industry Technology and Toxicology Centre, Brussels.
Kirk–Othmer ECT, 4th edn, Wiley, New York, 1993, Vol. 5, 35–53.
SI 1991 No. 472 The Environmental Protection (Prescribed Processes and Substances) Regulations 1991, HMSO, London, 1991.

CARBON DISULPHIDE CAS No. 75-15-0

CS_2 $\qquad\qquad\qquad\qquad\qquad M = 76.1$

B.p. 46°; dens. 1.263 g cm^{-3} (20°) i.e. 2.62 × air; v.p. (28°) 53.3 kPa (400 Torr). Highly flammable with a flash point of $< -30°$ and with explosion limits in air between 1% and 5% by volume.

When pure, carbon disulphide has a sweetish odour similar to that of chloroform but typically it resembles that of decaying vegetation. It is a solvent for fats, lipids, resins, rubber and white phosphorus. It first came into use for making matches in 1851.

Is obtained by the reaction of methane with sulphur in a process which also provides a route to carbon tetrachloride:

$$CH_4 + S(vapour) \rightarrow CS_2 + 2H_2S \qquad (S = S_2, S_6, S_8)$$

Levels down to $5\,mg\,m^{-3}$ can be determined by spectroscopy of the yellow copper complex ($\lambda = 420\,nm$) formed with the product of the reaction between carbon disulphide and diethylamine:

$$CS_2 + HN.Et_2 \longrightarrow Et_2N-C\overset{\displaystyle O}{\underset{\displaystyle SH}{\big<}}$$

N,N-diethyl dithiocarbamate

A major use is in the manufacture of Cellophane and viscose rayon, although this product now has a reduced share in the market. Minor uses are as an insecticide and soil disinfectant. CS_2 reacts with alkaline cellulose to produce colourless cellulose xanthate and orange sodium trithiocarbonate (Na_2CS_3). This solution in caustic soda is injected through spinnerets into dilute sulphuric acid to form filaments of rayon yarn:

$$ROC(S)S^-Na^+ + H_2SO_4 \rightarrow ROH + NaHSO_4 + CS_2$$
cellulose xanthate   cellulose

with decomposition of the co-produced thiocarbonate:

$$Na_2CS_3 + H_2SO_4 \rightarrow H_2S + CS_2 + NaHSO_4$$

A 1 kg amount of viscose gives rise to 20–30 g of CS_2 and 4–6 g of H_2S.

Carbon disulphide reacts with the amino groups of amino acids and proteins to give thiocarbonates, which themselves chelate with metals which are essential to enzyme function, such as copper and zinc, and this may be responsible for the neurotoxicity of CS_2. Through the destruction of cytochrome P450 it damages the liver, a toxic effect observed on the highly exposed group of workers in the viscose rayon industry. The consequences were loss of nervous reactions, disturbance of the gastrointestinal tract and optic neuritis. Workers exposed to CS_2 for 5–15 years at concentrations in the range $2–150\,mg\,m^{-3}$ showed effects on the CNS, kidneys, liver, and cardiovascular and respiratory systems.

Further reading

ECETOX (1991) *Technical Report 30(4)*, European Chemical Industry Technology and Toxicology Centre, Brussels.

Environ. Health Criteria, World Health Organization, Geneva, 1979, Vol. 10.

CARBON MONOXIDE CAS No. 630-08-0

CO $M = 28$

Carbon monoxide is an odourless, colourless gas with a melting point of $-207°$, and a boiling point of $-191.3°$. Produced by the partial oxidation of carbon compounds, mainly coal, coke and hydrocarbons, including petrol. CO is a major component of car exhausts and a common air contaminant, and becomes toxic at levels of 50 ppm. Residence time in the atmosphere is about 10 weeks. Global emissions ca 60×10^6 t year^{-1}. Annual rate of increase about 1%. Most common cause of human poisoning, both in industry and at home, where death occurs by asphyxiation. With the interruption of the normal oxygen supply and replacement by CO, haemoglobin forms carboxyhaemoglobin,where the affinity for CO is more than 200 times than that of oxygen. Brain and heart may be damaged by lack of oxygen. These are many well documented reports of the toxic effects on humans. Pregnant women and children are particularly at risk. Tobacco smoking is a possible source of CO in the home. Poorly maintained domestic gas boilers or oil heaters in under-ventilated rooms produce several fatalities per year in the UK. A common form of suicide is the feeding of the exhaust fumes from a car via a hosepipe into the seating area. Concentrations of 10 000–40 000 ppm lead to death within minutes. The concentration and length of exposure give rise to various symptoms. Concentrations in the range 500–1000 ppm in humans cause headaches, nausea, weakness, dizziness, mental confusion and hallucination.

At high altitudes, carbon dioxide is converted to carbon monoxide:

$$CO_2 + h\nu \rightarrow CO + O.$$

Principal removal of tropospheric CO is by bacterial and fungal processes in soil, e.g. by *Bacillus oligocarbophilus.*

Uses.   Reducing agent. Organic synthesis. Manufacture of metals.

Occupational exposure limits.   US TLV (TWA) 25 ppm (29 mg m^{-3}). UK short-term limit 300 ppm (330 mg m^{-3}).

Toxicity. Toxic by inhalation.   LC$_{50}$ (4 h) inhalation rat, guinea pig 1810–5720 ppm; LC$_{Lo}$ (30 min) inhalation human 4000 ppm.

Further reading
Kirk–Othmer ECT, 4th edn, Wiley, New York, 1993, Vol. 5, pp. 97–122.

CARBON TETRACHLORIDE CAS No. 56-23-5
LD$_{50}$ rat 2.92 g kg^{-1}; LC$_{50}$ (96 h) bluegill sunfish 125 ppm

CCl$_4$ $M = 153.8$

Tetrachloromethane, Freon 10; b.p. 77°; dens. 1.59 (25); v.p. 100 mmHg (23°); v. dens. 5.5; logP_{ow} 2.64 (20°). Solubility in water 1160 mg l^{-1}, miscible with most organic solvents.

CCl$_4$ can be obtained by the direct chlorination of methane. The industrial practice is linked with the demand for perchloroethylene, the common product of the thermal degradation/chlorination of mixtures of propane and propylene:

$$CH_2 = CHCH_3 + 7Cl_2 \xrightarrow{500°} CCl_4 + Cl_2C = CCl_2 + 6HCl$$
$$CH_3CH_2-CH_3 + 8Cl_2 \longrightarrow CCl_4 + Cl_2C = CCl_2 + 8HCl$$

Carbon tetrachloride is also obtained from the chlorination of carbon disulphide.

It is no longer used in Western Europe or the USA for the dry cleaning of clothes, but this use continues in Eastern Europe and S.E. Asia. In 1988 in the USA 283 $\times 10^3$ t were produced as a feedstock for CFC-11 (CCl$_3$F) and CFC-12 (CCl$_2$F$_2$), but with the phasing out of these two fluoro compounds production of carbon tetrachloride has declined steeply.

A fatality from exposure to carbon tetrachloride was reported as early as 1909 and occasional deaths occur as a result of acute liver and/or kidney necrosis following ingestion or inhalation – the risk of poisoning is highest in heavy drinkers. Carbon tetrachloride depresses the CNS and was once used as a surgical anaesthetic; symptoms of exposure are headache and giddiness. Humans exposed for 3 h at 10 ppm showed no adverse liver functions, but cases of liver cancer in animals and humans have been reported. Carbon tetrachloride is metabolized in the liver via the trichloromethyl radical, which reacts further to give chloroform and phosgene:

$$\overset{\bullet}{C}Cl_3 \xrightarrow{\ H-R\ } HCCl_3 + R^{\bullet}$$

$$\overset{\bullet}{O}-O-CCl_3 \qquad \left[HO.CCl_3\right] \text{ trichloromethanol}$$

$$\downarrow -HCl$$

$$O=C\overset{Cl}{\underset{Cl}{\big\langle}} \text{ phosgene}$$

The US TLV (TWA) is set at 5 ppm (31 mg m^{-3}), the UK long-term limit is 2 ppm and the MAK value is 10 ppm (65 mg m^{-3}).

As a carcinogen carbon tetrachloride is rated by the IARC as group 2B – sufficient evidence of activity in animals.

CARBONYLS

Most metal carbonyls, e.g. $Ni(CO)_4$, are extremely poisonous since they readily decompose to give **carbon monoxide**.

CATECHOL CAS No. 120-80-9
LD$_{50}$ rat 260 mg kg^{-1}

$C_6H_6O_2$ $M = 110.1$

1,2-Benzenediol, 1,2-dihydroxybenzene, pyrocatechol; m.p. 105° (sublimes); b.p. 245°; v.p. 10 Torr (15°); dens. 1.34; v. dens. 3.79. Soluble in water, EtOH, CHCl$_3$, benzene.

First prepared by Reinsch in 1839. Production worldwide in 1973 was 1.4–2.3 ×10^6 kg. It is used in photography as a developer, in dyeing fur and as a precursor of 4-t-butylcatechol, which is used to inhibit polymerization of stored styrene.

The acute oral toxicity of catechol is up to twice that of phenol, but its inhalation toxicity is less than that of phenol. Like phenol it produces a rise in blood pressure and dermatitis may follow skin contact. Catechol is absorbed from the gastrointestinal tract and through the skin. Part is excreted as a conjugate of sulphuric acid and part is metabolized to *o*-benzoquinone, which leads to dark discoloration of the urine. Catechol is an experimental teratogen.

The US TLV (TWA) is set at 5 ppm (20 mg m^{-3}) as for phenol.

CERIUM CAS No. 7440-45-1

Ce

Cerium is the second metal of the lanthanide (rare-earth) series. It has an atomic number of 58 , an atomic weight of 140.12, a specific gravity of 6.90 at 20°, a melting point of 799° and a boiling point of 3424°. The metal consists of cubic or hexagonal steel-grey crystals and forms compounds in the +3 or +4 oxidation state. The four natural isotopes are ^{140}Ce (88.48%), ^{142}Ce (11.08%, half-life > 5 × 10^{16} years), ^{138}Ce (0.25%) and ^{136}Ce (0.19%)

Cerium is the most abundant metal of the so-called rare earths. The average concentration in the earth's crust is $\sim$ 65 ppm. It is found principally in the minerals *monazite, bastnasite* and *cerite*. Metallic cerium is obtained from the reduction of cerium(III) fluoride with calcium. Reported levels in humans include blood 0.0002 mg l^{-1}, bone 2.7 ppm and liver 0.29 ppm. Daily intake is not available but is very low. Total body content for 70 kg adult is 40 mg.

Uses. Catalyst in manufacture of ammonia. Misch metal, pyrophoric alloys used for cigarette lighters. Catalyst in petroleum refining. Nuclear and metallurgical applications. Cerium sulphide has been suggested as a non-toxic replacement pigment for cadmium sulphide.

Cerium resembles aluminium in its pharmacological actions as well as in its chemical properties. Compounds of cerium are generally considered to be of low toxicity, except in concentrated form. Cerium dioxide dust can give rise to pneumoconiosis. The insoluble oxalate was used to prevent vomiting in pregnancy. The average dose is 0.05–0.5 g. The toxicity of cerium is dependent on it being in a biochemical form. Soluble salts can be bioconcentrated: the factor for freshwater fish is 500 and for marine fish 100. In humans, cerium has a variety of effects. Inhalation of cerium may affect the nervous system. Also reported to produce polycythaemia. Cerium salts increase the blood coagulation rate. Most of the health cases arise from exposure to workers during manufacture of cerium, cerium dioxide and some cerium salts. Large doses to experimental animals have caused writhing, ataxia, laboured respiration, sedation, hypotension and death. The metal dust presents a fire and explosion hazard.

Further reading

Gschneidner, Jr, K.A., and LeRoy, K.A. (eds) (1988) *Handbook on the Physics and Chemistry of the Rare Earths*, North-Holland, Amsterdam.

Kirk–Othmer, ECT, 4th edn, Wiley, New York, 1993, Vol. 5, pp. 729–749.

SI 1991 No. 472 The Environmental Protection (Prescribed Processes and Substances) Regulations 1991, HMSO, London, 1991.

Cerium (IV) ammonium nitrate CAS No. 16774-21-3

Ce(NH$_4$)$_2$(NO$_3$)$_6$

Uses. Volumetric analysis

Toxicity. Irritant.

Cerium(IV) ammonium sulphate dihydrate CAS No. 10378-47-9

$Ce(NH_4)_4(SO_4)_4.2H_2O$

Uses. Agent in volumetric quantitative analysis.

Toxicity. Irritant.

Cerium(III) nitrate hexahydrate CAS No. 10294-41-4

$Ce(NO_3)_3.6H_2O$

Uses. Preparation of cerium compounds.

Toxicity. LD_{50} oral rat 3154– 4200 mg kg^{-1}; intraperitoneal rat 290 mg kg^{-1}.

Cerium(IV) oxide CAS No. 1306-38-3

CeO_2

Uses. Manufacture of incandescent gas mantles. Hydrocarbon catalyst in 'self-cleaning' ovens. Glass polishing agent.

Toxicity. LD_{50} oral rat > 5 g kg^{-1}.

Cerium(IV) sulphate CAS No. 13454-94-9

$Ce(SO_4)_2$

Uses. Oxidizing agent in volumetric analysis.

Toxicity. Harmful by inhalation, contact with skin or if swallowed. Target organs are the testes.

CHEMICAL OXYGEN DEMAND (COD)

A measure of the oxidizable components present in water. Carbon and hydrogen in organic matter are oxidized but oxygen consumption depends upon the oxidant (often dichromate), the structure of the compound(s) and the procedure employed. COD values do not differentiate between stable and unstable matter in the environment and therefore it does not correlate with BOD.

Further reading
Watson, C. (1994) *Official and Standardised Methods of Analysis*, Royal Society of Chemistry, Cambridge, pp. 590, 623.

CHLORATES CAS No. 14866-68-3

ClO_3^-

The principal toxic effect of chlorates is the production of methaemoglobin in the blood and the destruction of red blood corpuscles.

Further reading
Kirk–Othmer ECT, 4th edn, Wiley, New York, 1993, Vol. 5, pp. 998–1016.

CHLORDANE CAS No. 57-74-9
(Velsicol)
Insecticide, non-systemic, LD_{50} rat, rabbit, mouse 300–400 mg kg^{-1}; LC_{50} fish 70–90 µg l^{-1}

$C_{10}H_6Cl_8$ $M = 409.8$

1,2,4,5,6,7,8,8-Octachloro-2,3,3a,4,7,7a-hexahydro-4,7-methano-1H-indene;
m.p.(*trans*)105° water solubility 0.1 mg l^{-1} (25°).

Technical chlordane is a mixture of 26 compounds; of these the *trans* (gamma) isomer accounts for 25%, and this is also a significant component in technical heptaclor (see **Heptachlor**)

It is harmful on skin contact and if swallowed. Chlordane is carcinogenic in animals and its cumulative toxicity also causes kidney and liver damage.

In 1984 Brazil imported 226 t but it was banned in Ecuador in 1991. In the USA an occupational exposure level of a TWA of 0.5 mg m^{-3} was set but it has been banned there since 1989. In that year 400 000 chickens were destroyed after they had been fed sorghum contaminated with chlordane. It is prohibited in the EC and FAO/WHO countries.

CHLORIC ACID CAS No. 7790-93-4

$HClO_3$ $\hspace{4cm}$ $M = 84.5$

Uses. Preparation of chlorates.

Toxicity. Poison Strong irritant. Powerful oxidizing agent.

CHLORIDES CAS No. 16887-00-6

Cl^-

Toxicity. Varies widely from sodium chloride (NaCl) of low toxicity to highly toxic phosgene, $COCl_2$.

CHLORINE CAS No. 7782-50-5

Cl (stable as Cl_2)

Chlorine, a member of Group 17 (the halogens) of the Periodic Table, has an atomic number of 17, an atomic weight of 35.45, a specific gravity of 1.47 at 20°, a melting point of −101° and a boiling point of −34.5°. Chlorine gas is made by electrolysis of NaCl solution. Natural isotopes are ^{35}Cl (75.77 %) and ^{37}Cl (24.23%). Oxidation states of −1, +1, +3, +5 and +7 exist.

In nature chlorine is found as *rock salt*, NaCl, *carnallite* $KCl.MgCl_2.6H_2O$, and *sylvite*, KCl. It is the 20th most abundant element in the earth's crust and occurs to the extent of 126 ppm. Chlorides of sodium, potassium and magnesium together constitute 0.03% of the earth's crust; the chloride content of the oceans' waters is 2%. Total volatile chlorine in the atmosphere is ∼ 1 ppb. Total body content (adult, 70 kg) 105 g. Chlorine is produced by the electrolysis of metal chlorides (usually NaCl); the mercury cathode process is now being supplanted by the diaphragm process.

Uses. Chlorine has many uses. Manufacture of hydrochloric acid and metal/non-metal chlorides. Reagent in preparative chemistry. Purifying water, as bleach and disinfectant, it reacts with water to give hypochlorite. Used in plastics manufacture, e.g. PVC. Chlorine is used in the preparation of organochlorine compounds. Some are highly toxic and may be used as pesticides (see **Pesticides**). Chlorinated fluorocarbons, used in refrigeration and aerosol propellants, are volatile and enter the atmosphere. These compounds are gradually being phased out because of their ability to damage the ozone layer (see **Ozone**)

Chlorine spillages and leakages are amongst the most common industrial accidents. Because of its irritating properties even at low concentrations, severe industrial exposure seldom occurs as workers are forced to leave the area. Thus, chlorine fumes from a fire at a chemical company in Minneapolis, USA, resulted in evacuation of the area, including 200 residents living nearby.

Chlorine is a very toxic gas and, when inhaled, affects the human respiratory system and can cause emphysema, chronic pulmonary oedema or congestion. Repeated exposure to chlorine can result in asthma (e.g. in workers in paper and pulp mills). A concentration of 3.5 ppm ($10.5\,mg\,m^{-3}$) produces a detectable odour, 15 ppm ($45\,mg\,m^{-3}$) causes irritation of the throat, concentrations of 50 ppm ($150\,mg\,m^{-3}$) are dangerous and 100 ppm ($300\,mg\,m^{-3}$) may be fatal. Exposure to a concentration of $110\,mg\,m^{-3}$ for 30 min proved fatal. A survey of workers in 22 chlorine-producing plants in the USA and Canada with concentrations ranging between 0.017 and $5\,mg\,m^{-3}$ showed that no long term damage ensued when compared with a control group. In 1915, during the first World War, 700 Canadian soldiers were exposed to chlorine in the field, of these one-third returned to duty after minimal treatment, and of the remainder one-third went to base hospitals. In the whole group there were six deaths, and four years later eight cases of bronchitis, eight of asthma and 18 neuroses were diagnosed.

A number of tanker spills have led to exposure of the general public. In 1947 40 kg of liquid chlorine leaked into a Brooklyn subway, causing 418 casualties; there were no deaths but 201 people were hospitalized. Cyanosis was frequently observed in these patients, and also increased respiration, heart rates and body temperature; there was no evidence of permanent pulmonary disease.

Occupational exposure limit.   US TLV (TWA) 0.5 ppm ($1.5\,mg\,m^{-3}$). UK short-term limit 1 ppm ($3\,mg\,m^{-3}$).

Toxicity. LC_{50} (96 h) rainbow trout $0.17\,mg\,l^{-1}$. LC_{50} (1 h) inhalation rat 293 ppm. LC_{Lo} (30 min) inhalation human $2530\,mg\,m^{-3}$

Further reading
Das, R. and Blanc, P.D. (1993) Chlorine gas exposure and the lung. *Toxicol. Indus. Health*, **9**, 439–55.
IARC Monogr. (1991) **52**, 45–141.
Kirk–Othmer ECT, 4th edn, Wiley, New York, 1991, Vol. 1, pp. 939–1039.
S.I. 1991 No.472 The Environmental Protection (Prescribed Processes and Substances) Regulations 1991, HMSO, London, 1991.

CHLORINE DIOXIDE CAS No. 10049-04-4

ClO_2 $M = 67.5$

Uses. Improvement of water quality. Bleaching agent, particularly wood pulp. Bactericide. Disinfectant. In aqueous solution it is often present as the chlorite ion, ClO_2^-.

Occupational exposures limits. US TLV (TWA) 0.1 ppm $(0.28\,mg\,m^{-3})$. UK short-term limit 0.3 ppm $(0.9\,mg\,m^{-3})$. MAC drinking water UK $0.5\,mg\,ClO_2\,l^{-1}$.

Toxicity. Irritant to laboratory animals and humans. Repeated exposures of workers caused cough, sneezing, sore throat and in severe cases bronchitis and pulmonary oedema. At least one fatality has occurred. Eye and throat irritant. LD_{50} oral rat $292\,mg\,kg^{-1}$.

Further reading

ECETOX (1991) *Technical Report No 30(4)*, European Chemical Industry Ecology and Toxicology Centre, Brussels.

Lykins, B.W., Goodrich, J.A. and Hoff, J.C. (1990) Concerns with using chlorine dioxide disinfection in the USA, *Aqua (London)* **39**, 376–386.

Kirk–Othmer ECT, 4th edn, Wiley, New York, 1993, Vol. 5, pp. 968–997.

CHLORINE TRIFLUORIDE CAS No. 7790-91-2

ClF_3 $M = 92.5$

Uses. Fluorinating agent. Propellant additive for rocket fuel. Silicon etching.

Occupational exposure limits. US TLV (TWA) 0.1 ppm $(0.38\,mg\,m^{-3})$.

Toxicity. Inhalation causes inflammation of skin, eyes and mucous membranes. Prolonged exposure leads to lung damage. LD_{50} (1 h) inhalation rat 299 ppm.

Further reading
See **Fluorides**.

CHLORMEQUAT CAS No. 999-81-5
(Cycogan)
Plant growth regulator. Toxicity class III. LD_{50} rat $600\,mg\,kg^{-1}$; LD_{50} mouse $54\,mg\,kg^{-1}$. No ill effects on rats fed on a 2 year diet at $1000\,mg\,kg^{-1}$. LC_{50} trout $4500\,mg\,l^{-1}$.

$$Cl{-}H_2C{-}H_2C{-}\overset{\text{Me}}{\underset{\text{Me}}{\overset{|}{\underset{|}{N^+}}}}{-}Me \quad Cl^- \qquad\qquad C_5H_{13}Cl_2N \quad M = 158.1$$

2-Chloroethyltrimethylammonium chloride; m.p. 245° (dec.); v.p. 0.75 Torr (20°); very soluble in water.

Usually employed as the chloride and is rapidly excreted by mammals in urine; used to control growth of cereals. Rapidly degraded in soil through enzymatic activity. The UK is cutting back its use on spring barley.

Codex Alimentarius acceptable daily intake $0.05\,mg\,kg^{-1}$. Maximum residue levels: wheat 5.0, milk 0.1 and grapes $1.0\,mg\,kg^{-1}$. EC limit in drinking water $0.1\,\mu g\,l^{-1}$.

CHLOROBENZENE CAS No. 108-90-7

LD_{50} rat 2.29 g kg^{-1}; LD_{Lo} mouse 250 mg kg^{-1}; LC_{50} fish (24–96 h) 24–73 mg l^{-1}

C_6H_5Cl $M = 112.6$

Benzene chloride, phenyl chloride, MCB; b.p. 131°; v.p. 8.8 mmHg; flash point 28°; dens. 1.104; v. dens. 3.88; log P_{ow} 2.84. Solubility in water 295 mg l^{-1}, soluble in EtOH, diethyl ether, benzene, CCl$_4$.

First produced in the middle of the 19th century. Obtained by the direct chlorination of benzene by Cohen and Hartley in 1905, the process used today with FeCl$_3$ as catalyst. The commercial product will inevitably contain more highly chlorinated compounds, especially the dichlorobenzenes, a typical mix contains 85% of monochlorobenzene and 15% of dichlorobenzenes. This byproduct is an important source of dichlorobenzenes, in which the *para*-isomer is at twice the concentration of the *ortho*-isomer.

In 1988 world production of chlorobenzene was 400 × 10^3 t divided between the USA (46%), Western Europe (34%) and Japan (20%). Three quarters of this goes into the manufacture of chloronitrobenzenes; it is also used as a paint solvent, in the preparation of diphenyl ether and sulphone polymers, as a heat-transfer medium and as a phase-transfer solvent.

Chlorobenzene is moderately toxic by inhalation and by skin penetration; it is strongly narcotic towards animals at 1200 ppm. In humans repeated exposure can damage the liver and kidneys and the presence of the metabolites 4-chlorocatechol and 4-chlorophenol in the urine may be evidence of this.

The original TLV set in the USA in 1945 was 75 ppm, but this has now been reduced to 10 ppm (46 mg m^{-3}) in line with the MAK value.

Further reading
Kirk–Othmer ECT, 4th edn, Wiley, New York, 1993, Vol. 6, pp. 87–100.

CHLOROETHYL ETHER
See **Bis (2-Chloroethyl) ether**.

CHLOROFLUOROCARBONS
The first of these compounds, known as CFCs, was synthesized by Midgely in 1928 and they came into industrial use in 1941. Their range of composition is as follows:

		ODP	Atmos. lifetime (years)
CFCl$_3$	CFC-11	1.0	77
CF$_2$Cl$_2$	CFC-12	1.0	139
C$_2$F$_3$Cl$_3$	CFC-113	0.8	92
C$_2$F$_4$Cl$_2$	CFC-114	1.0	180
C$_2$F$_5$Cl	CFC-115	0.6	380
CF$_2$BrCl	Halon 1211	2.7	12.5
CF$_3$Br	Halon 1301	11.4	77
C$_2$F$_4$Br$_2$	Halon 2402	5.6	Unknown

Owing to their inert behaviour and volatility they were used extensively as refrigerants and blowing agents, before it was shown by Rowland and Molina (1974) that their very stability ensured that they escaped into the stratosphere, where they undergo photolytic reactions which damage the ozone layer. In outline, this ozone depletion depends upon the release of chlorine atoms by photolytic cleavage of the CFC, followed by destruction of ozone and also of a molecule of odd oxygen, which generates the ozone in the first place:

$$CF_2Cl_2 + h\nu \longrightarrow CF_2Cl + Cl$$

or

$$CFCl_3 + h\nu \longrightarrow CFCl_2 + Cl$$

then

$$Cl + O_3 \longrightarrow ClO + O_2$$

and

$$ClO + O \longrightarrow Cl + O_2$$

adding

$$O + O_3 \longrightarrow 2O_2$$

The net reaction is the loss of two ozone molecules and the chlorine atoms which initiate the sequence continue to deplete the ozone in a chain reaction. Their potency is expressed as an ozone depletion potential (ODP) relative to that of CFC-11 as unity.

For brevity, a system of nomenclature is used which is based on the form CFC-XYZ where X = number of carbon atoms minus 1 (omitted if $X = O$); Y = number of hydrogen atoms plus 1; Z = number of fluorine atoms. Hence trichlorofluoromethane becomes CFC-11. The halon bromochlorodifluoromethane becomes Halon 1211, corresponding to carbon one, fluorine two, chlorine one and bromine one.

CFC	CAS No.	V.p., Torr (°)	M	Applications
CFC-11	75-72-9	400 (– 82°)	104.5	Fire extinguisher, refrigerant, aerosol propellant, etching printed circuits
CFC-12	75-71-8	4250 (20°)	120.9	As above, also for blowing foam and as local anaesthetic
CFC-113	76-13-1	360 (20°)	187.4	Fire extinguisher, refrigerant, dry cleaning
CFC-114	76-14-2	2014 (25°)	170.9	Propellant, blowing agent, refrigerant
CFC-115	76-15-3	–	142.5	Refrigerant and aerosols for food preparations

Samples from spray cans were taken in solvent-resistant Tedler bags and analysed by GC–MS. This showed that 54 of a total of 448 contained over 1% of CFCs, although many were claimed to be 'safe to the environment.'

These compounds are only slightly soluble in water, that of CFC-113 being $100 \, mg \, l^{-1}$, and their acute toxicity is low, although CFCs 11 and 12 are weakly narcotic. The TLV (TWA) in the USA for CFCs 113, 114 and 115 is 1000 ppm and the UK level for long term exposure in the workplace is also 1000 ppm.

Ozone depletion was first confirmed in the polar vortex over Antarctica – a region of very cold air ($< -190°$) supported by westerly winds. Here light-induced chemical change occurs independently of the large volume of surrounding air, assisted by ice crystals on the polar stratospheric clouds. A number of intermediate species interact, for example:

$$M + ClO\bullet + NO_2 \longrightarrow ClO\bullet NO_2$$

and these are stored on ice crystals, yet generate light sensitive products on hydrolysis:

$$H_2O + Cl\bullet ONO_2 \xrightarrow{H_2O} HNO_3 + HO\bullet Cl \xrightarrow{hv} HO\bullet + Cl\bullet$$

when ozone depletion recommences.

Once the link between depletion and the CFCs was accepted, the USA proposed a ban on non-essential use in 1979, while the EC countries proposed a limit on total production capacity. The first direct evidence of ozone depletion over Antarctica came in 1985 and restrictions on CFCs were then proposed at the Vienna Convention and confirmed by the Montreal Protocol of 1987, which came into force in January 1989. The two main provisions were based on the EC approach, namely: (1) consumption of the five CFCs listed above to be frozen at 1981 levels and (2) production to be reduced to 50% of this by 1999. With a worsening situation the Montreal Protocol was re-negotiated at Copenhagen in 1992 and the following restrictions on production of CFCs were agreed: 1996, freeze at present level; 2004, cut by 35%; 2010, cut by 65%.

In the Antarctic spring of August 1993 the ozone level there fell by 30%, greater than in 1992. The ozone-poor air is transported in a complex way within atmospheric pressure systems and it is not yet clear whether Arctic ozone is being affected, however, there has recently been a 4% rise in UVB ($\lambda = 280$–320 nm) in the UK as a result of greater penetration of the layer.

The current EC restrictions are more stringent than those set out in the amended Montreal Protocol and require: (1) no production of CFCs after 30 June 1997; (2) no production of Halons after 31 December 1997; (3) no production of carbon tetrachloride after 31 December 1997; (4) no production of 1,1,1-trichloroethane after 31 December 2004.

Halon 1211, presently produced at 7000 t year^{-1}, is to be phased out by 2005; its effect will then be negligible by 2050 as the half-life is only 12.5 years. It generates bromine atoms by the reaction

$$CF_2BrCl \xrightarrow{hv} CF_2Cl\bullet + Br$$

These are 50 times as damaging to the ozone layer as chlorine atoms because they are released more readily from reservoir compounds such as $BrO\bullet NO_2$.

Halon 1301, presently produced at 3500 t year^{-1}, is to be phased out by 2070; owing to its long half-life of 77 years, a level of 1 pptv will persist well into the next century.

It was formerly believed that the related HCFCs were satisfactory substitutes because the C-H bond is open to attack in the troposphere by radicals such as $\bullet$OH, so reducing their concentration in the stratosphere. Typical of this group are the following:

HCFC-123	CF_3CHCl_2	for blowing foam	ODP 0.013
HCFC-134	CF_3CH_2F	a refrigerant	ODP 0.0
HCFC-132	CH_2ClCF_2Cl	cleaning electronics	–

However, it is now realised that they persist long enough to reach the stratosphere and they also contribute to the 'greenhouse effect,' and consequently they too will be phased out completely by 2030. In 1996 the production of HCFCs in the world was 791×10^3 t and in Europe 143×10^3 t.

Reference

Molina, M.J. and Rowland, F.S. (1974) Stratospheric sink for CFCs, chlorine atom catalysed destruction of ozone, *Nature (London)*, **249**, 810.

Further reading

Haigh, N. (1991) *Manual of Environmental Policy in the EC and Britain*, Longman, Harlow.

Rastogi, M.C. (1990) A routine GC method for the analysis of CFCs in aerosol cans, *Chromatographia*, **29**, 152–154.

UK Strategic Ozone Research Group (1993) *Stratospheric Ozone*, HMSO, London.

CHLOROFORM CAS No. 67-66-3
LD_{50} rat $0.45–2.0\,g\,kg^{-1}$; LC_{50} fish $19–190\,mg\,l^{-1}$

$CHCl_3$ $\qquad\qquad\qquad\qquad\qquad M = 119.4$

Trichloromethane; b.p. 62°; sp. gr. 1.49 (20°) v.p. 21.3 kPa (160 mmHg, 20°) v. dens. $4.36\,kg\,m^{-3}$. Solubility in water $8\,g\,l^{-1}(25°)$; soluble in most organic solvents, the technical grade includes 0.5–1.0% EtOH. It is non-flammable but decomposes on heating to emit HCl and phosgene; it reacts violently with some metals, including sodium, potassium and aluminium.

Chloroform has a sweetish odour. The GC limit of detection in air is $0.01\,\mu g\,l^{-1}$ and in water $1\,\mu g\,l^{-1}$. It is manufactured by the action of hydrogen chloride on methanol or by chlorination of methane; in 1987 world production was 440 kt.

Chloroform was formerly used as an anaesthetic. It is now used in formulating pesticides, drugs and flavourings but is banned from human medication and cosmetics by the US FDA (1976).

The residence time in air is several months. Typical human intake is $2\,\mu g\,day^{-1}$ from food, indoor air and water; 60–80% of that inhaled is exhaled as CO_2. The remainder is distributed throughout the body, with the highest levels in fat, blood, the liver, kidneys, lungs and the nervous system. Chloroform is metabolized by the following route:

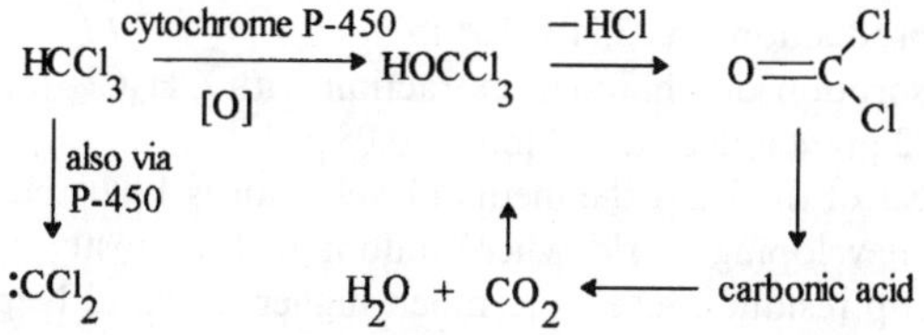

Chloroform is formed in unacceptable concentrations in water when chlorine disinfection is not properly controlled. A survey of 50 US cities showed that THMs ranged from 2 to $268\,\mu g\,l^{-1}$ and accounted for 50% of total organic compounds.

Humans occupationally exposed have suffered liver damage at 80–160 mg m^{-3} and the estimated lethal dose for an adult is 45 g. Chloroform induces liver tumours in mice and rats on a daily dose of ca 200 mg kg^{-1} but presents a low carcinogenic risk to humans.

Further reading
Environ. Health Criteria, World Health Organization, Geneva, 1994, vol. 163.

CHLOROMETHANE CAS No. 74-87-3
LD$_{50}$ rat 1.8 g kg^{-1}, LC$_{50}$ (96 h) bluegill sunfish 550 ppm

CH$_3$Cl $M = 50.5$

Methyl chloride; b.p. −24°; flash point −24° v.p. 3.8 × 10^3 mmHg (20°); density 1.78; logP$_{ow}$ 0.91. Water solubility 3.03 g l^{-1} (20°), soluble in EtOH, diethyl ether, CHCl$_3$.

A sweet-smelling, highly flammable gas. Produced naturally in oceans, possibly by the action of chloride ions on methyl iodide, also from forest fires, volcanoes and cigarette smoke.

Formerly used as a local anaesthetic and refrigerant, but its use has been discontinued owing to the fire hazard. Currently used in the synthesis of silicones, based on the sequence.

$$2CH_3Cl \;+\; Si(Cu) \;\longrightarrow\; (CH_3)_2SiCl_2 + Cu$$
$$Si/Cu\,alloy$$

followed by hydrolysis to $(CH_3)_2Si(OH)_2$, which polymerizes to

```
    Me      Me
    |       |
—(Si—O—Si—O)ₙ—
    |       |
    Me      Me
```

It is also used in the synthesis of methyl cellulose and quaternary methylamines and as a blowing agent for polystyrene foam and butyl rubber. It has been estimated that 40 000 US workers are exposed to it.

An important use, which is now declining, is the manufacture of methylene chloride:

$$CH_3Cl + Cl_2 \longrightarrow CH_2Cl_2 + 2HCl$$

In 1993 world production was 354 × 10^6 lb.

Analysis by sorption on charcoal, extraction with CH$_2$Cl$_2$ followed by GC with Chromosorb 102 gives a detection limit of 0.08 μg l^{-1}.

In remote areas of the USA the median level in air is 1300 ppt, in the Netherlands 700 ppt. In the developing world where indoor fuel is mostly wood, animal dung, charcoal and crop residues, levels are much higher. In rural Nepal indoor levels of chloromethane reached 6950 pptv.

Chloromethane penetrates the skin and acts as a narcotic, but it is weaker than chloroform. Exposure to concentrations of ca 500 ppm leads to dizziness, nausea, vomiting and abdominal pain; convulsions and death have been reported. In 1993 world production was 849×10^6 lb.

It is lost from water bodies by volatilization $(t_{1/2} = 2.4 \text{ h})$; in air chloromethane diffuses upward and is photodissociated to the methyl radical and chlorine atoms, it also reacts with OH radicals:

$$\bullet OH + CH_3Cl \longrightarrow H_2O + \bullet CH_2Cl$$

The low value of its $\log P_{ow}$ indicates that there is no tendency for bioconcentration by aquatic organisms.

Chloromethane is teratogenic in mice at 500 ppm and there is limited evidence of its carcinogenicity; it is placed in the IARC group 3. In the USA the TLV (TWA) is 50 ppm (103 mg m^{-3}); the STEL is 100 ppm (207 mg m^{-3})

Further reading

Howard, R.E. (ed.) (1989) *Handbook of Environmental Fate and Exposure Data for Organic Compounds*, Lewis, Chelsea, MI, Vol. 1, pp. 394–401.

Wagner, R.E., Kotas, W. and Yogis, G.A. (1992) *Environmental Analytical Methods*, 2nd edn, Genium, Schenectady, NY.

CHLOROMETHYL ETHER
See **Bis(chloromethyl) ether**.

CHLOROMETHYL METHYL ETHER CAS No. 107-30-2
LD_{50} rat 817 mg kg^{-1}; LC_{50} inhalation, hamster (7 h) 65 ppm

$$ClCH_2—O—CH_3 \qquad\qquad C_2H_5ClO \quad M = 80.5$$

Chloromethoxymethane, methoxymethyl chloride; b.p. 59°; v.p. 260 Torr (20°), dens. 1.062. Rapidly hydrolysed to give hydrogen chloride, formaldehyde and methanol. It is soluble in ethanol, diethyl ether and acetone.

Chloromethyl methyl ether is manufactured by passing HCl through a mixture of formalin and methanol. It is used in the synthesis of chloromethylated compounds and for the manufacture of ion-exchange resins.

It is highly flammable and may explode in contact with oxidizing agents. A severe eye and lung irritant; acute exposure can cause pneumonia. Four cases of lung cancer were found in a group of 111 workers exposed over 4 years; a significant increase of respiratory tract cancer was found in another study of 2850 workers over a 23 year period.

Chloromethyl methyl ether has been classed by the ACGIH in group A2 – a suspect human carcinogen. The still more dangerous bis-(chloromethyl) ether is a normal contaminant of this monochloro compound.

CHLORONITROBENZENES

$$C_6H_4ClNO_2 \quad M = 157.6$$

Compound	CAS No.	M.p. (°)	B.p. (°)	Dens.	v.p. (20°) (Torr)	Log P_{ow}
2-Chloronitrobenzene	88-73-3	33	246	1.368	3×10^{-2}	2.24
3-Chloronitrobenzene	121-73-3	46	236	1.534	–	2.41
4-Chloronitrobenzene	100-00-5	83	242	1.520	9×10^{-3}	2.39

Water solubility 199 mg l^{-1} (25°, 2-chloro); all are soluble in ethanol, acetone, benzene, diethyl ether.

The corresponding nitrophenols are formed from them by treatment with caustic alkali and nitroanilines by reaction with ammonia. Reaction with sodium alkoxides or sodium phenoxide affords the nitro ethers. 2-Nitrophenol is used to prepare the carbamate carbofuran, an insecticide and acaricide.

2-Nitroaniline is an intermediate for the fungicide benomyl. 3-Nitroanisole and 3-chloronitrobenzene are intermediates for 3,3'-dimethoxybenzidine and 3,3'-dichlorobenzidine, respectively. 4-Chloronitrobenzene is a precursor of parathion and also leads to 4-nitroaniline, which gives p-phenylenediamine. The latter is used as an antioxidant in petrol and as an intermediate for dyestuffs and rubber additives.

2-Chloro- and 3-chloronitrobenzene are classified in the UK as 'toxic substances,' but no occupational exposure limits have been set. 2-Chloronitrobenzene had a half-life in the river Rhine of 3.2 days but survived transport down the Mississippi. It was only 30% degraded by activated sludge after 14 days. Dyestuffs workers are at risk from it as are members of the public when drinking water is polluted.

The acute toxicities are as follows:

Compound	LD$_{50}$ rat (mg kg^{-1})	LC$_{50}$ guppy (mg l^{-1})
2-Chloronitrobenzene	135–288	30 (14 day)
3-Chloronitrobenzene	390–470	20 (96 h)
4-Chloronitrobenzene	420–650	13

4-Chloronitrobenzene (PNCB): workmen involved in the nitration of chlorobenzene were hospitalized suffering from cyanosis of the blood and anaemia. Methaemoglobinaemia was also found in rabbits and at a level of 3.6 mg m^{-3} there was an adverse response in humans, including eczema. In view of the volatility of PNCB, the TLV was revised to 0.1 ppm (0.64 mg m^{-3}) in the USA in 1988. The STEL in the UK is 2 mg m^{-3} (0.3 ppm).

Further reading

Dawson, G.W., Jennings, A.L., Drozdowski, D. and Rider, E. (1975) The acute toxicity of 47 industrial chemicals to fresh and saltwater fishes, *J. Hazardous Mater..*, **1**, 303–318.

Howard, P.H. (ed.) (1989) *Handbook of the Environmental Fate and Exposure data for Organic Compounds*, Lewis, Chelsea, MI, Vol. 1, pp. 146–160.

CHLOROPHENOLS

Cl

C_6H_5ClO $M = 128.6$

OH

Compound	CAS No.	M.p. (°)	B.p. (°)	Dens. (20°)	V.p. (25°) (Torr)	$\text{Log} P_{ow}$	Water solulibility (25°) (g l^{-1})
2-Chlorophenol	95-57-8	9	175	1.263	1.42	2.15	28
3-Chlorophenol	108-43-0	33	214	1.245	0.12	2.50	26
4-Chlorophenol	106-48-9	43	220	1.306	0.087	2.39	27

The LD_{50} values for the rat are in the range 570–670 mg kg^{-1} and the LC_{50} for guppy and rainbow trout are in the range 16–26 ppm.

All three compounds are used in synthesis but as the 2- and 4-isomers are the normal products of the chlorination of phenol they are more significant. They are used in the manufacture of dyestuffs, while 4-chlorophenol additionally is used to denature alcohol and as an antiseptic.

The mixed isomers can be adsorbed from air by passage through a tube of silica gel, extracted with acetonitrile and analysed by HPLC when UV will detect a sample of 2.5 μg. A 3 l sample of air suffices for concentrations in the range 1–23 mg m^{-3} and 2-chlorophenol can be separated from a mixture including dichlorophenols.

The chlorophenols severely irritate the skin and are dangerous to inhale. They do not readily hydrolyse and so bioconcentrate in water. Their evaporation rate is slow – the half-life in rivers is about 73 days. In air they are destroyed by ozonolysis with a half-life of 2.0 days. They may be formed from phenol during the chlorination of water. 2-Chlorophenol was found at 39 ppt in Canadian drinking water; in 1974, 200 samples from the Rhine had levels between 3 and 20 ppb. The 4-chloro-isomer was found in water at Love Canal.

In 1983 NIOSH estimated that 2500 US employees were at risk of contact with chlorophenols, but no TLV has been set. A limit of 10 μg l^{-1} has been set for 2-chlorophenol in groundwater and the limit in soil is 660 μg kg^{-1}.

Further reading
Environmental Health Criteria. Chlorophenols Other Than Pentachlorophenol, World Health Organization, Geneva, 1989.

CHLOROTHALONIL CAS No. 1897-45-6
(Daconil)
Wide-spectrum fungicide, bacteriocide.
LD_{50} rat 10 g kg^{-1}; LC_{50} trout 0.25 mg kg^{-1}

Cl

Cl   CN

Cl   Cl

CN

$C_8Cl_4N_2$ $M = 265.9$

2,4,5,6-Tetrachloro-1,3-benzenedicarbonitrile; v.p. < 0.01 mmHg (40°); water solubility 0.6 mg l^{-1}.

Used for the control of potato blight, mould on vines fruit and vegetables and also diseases of wheat. Greenhouse workers and those using wood preservatives containing chlorothalonil have become infected by dermatitis. There is also limited evidence of its inducing cancer in animals. The half-life in soil is 1.5–3 months, degraded in plants to 4-hydroxy-2,5,6-trichloroisophthalonitrile.

Codex Alimentarius acceptable daily intake 0.03 mg kg^{-1}. Maximum residue limits: vegetables 1–15, potato 0.1 and banana 0.2 mg kg^{-1}

CHLORPYRIFOS CAS No. 2921-88-2
(Dursban, Lorsban)
Insecticide, acaricide. Toxicity class II, moderately toxic, 10 mg < LD$_{50}$ < 50 mg kg^{-1}. LD$_{50}$ rat 135–160 mg kg^{-1}; LC$_{50}$ trout 3 µg l^{-1}; LD$_{50}$ contact bee 59 ng per bee

$C_9H_{11}Cl_3NO_3PS$ $M = 350.6$

O,O-Diethyl-*O*-(3,5,6-trichloro-2-pyridyl)phosphorothioate; m.p.42°; v.p. 1.5 mPa (25°); solubility in water 2 mg l^{-1}.

Used to control mosquitoes, animal flies, aphids in cereals, root flies in brassicas and also on fruit and vines. In Californian fogwater levels were found in the range 0.39–7.7 µg l^{-1} (January 1986). The hydrolytic $t_{1/2}$ of chlorpyrifos is 35–78 days and it is rapidly metabolized by mammals to its monoethyl ester and 3,5,6-trichloro-2-pyridinol.It exhibits anticholinesterase activity in humans.

Chlorpyrifos is a major import into South America. Uncontrolled use in banana plantations has led to river pollution; in Costa Rica cucumbers grown downstream included 8 ppm and the International Water Tribunal ruled that proper controls be introduced.

Codex Alimentarius acceptable daily intake 0.01 mg kg^{-1} (1982). Maximum residue limits rice 0.1, milk 0.01, potatoes 0.05, citrus fruit 0.3 mg kg^{-1}. In the workplace an exposure level of TWA 0.2 mg m^{-3} has been proposed in the USA (ACGIH) and in the UK.

CHROMATES CAS No. 11104-59-9

CrO_4^{2-}

Uses. Oxidising agents. Pigments.

Toxicity. Highly toxic and carcinogenic.

CHROMIC ACID CAS No. 7738-94-5

H_2CrO_4 $M = 118.0$

Uses. Oxidizing agent in organic chemistry. Corrosion inhibitor.

Occupational exposure limit. US TLV (TWA) 0.05 mg Cr m^{-3}

Toxicity. Poison. Mutation data. Inhalation can cause serious internal damage.

CHROMIUM CAS No. 7440-47-3

Cr

Chromium, a member of Group 6 of the Periodic Table, has an atomic number of 24, an atomic weight of 51.996, a specific gravity of 7.20 at 20°, a melting point of 1907°, and a boiling point of 2670°. It is a silvery, shiny, malleable metal and exists in oxidation states from -2 to $+6$. Naturally occurring isotopes are ^{50}Cr (4.35%), ^{52}Cr (83.79%), ^{53}Cr (9.50%), ^{54}Cr (2.37%). Chromium is a common element, the average crustal abundance is around 100 ppm, 21st in order of relative abundance. The main chromium ores are *chromite*, $FeCr_2O_4$, and *crocoite*, $PCrO_4$. The metal is obtained by the reduction of chromium(III) oxide with aluminium.

The main anthropogenic sources of chromium in the environment are the manufacture of chromium steels, the tanning industry, building industry, oil industry, textile industry and industrial effluents enriched in sewage sludge. Typical concentrations are soil 10–90 mg kg^{-1}; sea water 0.3 μg l^{-1}; fresh water 1–10 μg l^{-1}; beer and wine 300–450 μg l^{-1}, air 10 ng m^{-3} rural and up to 70 ng m^{-3} industrial urban, plants 0.02–14 mg kg^{-1} dry weight, sea fish 0.03–2 mg kg^{-1} dry weight, mammalian muscle 0.002–0.8 mg kg^{-1} and mammalian bones 0.1–30 mg kg^{-1}. Adult humans store around 5–10 mg of chromium, mainly in the spleen and liver. The dietary requirement of humans is 0.01–0.3 mg day^{-1}.

Uses. Hardened steel. Manufacture of stainless steel. Alloys. Chrome plating, anticorrosion.

Chromium is considered to be essential to the health of living organisms, and a deficiency of chromium in animals can produce diabetes, arterioscleriosis, growth problems and eye cataracts. A deficiency in rats and monkeys has been shown to impair glucose tolerance, decrease glycogen reserve and inhibit the utilization of amino acids. A deficiency in Cr increases the toxicity of lead. Chromium is probably the least toxic of the trace elements and, generally, mammals can tolerate 100–200 times their total body content of chromium without harmful effects. Acute exposure to the metal causes acute nephrotoxicity, which may lead to renal failure and death. Only trivalent and hexavalent chromium are biologically significant. Chromium is required for normal carbohydrate and lipid metabolism. Certain Cr(III) compounds regulate the action of insulin. Trivalent chromium salts are usually not very toxic although prolonged contact can cause skin lesions and sensitization of the skin. The uptake of Cr is strongly influenced by its oxidation state. Between 0.5 and 3% of total intake of Cr(III) is absorbed in the body (taken up by transferrin); excretion is almost solely via urine. Plants seem to be able to take up relatively large amounts of Cr(III) without too much harm. Toxic to algae at concentrations ca 7 μg ml^{-1}.

It is usually assumed that hexavalent Cr is 100 to 1000 times more toxic than Cr(III), and carcinogenic. Cr(VI) easily crosses cell membranes. The specific toxic effects of hexavalent chromium are mainly a variety of DNA lesions. In the body, reductants of hexavalent chromium include thiol-containing substances such as glutathione and the enzyme proteins P450 and P450IIE1. Chromates and dichromates are severe irritants to skin (dermatitis), mucous membranes and lung (cancer). On the skin they cause deep ulcers. Hexavalent chromium sensitization is known to occur from exposure to cement. A daily intake of 30–100 µg of Cr(VI) compounds is irritating and corrosive when absorbed through the digestive tract, skin and alveoli of the lungs. Cr is found in lungs, muscle, fat and skin. Reducing the use of Cr in impregnation agents and paints has been highlighted as a possible action to reduce input into the North Sea. The toxicity of Cr(VI) to fish is fairly high: LC_{50} fresh water fish 250–400 mg l^{-1}; for sea fish 170–400 mg l^{-1}; and trout at 10 mg l^{-1} and crabs at 0.3–0.7 mg l^{-1} were badly affected. An oral dose of 0.5–1.0 g of potassium dichromate is fatal for humans; skin absorption of this compound is also very dangerous, resulting in diarrhoea, internal bleeding and serious kidney and liver damage.

Chromium(III) compounds are classified as IARC group 3 (not classified).

Occupational exposure limits. ACGIH (PEL) TWA Cr(0) and soluble Cr(III) 0.5 mg m^{-3}; Cr(VI) soluble compounds 50 µg m^{-3}; insoluble Cr(VI) 0.01 mg m^{-3}. MAC (EPA) drinking water 100 µg l^{-1}.

Toxicity. LD_{Lo} oral human 50–70 mg kg^{-1} (soluble salts).

Further reading
Alexander, J., (1993) Toxicity versus essentiality of chromium, *Scand. J. Work Environ. Health* **19**, 126–127.
ECETOX (1991) *Technical Report 30(4)*, European Chemical Industry Technology and Toxicology Centre, Brussels.
Environmental Health Criteria 61, Chromium, WHO/IPCS, Geneva, 1988.
Hamilton, J.W. and Wetterhahn, K.E. (1988) Chromium, in Seiler, H.G., Sigel, H. and Sigel, A. (eds) (1988) *Handbook on Toxicity of Inorganic Compounds*, Marcel Dekker, New York, pp. 239–250.
IARC Monogr. (1990) **49**, 49–256
Kirk–Othmer ECT, 4th edn, Wiley, New York, 1993, Vol. 6, pp. 228–311.
Nriagu, J.O. and Nieboer, E. (eds) (1988) *Chromium in the Natural and Human Environments*, Wiley, New York.
O'Flaherty, E.J. (1993) Chromium as an essential and toxic metal, *Scand. J. Work Environ. Health*, **19** (Suppl. 1), 124–125.
SI 1991 No. 472 The Environmental Protection (Prescribed Processes and Substances) Regulations 1991, HMSO, London, 1991.
ToxFAQs (1993) *Chromium*, ATSDR, Atlanta, GA.

Chromium(III) chloride CAS No. 10025-73-7

$CrCl_3$

Uses. Chromium plating. Preparation of chromium compounds. Polymerization catalyst. Textile mordant.

Occupational exposure limits. US TLV (TWA) 0.5 mg Cr m^{-3}.

Toxicity. LD_{50} oral rat 1870 mg kg^{-1}; LC_{50} (2 h) inhalation mouse 31.5 mg m^{-3}. IARC group 3.

Chromium(III) fluoride CAS No. 7788-97-8

CrF_3

Uses. Printing and dyeing woollens. Halogenation catalyst.

Occupational exposure limits. US TLV (TWA) $0.5\,mg\,Cr\,m^{-3}$.

Toxicity. LD_{Lo} oral rat $150\,mg\,kg^{-1}$.

Chromium(III) nitrate CAS No. 13548-38-4

$Cr(NO_3)_3$

Uses. Corrosion inhibitor. Textiles. Manufacture of chromium(III) oxide.

Occupational exposure limits. US TLV (TWA) $0.5\,mg\,Cr\,m^{-3}$.

Toxicity. LD_{50} oral mouse $2976\,mg\,kg^{-1}$.

Chromium (III) oxide CAS No. 1308-38-9

Cr_2O_3

Uses. Abrasive. Catalyst. Protective coatings. Pigments.

Occupational exposure limits. N.a.

Toxicity. Carcinogenic. Extremely hazardous chemical.

Chromium(VI) oxide CAS No. 1333-82-0

CrO_3

Uses. Oxidizing agent in organic chemistry. Chromium plating. Aluminium anodizing. Corrosion inhibitor.

Occupational exposure limits. US TLV (TWA) $0.05\,mg\,Cr\,m^{-3}$.

Toxicity. Corrosive and oxidizing. Human carcinogen. LD_{50} oral rat $80\,mg\,kg^{-1}$.

Chromium(III) sulphate CAS No. 10101-53-8

$Cr_2(SO_4)_3$

Uses. Tanning industry. Mordant in textiles. Chrome plating. Paints, inks and glazes for porcelain.

Occupational exposure limits. US TLV (TWA) $0.5\,mg\,Cr\,m^{-3}$.

Toxicity. LD_{50} intraperitoneal mouse $28.1\,mg\,Cr\,kg^{-1}$

COBALT CAS No. 7440-48-4

Co

Cobalt, a member of Group 9 of the Periodic Table, has an atomic number of 27, an atomic weight of 58.93, a specific gravity of 8.9 at 20°, a melting point of 1495°, a boiling point of 2927° and one natural isotope, ^{59}Co. Cobalt- 57 ($t_{1/2} = 271$ days) is used in vitamin B_{12} diagnosis. Metallic cobalt forms a lustrous, grey, strongly ferromagnetic solid. The main oxidation states of cobalt are $+2$, $+3$, and $+1$ in vitamin B_{12} (which is required by all animals.)

Cobalt is found in the earth's crust at a concentration of $\sim$ 20 ppm, placing it 32nd in order of terrestrial abundance. Commercial sources of cobalt are *smaltite*, $CoAs_2$, *cobaltite*, CoAsS, and *linnaeite*, Co_3S_4. Cobalt is also obtained as a by-product from the extraction of other metals, in particular from nickel and copper ores. The metal is obtained by reduction of cobalt(III) oxide with charcoal. Average cobalt contents: soils ~ 8 ppm, sea water ~ 0.01 ppb, river water 0.2 ppb. Normal intake of cobalt (humans) is ~ 10–$80\,\mu\mathrm{g\,day}^{-1}$ and total cobalt (70 kg adult) ~ 3 mg.

Uses. Main uses of cobalt and its compounds are in the ceramics and paint industries, as catalysts in industrial organic syntheses and (cobalt metal) in the manufacture of high-grade steel and magnetic alloys. Bombardment of ^{59}Co by thermal neutrons converts this to the radioactive ^{60}Co($t_{1/2} = 5.271$ years). It undergoes β decay and at the same time gives off intense γ-radiation which is used in hospitals for radiotherapy of cancerous tissues. It is also used for irradiation of food. Cobalt-60 is a by-product of nuclear explosions.

Inorganic cobalt salts are moderately toxic to higher animals and humans and can be carcinogenic. Cobalt is an essential trace element and is mostly found in humans and animals as vitamin B_{12}. Cobalt is also essential for N_2 fixation by bacteria or blue–green algae. A deficiency of Co is of far greater concern than potential toxic levels in plants. Low levels of Co in feedstuffs can cause nutritional diseases in ruminants and pernicious anaemia in humans. Treatment is by the addition of cobalt sulphate to fertilizers. Chronic cobalt toxicity symptoms in humans include anorexia, nausea, vomiting, diarrhoea, skin rashes, tinitus and deafness. Cobalt inhibits enzymes of the citric acid cycle, which results in an increase of the lipoprotein, erythropoietin, which can produce polycythaemia (increase of red blood cells) in bone marrow. There are several reported incidents of cobalt toxicity in work environments. Mixtures of tungsten, titanium and tantalum carbides with cobalt as a binding agent are known as hard metals and are used as abrasives. The dusts are toxic on inhalation due to their cobalt contents. The pulmonary fibrosis of hard metal grinders has been recognized as an occupational disease.

A patient treated with $50\,\mathrm{mg\,day}^{-1}$ of cobalt(II) chloride for 3 months died; autopsy revealed the myocardial concentration of cobalt was $1.65\,\mathrm{mg\,kg}^{-1}$ wet weight, 25–80 times higher than normal. Patients were administered cobalt chloride three times a day for several weeks (total dose 100 mg) for the treatment of anaemia. Adverse effects to the alimentary canal were observed and included vomiting and anorexia in some patients. Occupational exposure to cobalt chloride aeresols can give rise to a number of diseases: rhinitis, pharyngitis, laryngitis and irritation of the eyes. Several hundred cases of cobalt toxicity occurred in the mid-1960s in the USA, Canada and Belgium among consumers of a certain brand of beer to which cobalt sulphate at $1\,\mathrm{mg\,l}^{-1}$ was added as a foam stabilizer. A large percentage of heavy drinkers with cobalt intakes of $> 6\,\mathrm{mg\,day}^{-1}$ died from congestive heart failure. Moderate drinkers of this brand of beer suffered from thyroid dysfunction and non-fatal heart disease. Workers exposed to cobalt containing dusts developed pulmonary fibrosis and other forms of chronic lung damage. One fatality has been reported for a worker employed as a hard metal tool grinder. Cobalt allergy produced lung diseases among diamond polishers in Antwerp, Belgium. It is a suspect cancer and tumour agent for recipients of hip replacements still in use after 10 years. In workers acutely exposed to cobalt carbonyl vapour, headaches, weakness, irritability, changes in reflexes and changes in brain activity were observed.

Occupational exposure limits. US TLV (TWA) $0.02\,\mathrm{mg\,m}^{-3}$ UK long-term limit $0.1\,\mathrm{mg\,m}^{-3}$. Cobalt dust is a fire hazard.

Toxicity. Cobalt compounds are placed in IARC group 2B (sufficient animal data for carcinogenicity).

Further reading

Control of Substances Hazardous to Health and Control of Hazardous Substances; Control of Substances Hazard to Health Regulations 1988 Approved Codes of Practice, 3rd edn HMSO, London, 1991.

ECETOX (1991) *Technical Report 30(4)*, European Chemical Industry Ecology and Toxicology Centre, Brussels.

HSE (1993) *Toxicity Review 29: Cobalt and Its Compounds*, HMSO, London.

IARC Monogr. (1991) **52**, 363–472.

Kirk–Othmer ECT, 4th edn, Wiley, New York, 1993, Vol. 6, pp. 760–793.

SI 1991 No. 472 The Environmental Protection (Prescribed Processes and Substances) Regulations 1991, HMSO, London, 1991.

Cobalt(II) bromide CAS No. 7789-43-7

$CoBr_2$

Uses. Catalyst for organic reactions.

Toxicity. LD_{50} oral rat $406\,mg\,kg^{-1}$.

Cobalt(II) carbonate hydrate CAS 513-78-1

$CoCO_3.H_2O$

Uses. Ceramics. Trace element added to soils and animal feed. Catalyst. Pigment.

Toxicity. LD_{50} oral rat $640\,mg\,kg^{-1}$.

Cobalt carbonyl CAS No. 10210-68-1

$Co_2(CO)_8$

Uses. Catalyst. Anti-knock agent in petrol.

Toxicity. Potentially harmful due to release of CO.

Cobalt(II) chloride CAS No. 7646-79-9; 7791-13-1 (hexahydrate)

$CoCl_2$

Uses. In glass and porcelain painting. Electroplating. Fertiliser and feed supplement. Catalyst. Manufacture of vitamin B_{12}

Toxicity. Poison. LC_{50} (96 h) fathead minnow $48\,mg\,l^{-1}$. LD_{50} oral rat 150–$500\,mg\,kg^{-1}$. Possible human carcinogen.

Cobalt(II) oxide CAS No. 1307-96-6

CoO

Uses. Glass manufacture. Ceramics. Catalyst preparations.

Toxicity. Harmful if swallowed, inhaled or absorbed through skin. LD_{50} oral rat $202\,mg\,kg^{-1}$; subcutaneous mouse $125\,mg\,kg^{-1}$.

Cobalt(II) sulphate CAS No. 10124-43-3; 60459-08-7 (heptahydrate)

$CoSO_4$

Uses. Pigments, ceramics, enamels and glazes. Batteries and electroplating baths. Manufacture of vitamin B12. Addition to fertilizers to supplement low cobalt content in soils.

Toxicity. Possible human carcinogen. LD_{50} oral rat 582–768 mg kg^{-1}.

Cobalt(II) sulphide, CAS No. 1317-42-6

CoS

Uses. Catalyst for hydrogenation or hydrosulphurization in petroleum refining.

Toxicity. Possible human carcinogen.

COPPER CAS No. 7440-50-8

Cu

Copper, a reddish coloured metal, is a member of Group 11 of the Periodic Table, has an atomic number of 29, an atomic weight of 63.54, a specific gravity of 8.92 at 20°, a melting point of 1084°, a boiling point of 2562° and two natural isotopes with the following percentages abundance. ^{63}Cu 69.2 % and ^{65}Cu 30.8%. Two oxidation states, +1 and +2

Widely distributed in all continents. 47–55 $\times 10^{-4}$% ($\sim$ 50 ppm), 26th in relative abundance. Occurs as *chalcopyrite*, $CuFeS_2$, *bornite* Cu_5FeS_4, *malachite*, $CuCO_3.Cu(OH)_2$, CuS, Cu_2S and native copper. Associated with Zn, Cd, Pb, As, Sb, Se, Ni, Pt, Mo, Au and Te ores. Metal produced from sulphide, oxide or carbonate ores by pyro- or hydrometallurgical processes. Anthropogenic sources of copper are copper mining and smelting/refining works and non-ferrous industries. Considerable amounts of copper are introduced into soils via fertilizers, pesticides and sewage. Owing to the high solubility of common copper salts, copper very easily enters the food chain. The following are typical concentrations: sea water $\sim$ 0.3µg l^{-1}, river water $\sim$ 7µg l^{-1}, drinking water 0.05–1.5 ppm (UK MAC in drinking water is 3000 µg l^{-1}; EC advisory level for drinking water 100 µl l^{-1} at the source of supply. WHO guide 2 mg l^{-1}. UK National Environmental Quality Standards, fresh water 20 µg Cu l^{-1}; abstracted to potable supply 50 µg l^{-1}); air 10–100 ng m^{-3}; soil 1–50 ppm (15–60 mg kg^{-1}); plant tissues 6–40 ppm, certain birds 3.6–14 mg kg^{-1} dry weight plumage, blue–green algae and molluscs 6.8 mg Cu kg^{-1} dry weight. Under ordinary conditions the human body receives 2–5 mg Cu day^{-1}. A 70 kg human carries 50–120 mg of Cu, with 1200 µg l^{-1} copper in the blood. Maximum permissible levels of copper in food: sugar 1, tea 100, meat/poultry 5 and infant food milk 1 mg kg^{-1}.

Uses. Electrical, plumbing, wire, alloys – bronze, brass, cupronickel – many manufacturing industries including machine and ship building industries. ^{67}Cu($t_{1/2}$ = 62 h) is used in radio(immuno)therapy. ^{64}Cu($t_{1/2}$ = 12.70 h) is used in medical diagnosis.

Copper is one of the 27 elements known to be essential to humans. It is present in many proteins, such as haemocyanin, galactose oxidase, ceruloplasmin and superoxide dismutase. Copper is an important oxidation catalyst and oxygen carrier in humans. Copper aids photosynthesis and other oxidative processes in plants. It can be acutely toxic, in excess, to all species, including humans. Greater than 60 mg kg^{-1} is considered excessive for most plants and can lead to chlorosis. The acute oral toxicity in humans, LD_{Lo}, is about $100\,mg\,kg^{-1}$. Generally, Cu toxicity is increased by low Mo, Zn and sulphate intake. Application of calcium blocks copper entry. Metallic copper is moderately toxic to fish, whereas copper chloride, sulphate and nitrate are very toxic even at low concentration, $\sim 0.02\,mg\,l^{-1}$. In sea water, a number of creatures, e.g. fish, mussels and crabs, bioaccumulate copper; the copper-containing haemocyanin functions as the dioxygen carrier for anthropods and molluscs.

In humans, the biological function of copper is related to the enzymatic action of specific essential copper proteins. A critical factor in the toxicity is the ability of Cu^{2+} to block sulphydryl groups in these proteins. In chronic intoxication of man by copper or its salts, functional disturbances of the nervous system, kidneys and liver may result, as well as ulceration and perforation of the nasal septum. Copper compounds are hepatotoxic. Excess copper accumulates in the liver.

Abnormalities are observed from workers in the manufacture and handling of Cu and Cu alloys, electrolytic refining of copper and handlers of Cu powders. Pneumoconiosis is a common disease of copper miners. Smelter workers are affected variously: enlarged livers, bilirubinaemia and gastric disorders. Allergic dermatitis in copper metalworkers results in eczema and vitiligo. Higher than usual copper levels in hair are an indication of copper working. Welding and cutting of copper materials by metal workers can lead to headaches, weak legs, nausea, muscle pain, chills and temperature rise. Inhalation of copper fumes can cause metal fume fever and haemolysis of red blood cells. Also, there is a greater risk of cancer for workers in the copper smelting industry. Copper compounds are probably only mildly carcinogenic although there is evidence of teratogenic and reproductive effects. Among male Japanese copper smelter workers, squamous cell carcinomas have been found to be very frequent.

In humans, accidental ingestion of large amounts of copper salts from foods or beverages contaminated by copper can cause gastrointestinal disturbances. Amounts of 1–2 g of copper sulphate have caused vomiting, gastric pains, diarrhoea, exhaustion, cramps, convulsion, shock and coma. There are a few cases of accidental/suicidal oral poisoning by copper sulphate. A lethal dose can occur between 27 and 120 g. Death occurred by renal failure.

The incidence of cardiovascular disease has been correlated with the drinking water concentrations of copper (and titanium). The copper ion is more toxic in soft water ($3–80\,mg\,l^{-1}$, species dependent); in hard water the copper is precipitated as the basic carbonate, malachite. Absorption of metal from cooking utensils and during dialysis caused hepatotoxicity. Poisoning from a copper container of fruit jelly with $224\,mg\,Cu$ per 100 g has been reported. Chronic exposure of domestic animals to copper varies according to species. Ruminants, particularly sheep, are very sensitive to copper dosage. A $100\,mg\,Cu\,kg^{-1}$ feed for 5 days resulted in weak, sick animals and some died. To avoid poisoning, the maximum permissible copper levels proposed for feedstuffs are $80\,mg\,kg^{-1}$ in mixed feeds for pigs/poultry and $30\,mg\,kg^{-1}$ in mixed feeds for cattle.

Two rare genetic disorder diseases are associated with copper. The symptoms of Wilson's disease (hepatolenticular degeneration) are due to the accumulation of copper and arise from the body's failure to excrete copper in the bile; it may be fatal, but treatment with a copper-chelating agent is invariably successful. In contrast, the genetic disorder Menkes' disease, which arises when there is insufficient nutritional copper, is fatal, usually at 6 months–3 years of age.

Occupational exposure limits. OSHA PEL (TWA) (dust and mists) $1\,mg\,Cu\,m^{-3}$; fume $0.1\,mg\,m^{-3}$. Copper dust is flammable.

Further reading
ECETOX (1991) Technical Report 30(4), European Chemical Industry Ecology and Toxicology Centre, Brussels.
Flemming, H.C. (1993) Copper and the Environment, *Metall*, **47**, 1020–1027.
King, R.B. (ed.) (1994) *Encyclopedia of Inorganic Chemistry*, Wiley, Chicester, Vol. 2, pp. 829–849.
Kirk–Othmer ECT, 4th edn, Wiley, NewYork, 1993, Vol. 7, 381–520.
SI 1991 No. 472 The Environmental Protection (Prescribed Processes and Substances) Regulations 1991, HMSO, London, 1991.

Copper(II) acetate CAS No. 142-71-2

$(CH_3CO_2)_2Cu$

Uses. Catalyst. Pigment for ceramics. Manufacture of Paris Green. Fungicide.

Occupational exposure limits. N.a.

Toxicity. LD_{50} oral rat $595\,mg\,kg^{-1}$; LC_{50} (96 h) fathead minnow $0.39\,mg\,l^{-1}$; LC_{50} ip mouse $2.5\,mg\,kg^{-1}$.

Copper(II) acetoarsenite CAS No. 12310-22-4
(Paris Green)

$(CH_3CO_2)Cu_2(AsO_2)_3$

Uses. Wood preservative. Insecticide. Pigment in paints.

Occupational exposure limits. US TLV (TWA) $0.1\,mg\,As\,m^{-3}$. UK long-term limit MEL $0.1\,mg\,As\,m^{-3}$.

Toxicity. Irritant. Toxic by injection. LD_{50} oral rabbit, rat 13–$100\,mg\,kg^{-1}$.

Copper(I) chloride CAS No. 7758-89-6

CuCl

Uses. Absorbent of carbon monoxide. Catalyst – petroleum industry and organic reactions. Catalyst in acrylonitrile production. Converts ethylene dichloride to vinyl chloride. Condensing agents for soaps, fats and oils.

Occupational exposure limits. US TLV (TWA) $1\,mg\,Cu\,m^{-3}$. UK long-term limit $1\,mg\,Cu\,m^{-3}$.

Toxicity. LD_{50} oral rat $140\,mg\,kg^{-1}$. LC_{50} oral mouse $347\,mg\,m^{-3}$; inhalation mouse $1008\,mg\,m^{-3}$.

Copper(II) chloride dihydrate CAS No. 10125-13-0

$CuCl_2.2H_2O$

Uses. Wood preservative and anti-mildew agent. Feed additive. Electroplating. Photography. Mordant dyeing and printing textiles. Indelible inks. Adsorbent of carbon monoxide. Mineral paints.

Occupational exposure limits. US TLV (TWA) $1\,mg\,Cu\,m^{-3}$. UK long-term limit $1\,mg\,Cu\,m^{-3}$.

Toxicity. Skin irritant. LD_{50} (16–24 h) stickleback $2\,mg\,l^{-1}$. LD_{50} intraperitoneal mouse $7.4\,mg\,kg^{-1}$. LD_{50} oral rat $163\,mg\,kg^{-1}$.

Copper(II) chloride oxide hydrate CAS No. 1332-40-7

$Cu_2Cl(OH)_3$

Uses. Foliar fungicide (Bordeaux A and Z) and bactericide. Catalyst. Antifouling agent.

Occupational exposure limits. US TLV (TWA) $1\,mg\,Cu\,m^{-3}$. UK long-term limit $1\,mg\,Cu\,m^{-3}$.

Toxicity. LC_{50} (48 h) carp $2.2\,mg\,l^{-1}$. LD_{50} oral rat $1440\,mg\,kg^{-1}$.

Copper(I) cyanide CAS No. 544-92-3

CuCN

Uses. Insecticide, Fungicide. Antifouling agent in marine paints. Catalyst. Electroplating.

Occupational exposure limits. US TLV (TWA) $5\,mg\,Cu\,m^{-3}$.

Toxicity. Poison.

Copper (II) hydroxide CAS No. 20427-59-2

$Cu(OH)_2$

Uses. Fungicide and bactericide. Rayon manufacture. Catalyst. Electroplating.

Toxicity. Harmful by inhalation, contact with skin or if swallowed. Target organ, kidney.

Copper(II) nitrate CAS No. 3251-23-8

$Cu(NO_3)_2$

Uses. Wood preservative. Fungicide. Herbicide. Pyrotechnics. Electroplating. Cotton printing. Enamels.

Occupational exposure limits. US TLV (TWA) $1\,mg\,Cu\,m^{-3}$. UK long-term limit $1\,mg\,Cu\,m^{-3}$.

Toxicity. LD_{Lo} (16–24 h) steelhead trout $2\,mg\,l^{-1}$. LD_{50} oral rat $940\,mg\,kg^{-1}$.

Copper(I) oxide CAS No. 1317-39-1

Cu_2O

Uses. Pigments in glass and enamels.

Occupational exposure limits. US TLV(TWA) $1\,mg\,Cu\,m^{-3}$. UK long-term limit $1\,mg\,Cu\,m^{-3}$

Toxicity. LD_{50} oral rat $470\,mg\,m^{-3}$.

Copper(II) oxide CAS No. 1317-38-0

CuO

Uses. Catalyst. Manufacture of glass, enamels, ceramics and porcelain.

Occupational exposure limits. US TLV (TWA) $1\,mg\,Cu\,m^{-3}$. UK long-term limit $1\,mg\,Cu\,m^{-3}$.

Toxicity. Inhalation of sublimed CuO can cause metal fume fever in humans.

Copper(II) selenate CAS No. 15123-69-0

CuSeO$_4$

Uses. Colouring copper and copper alloys.

Toxicity. Toxic.

Copper(II) sulphate, CAS No. 7758-99-8

CuSO$_4$.5H$_2$O

Uses. Antifungal agent. Agricultural fungicide, herbicide, algicide and bactericide. Copper sulphate–copper hydroxide–lime–Bordeaux Mixtures, used in spraying fruit trees. Copper deficiency/growth promoter additive in livestock and poultry feeds and as an agent for disease control in livestock and poultry production. Tanning leather. Battery electrolyte. Photography. Wood preservative, in mineral paints, in dyeing and printing and in electroplating

Occupational exposure limits. US TLV(TWA) $1\,mg\,Cu\,m^{-3}$. UK long-term limit $1\,mg\,Cu\,m^{-3}$.

Toxicity. Skin irritant. LC$_{50}$ (96 h) rainbow trout 0.1–$2.5\,mg\,l^{-1}$. LD$_{50}$ oral rat $300\,mg\,kg^{-1}$; LD$_{Lo}$ oral human $50\,mg\,kg^{-1}$.

CRESOLS

LD$_{50}$ rat *ortho* 1470, *meta* 2010, *para* $1430\,mg\,kg^{-1}$; LC$_{50}$ rainbow trout 7.9–$8.4\,mg\,l^{-1}$

Hydroxytoluenes.

Cresol	CAS No.	M.p. (°)	B.p.(°) (1 atm)	V.p. (25°) (Torr)	Water solubility (25°) (g l^{-1})
ortho	95-48-7	31	191	0.31	25.9
meta	108-39-4	12	202	0.14	22.7
para	106-44-5	35	202	0.13	21.5

The cresols are all freely soluble in ethanol, diethyl ether, acetone, benzene and aqueous alkalis. The log (n-octanol–water partition coefficient) is in the range 1.94–1.96 with bioconcentration factors between 14 and 20.

Commercial cresol or cresylic acid contains all three isomers and also xylenols. A typical technical grade contains *ortho* 20%, *meta* 40%, and *para* 30%, while the remaining 10% consists of phenol itself and the xylenols. The chemical reactions are similar to those of phenol and include halogenation, nitration and carbonyl condensations.

Cresols and their derivatives occur naturally in plants, e.g. oil of jasmine and those of conifers, oaks and sandalwood – also in mammalian urine and faeces and in poultry manure. They occur in crude petroleum and coal tar and may be released during processing; another source is the reaction between toluene and hydroxyl radicals. The USEPA estimated that in 1987 52 t of cresols were released to air and 172 t to waste water.

Demand for cresol has outstripped the supply from coal tar and petroleum and recent methods include the conversion cymene $\longrightarrow$ cymene hydroperoxide $\longrightarrow$ cresols + acetone. In 1990 US production was 38 300 t.

Uses. The *ortho*-isomer is used as a solvent and insecticide and in the formulation of novolak resins for sealing integrated circuits also in phenol–formaldehyde resins. The *meta*-isomer is an intermediate for contact pesticides and tricresyl metaphosphate is a flame-retarding plasticizer for PVC and other plastics. *para*-Cresol is used for the manufacture of antioxidants such as di-*t*-butylphenol for addition to rubber, food-stuffs and to lubricating and motor oils.

The cresols are short-lived in air but high concentrations have been detected near manufacturing plants, e.g. in 1978 o-cresol was found at 179 $\mu g\,m^{-3}$ (40 ppb) near a phenolic resin factory. In a room with an open fire, indoor air sampling on a cellulose ester membrane followed by HPLC showed o-cresol 5.0 and m- + p-cresol at 4.9 $\mu g\,m^{-3}$.

The cresols are rapidly degraded in surface water but are more persistent in groundwater; the mean level in 315 groundwater samples was *ortho* 10.9 and *meta* + *para* 12.5 $\mu g\,l^{-1}$. Rainout on seven occasions in Portland, Oregon (1984) had o-cresol in the range 0.24–2.8 $\mu g\,l^{-1}$ and m- + p- cresol at 0.38–2.0 $\mu g\,l^{-1}$. They have been detected in foods, e.g. tomatoes, cheese, butter, coffee and tobacco; 1 cigarette releases ca 75 μg.

Cresols are strongly irritant towards humans, being corrosive to the skin and burning the mouth – the threshold for mucosal irritation is 6 $mg\,m^{-3}$. Two deaths occurred in a group of patients who ingested 4–120 ml of disinfectant containing 25–50 ml of cresols. General symptoms of poisoning include mouth/throat burns, abdominal pains, vomiting and coma. There are no human data on carcinogenesis but cresols may promote cancer in animals when it has been initiated by other agents, e.g. benzene. The USEPA rates them as extremely hazardous.

Further reading
Environ. Health Criteria, World Health Organization, Geneva, 1995, Vol. 168.

CUMENE CAS No. 98-82-8
LD$_{50}$ rat 1.5 g kg^{-1}; LC$_{50}$ (96 h) goldfish 2.7 mg l^{-1}

Pr^i

C_9H_{12} $M = 120.2$

Isopropylbenzene; flammable; b.p. 152°; v.p. 3.2 Torr (20°); dens. 0.86; v. dens. 4.13. Solubility in water $50\,mg\,l^{-1}$ (20°); soluble in most organic solvents.

Cumene is a constituent of petroleum. It is manufactured by the alkylation of benzene with $AlCl_3$, H_3PO_4 or Al_2O_3. It is a component of high-octane aviation fuel ($20\,mg\,ml^{-1}$) and is used as a thinner for cellulose paints. A major outlet is the manufacture of styrene and in the synthesis of acetone and phenol. During 1987–95 the annual UK production followed the demand for phenol and was estimated at 120×10^3 t and that in the USA in 1987 was 1.9×10^6 t.

Cumene reacts violently with nitric acid and also forms an explosive peroxide. The odour threshold is 5×10^{-2} $mg\,m^{-3}$ (10 ppb). The principal hazard from the vapour is irritation of the eyes and skin although it is a narcotic at high concentration. Cumene accumulates in adipose tissue and is known to cause lung damage in animals but this has not been seen in humans. It occurs in tobacco smoke but there is no evidence of its being mutagenic or carcinogenic.

Cumene is metabolized by hydroxylation of the side chain rather than by the production of phenols, for example to give $PhCH_2CH(OH)CH_3$ (25%), which is exhaled by humans and which is found in the urine before being excreted as its glucuronide. Also formed are $PhCH(CH_3)CH_2OH$ and 2-phenylpropionic acid [$PhCH(CH_3)CO_2H$]:

$$\underset{\overset{|}{CH_2OH}}{Ph-CH-CH_3} \longrightarrow \underset{\overset{|}{CO_2H}}{Ph-CH-CH_3}$$

2-phenylpropionic acid

In the UK in 1987 the total emission of non-methane hydrocarbons (NMHC) was 2×10^6 t and of this cumene accounted for 3540 t, including 1200 t from vehicle emissions. The following ocurrences in air have been recorded: Gatwick Airport 12, central London 5 and urban peak 41 $\mu g\,m^{-3}$ Cumene was found at levels up to $54\,mg\,l^{-1}$ in US groundwater but it is not significant in the UK, although found in the river Thames at $60\,ng\,l^{-1}$. It is a prescribed substance under release to land in the UK Environmental Protection Regulations, 1991. US workplace levels are restricted to a TLV (TWA) of 50 ppm ($246\,mg\,m^{-3}$) and in the UK the STEL is 75 ppm ($370\,mg\,m^{-3}$).

Further reading
Chem-Facts, Chemical Intelligence Services, Reed Telepublishing, Dunstable, 1987.
Chemical Safety Data Sheets Vol. 1, Solvents, Royal Society of Chemistry, Cambridge, 1989.
Nielsen, I.R., Diment, J. and Dobson, S. (1994), *Environmental Hazard Assessment, Cumene*, DOE, Watford.

CYANAMIDE CAS No. 420-04-2

H_2NCN

Calcium cyanamide is used in agriculture as a weed killer and defoliant. It is toxic to humans due to decomposition with water to give lime (CaO) and cyanamide (NH_2CN). Cyanamide exerts systemic effects on respiratory centres and on blood. Skin exposure causes severe dermatitis.

CYANATES CAS No. 661-20-1

OCN⁻

Sodium and potassium cyanate are used in organic chemistry to obtain urea. Patients receiving sodium cyanate for sickle cell anaemia developed cataracts.

CYANAZINE Cas No. 21725-46-2
(Bladex).
Herbicide. Toxicity class III. LD_{50} rat 149 mg kg^{-1}; LD_{50} rabbit 141 mg kg^{-1}

$C_9H_{13}ClN_6$ $M = 240.7$

2-(4-Chloro-6-ethylamino-1, 3, 5-triazin-2-ylamino)-2-methylpropionitrile; m.p.167°; v.p. 1.6×10^{-9} Torr (20°). Water solubility 170 μg kg^{-1}. $t_{1/2}$ in soil ca 14 days; degraded by hydrolysis of the CN group and replacement of Cl by OH; eliminated by animals within 4 days.

Introduced in 1972 by Shell and of major use on cereals and cotton in the 14 States of the US cornbelt where maximum levels are found in early summer following application in the spring. In the Charles river of Iowa maximum levels in June were 5 μg l^{-1} and in the Minnesota river 6.5 μg l^{-1}, falling rapidly to low levels by September. A typical level in the mainstream Mississippi was 300 ng l^{-1}.

In 1990 in the cornbelt of Iowa, cyanazine with alachlor, atrazine and metolachlor accounted for 72% of all herbicides amounting to 22 000 tons of active compounds.

The EC limit for drinking water is 0.1 μg l^{-1}. The threshold limit value in the USA is a TWA of 5 mg m^{-3}.

CYANIDES CAS No. 57-12-5

CN⁻

Uses. Extraction of gold and silver from ores. Chemical synthesis. Electroplating. Hardening of steel.

Cyanides are common chemicals and are used as rat/pest poisons. Cyanides have a bitter, almond smell. Sodium and potassium cyanide are used in the extraction of gold and in photography. Human poisoning by cyanides may occur from inhalation of the vapour, ingestion or absorption by the skin. Handling of cyanides and exposure to skin cause severe dermatitis. Cyanides are less toxic than hydrogen cyanide but become more toxic when ingested. Poisoning may occur from dusts or splashes from industrial baths. In humans, systemic effects include nausea, vomiting, giddiness, palpatations, bradycardia, unconsciousness, convulsions and death. Because sodium and potassium cyanide are highly soluble they are potent poisons; $2–3\,mg\,kg^{-1}$ is a fatal dose for humans. A blood cyanide level $> 0.02\,\mu g\,l^{-1}$ is considered toxic and concentrations $> 1\,\mu g\,ml^{-1}$ are lethal. It is possible to get cyanide poisoning from amygdalin, a cyanogenic glycoside found in apricots, peaches and sweet almonds. It is used as an anticancer drug.

Occupational exposure limits.   USA TLV (TWA) $5\,mg\,CN^-\,m^{-3}$. UK MAC drinking water $50\,\mu g\,CN^-\,l^{-1}$.

Toxicity.   Very poisonous. Potential human carcinogen. LC_{50} (20 days) rainbow trout $0.005–0.05\,mg\,HCN\,l^{-1}$. LD_{50} intraperitoneal mouse $3\,mg\,CN^-\,kg^{-1}$. Target organs are CNS, CVS, liver, kidneys and skin.

Further reading
ECETOX (1991) *Technical Report No 30(4)*, European Chemical Industry Techno-
   logy and Toxicology Centre, Brussels.
Kirk–Othmer ECT, 4th edn, Wiley, New York, 1993, Vol. 7, pp. 753–782.
ToxFAQS (1993) *Cyanide*, ATSDR, Atlanta, GA.

CYANOGEN CAS No. 460-19-5

NCCN

Uses.   Intermediate in chemical synthesis.

Occupational exposure limits.   US TLV (TWA) $10\,ppm$ ($21\,mg\,m^{-3}$).

Toxicity.   Irritant. Toxic by inhalation. LC_{50} (1 h) inhalation rat $350\,\mu g\,m^{-3}$.

Further reading
Merck Index, 12th edn, Chapman & Hall, London, 1996.

CYANOGEN BROMIDE CAS No. 506-68-3

CNBr

Uses.   Fumigant. Blocking of enzymes.

Occupational exposure limits.   US TLV (TWA) $5\,mg\,CN\,m^{-3}$.

Toxicity.   Toxic by inhalation and corrosive substance. LC_{50} (96 h) bluegill sunfish $0.24\,mg\,l^{-1}$. LC_{Lo} (10 min) inhalation humans $92\,ppm$.

CYANOGEN CHLORIDE CAS No. 506-77-4

CNCl

Uses.   Chemical intermediate. Insecticide. Military poison gas.

Occupational exposure limit.   US TLV (TWA) $0.3\,ppm$ ($0.75\,mg\,m^{-3}$).

Toxicity.   Toxic irritant gas. LD_{50} oral cat $6\,mg\,kg^{-1}$. LD_{100} (10 min) inhalation humans $159\,ppm$.

CYCLAMATE CAS No. 100-88-9

$C_6H_{13}NO_3S$ $M = 179$

Cyclohexylsulphamic acid; m.p. 170°. Water solubility 13 g/100 g; that of the sodium salt is 20 g/100 g.

First synthesized by Sveda in 1937. In 1958 it was classified as 'generally recognised as safe (GRAS)' in the USA and at a production cost of only $2 lb^{-1} was widely used as a non-calorific sweetener. This usage rose rapidly through the 1960s owing to increased consumption of sweetened soft drinks.

In 1969 cyclamate was claimed to induce bladder cancer in rats which were fed diets of sodium cyclamate (10 parts)–sodium saccharin (1 part) at doses of up to 2.5 g day^{-1}. It was subsequently shown that the tumours could have resulted from calcification of the urinary tract and there was evidence of an external factor. In 1970, in the light of the above study, cyclamate was banned in the USA by the FDA for use in foods, beverages and drugs. Four subsequent detailed studies at doses of 2.6 g kg^{-1} on mice, hamsters and rats and also a 9 year study in dogs failed to detect bladder cancer or any other cancers. Through the 1980s numerous other studies indicated that both cyclamate and saccharin were safe. Cyclamate is GRAS according to the WHO and the EC and is allowed as a food additive in over 40 other countries (Jones, 1992). It has been said that the continuance of the ban into the 1990s in the USA was a political rather than a scientific decision.

Reference
Jones, J.M. (1992) *Food Safety*, Eagan Press, St Paul, MN.

Further reading
Bopp, B.A. and Price, P. (1991) in Nabors, L.O'B. and Gelardi, R.C. (eds), *Alternative Sweeteners*, 2nd edn, Marcel Dekker, New York.

CYCLOHEXANE CAS No. 110-82-7
LD$_{50}$ rat 12.7 g kg^{-1}

C_6H_{12} $M = 84.2$

Hexahydrobenzene, hexamethylene, m.p. 6.5°; b.p. 81°; flash point −0.5°, dens. 0.775; v.p. 77 Torr (20°). Water solubility 55 mg l^{-1}, miscible with most organic solvents.

Cyclohexane is produced by the procedure first used by Sabatier in 1898, namely the hydrogenation of benzene catalysed by nickel or platinum. In order to protect the catalyst the sulphur content of the benzene must not exceed 1 ppm.

Cyclohexane occurs in crude petroleum. It is used as a solvent for resins and lacquers but the bulk of the production is converted into adipic acid and caprolactam, the precursors of nylon 6,6.

The route to adipic acid depends upon the pre-oxidation of cyclohexane to cyclohexanol/cyclohexanone with air/Co acetate at 150° followed by oxidative cleavage with nitric acid. Cyclohexanone affords caprolactam by oxidation followed by acid-catalysed rearrangement:

caprolactam   adipic acid

Nylon 6,6 is obtained by the combination of adipic acid with hexamethylene diamine. The latter can be made from adipic acid but the preferred route is from 1,3-butadiene. Alternatively, if 1 mol of water is added to caprolactam it is hydrolysed and then undergoes stepwise condensation with renewal of the required water:

nylon 6,6

The production of caprolactam in the USA in 1990 was 610×10^3 t.

The oxidation products cyclohexanol and cyclohexanone both have a TLV (TWA) of 50 ppm (200 mg m^{-3}). Cyclohexane is flammable and narcotic at high concentrations, although it does not exhibit the neurotoxicity of n-hexane. The TLV (TWA) is 300 ppm (1030 mg m^{-3}).

A leakage of cyclohexane was responsible for the disastrous incident at the ICI complex at Flixborough, UK (Meyer, 1977). The initial leakage was due to the collapse of a temporary pipe connection between two reactors, so releasing cyclohexane confined above its boiling point at 145° and 120 psi (p absolute = 14.7 psi) or 8 atm. The subsequent explosion and fire killed 28 and caused $100 million of damage.

Reference
Meyer, E. (1977) *Chemistry of Hazardous Materials*, Prentice-Hall, Englewood Cliffs, NJ.

Further reading
Wittcoff, H.A. and Reuben, B.G. (1996) *Industrial Organic Chemicals*, Wiley, New York, pp. 248–255.

CYPERMETHRIN CAS No. 67375-30-8 (α-form as shown)
(Ammo, Cymbush)   52315-07-8.
Broad-spectrum insecticide Toxicity class II. α-form: LD$_{50}$ rat 79 mg kg^{-1}. LC$_{50}$ brown trout 2 μg l^{-1}

$C_{22}H_{19}Cl_2NO_3 \quad M = 416.3$

(RS)

α-Cyano-3-phenoxybenzyl(1R, S)-cis,trans-3-(2, 2-dichlorovinyl)-2, 2-dimethyl-cyclopropane carboxylate.

It is widely used in Paraguay and has been identified as a source of pollution. Under proper control fish and amphibia are not affected in ponds subjected to doses in excess of spray drift; however, invertebrates are affected at these levels. Birds and bees are not at risk under field conditions. Cypermethrin is a possible carcinogen and the risk to humans has been estimated as 3.73 persons/million.

It is effective against insects in cereals, citrus fruit, cotton, other fruits, vegetables and coffee. The metabolites include 3-phenoxybenzoic acid.

Codex Alimentarius acceptable daily intake $0.05\,mg\,kg^{-1}$. Maximum residue limits: wheat 0.2, milk 0.05 (EC 0.02), meat 0.2 (EC 0.05), citrus fruit $2.0\,mg\,kg^{-1}$.

p, p′-DDT CAS No. 50-29-3
Insecticide. Toxicity class III, moderately toxic, $50\,mg < LD_{50} < 500\,mg$; TD_{Lo} oral human $16\,mg\,kg^{-1}$; LD_{50} oral monkey $200\,mg\,kg^{-1}$. Induces cancer in animals. LD_{50} TE oral rat $87\,mg\,kg^{-1}$

$C_{14}H_9Cl_5 \quad M = 354.5$

Dichlorodiphenyltrichloroethane 1,1′-(2,2,2-trichloroethylidene)bis(4-chloroben-zene); m.p. 109°; v.p. 3.3×10^{-5} Pa (25°).Water solubility $3.0\,\mu g\,l^{-1}$.

DDT is synthesized by the condensation of trichloroacetaldehyde with chloroben-zene in sulphuric acid, which is not a unique pathway and so significant amounts of the *ortho*-isomers are formed. The dichloroacetaldehyde is also present and gives rise to dichlorodiphenyldichloroethane (DDD).

DDT is not very toxic to humans and a test group ingested 35 mg daily for an extended period without ill effect. Owing to the development of resistant insect species and to overkill, large overdoses were used, leading to the death of fish and raptorial birds. For details see Alloway and Ayres (1997). As a consequence, DDT was banned in the USA in 1973 and in the UK in 1984 but as its $t_{1/2}$ is about 8 years significant residues remain today. This is all the more true in Third World countries which may not be able to afford the more expensive alternatives to organochlorines, i.e. organophosphates, carbamates and pyrethroids.

In 1991 India produced 7000 t and the average daily intake there then was 0.27 mg. In South Africa, intake by Kwa-Zulu babies was in the range 0.1–$0.3\,mg\,day^{-1}$ compared with the acceptable daily intake of only $0.02\,mg\,day^{-1}$. The following table provides evidence of the effect of continuing usage in tropical Asia as levels in air over water bodies are higher there, as also are levels in surface water.

Levels of p, p'-DDT in ocean air(pg m^{-3}, 1990; see Iwata *et al.*, 1993)

Location	Range	Mean
North Pacific	0.4–27	5.7
Bay of Bengal	19–590	140
North Atlantic	1.1–9.6	3.9
South China Sea	3.7–46	20

The workplace limit in the USA and UK is a TWA of 1 mg m^{-3}. The EPA limit for drinking water is 0.59 ng l^{-1} and the WHO gives 1 μg l^{-1}.

The principal metabolite of DDT is dichlorodiphenyldichloroethylene (DDE, CAS No. 72-55-9), which has an even longer life and may persist for over a decade. Its regulatory details are as follows: *Codex Alimentarius* acceptable daily intake 0.02 mg kg^{-1}; maximum residue limits, cereal grains 0.1, milk 0.05, eggs 0.5, meat 5.0 mg kg^{-1}.

References

Alloway, B.J. and Ayres, D.C. (1997) *Chemical Principles of Environmental Pollution*, 2nd edn, Blackie, London, pp. 289–290.

Iwata, H., Tanabe, S., Sakai, N. and Tatsukawa, R. (1993) Distribution of persistent organochlorines in ocean air and surface seawater and the role of oceans on their global transport and fate. *Environ.Sci. Technol.*, **27**, 1080–1098.

Further reading

Dinham, B. (1993) *The Pesticide Hazard*, Pesticide Trust, ZED Books.

DELTAMETHRIN CAS No. 52918-63-5
(Crackdown)
Insecticide. Toxicity class II. LD$_{50}$ rat 128 mg kg^{-1}; LC$_{50}$ trout 0.5 μg l^{-1}

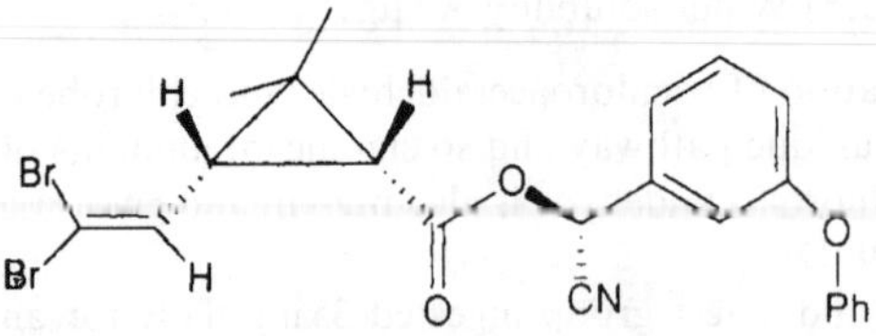

$C_{22}H_{19}Br_2NO_3$ M = 505.2

(S)-α-Cyano-3-phenoxybenzyl-(1R,3R)-3-(2,2-dibromovinyl)-2,2-dimethylcyclopropane-1-carboxylate; m.p. 101°; v.p. 0.015 mmHg (25°). Water solubility <0.1 mg l^{-1}.

There were 16 incidents of poisoning by deltamethrin in the Parana state of Brazil in 1990. It is dangerous to fish but less so towards bees, which are repelled by it.

Deltamethrin is widely used for the control of aphids and caterpillars in a range of vegetables, fruit and cereals and also against mosquitoes and animal flies. It is degraded by soil microbes in 1–2 weeks.

Codex Alimentarius acceptable daily intake 0.01 mg kg^{-1} (1982). Maximum residue levels cereal grains 1.0, milk 0.02, root vegetables 0.01, banana 0.05 mg kg^{-1}.

DIAZINON CAS No. 333-41-5
(**Basudin, Spectracide**). Non-systemic insecticide, acaricide. Toxicity class II or III.
LD_{50} rat 250–320 mg kg^{-1}; LC_{50} trout 2.6 mg kg^{-1}

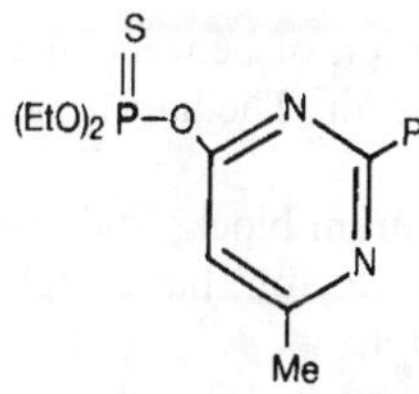

(EtO)$_2$ P—O [structure]

$C_{12}H_{21}N_2O_3PS$ $M = 304.3$

O,O-Diethyl-O-(2-isopropyl-6-methylpyrimidin-4-yl)phosphorothioate; yellow oil;
v.p. 1.4×10^{-4} Torr(20°). Water solubility 40 mg l^{-1}.

Used to control pests that have become resistant to organochlorines, for example, it is used in South Africa against locusts in place of BHC. It is also used for the control of sucking and chewing insects on fruit, vines, vegetables and cereals.

Diazinon is strongly adsorbed on soil. The principal metabolites are diethyl thiophosphate and diethyl phosphate; on storage it gives the more toxic Sulfotep.

The workplace limit in the UK and USA is a TWA of 0.1 mg m^{-3}. *Codex Alimentarius* acceptable daily intake 0.002 mg kg^{-1}. Maximum residue limits wheat 0.1, milk 0.02, olive oil 2.0, citrus fruit 0.7 mg kg^{-1}.

1,2-DIBROMOCHLOROPROPANE CAS No. 96-12-8
(**Nemafume, Nemagon**) Soil fumigant. LD_{50} rat 170 mg kg^{-1}; LD_{50} rabbit 180 mg kg^{-1}. Fish unaffected at 5 ppm

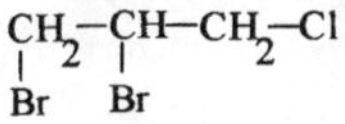

CH$_2$—CH—CH$_2$—Cl
   | |
  Br   Br $C_3H_5Br_2Cl$ $M = 235.4$

Rated as a cancer hazard by IARC and also by OSHA. A successful action for damages was brought in Texas against the manufacturers by exposed field workers. Limit in drinking water USA (EPA) 0.2 μg l^{-1}, WHO guideline 1 μg l^{-1}

DICHLOROBENZENES

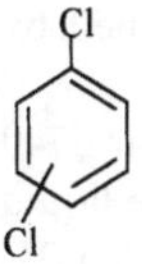

[structure]

$C_6H_4Cl_2$ $M = 146.1$

Compound	CAS No.	M.p. (°)	B.p. (°)	V.p. (25°) (Torr)	LogP_{ow}	Water solubility (25°) (mg l^{-1})
1,2-Dichlorobenzene	95–50–1	−17	180	1.50	3.38	145
1,3-Dichlorobenzene	541–73–1	−25	173	2.3	3.48	–
1,4-Dichlorobenzene	106–46–7	53	174	1.0	3.40	–

1,2-Dichlorobenzene is used as a solvent for the production of toluene diisocyanate, for degreasing, as an insecticide and soil fumigant, for the synthesis of herbicides and as a bacteriostat in deodorant soaps. In 1987 the US demand was 23×10^3 t.

1,3-Dichlorobenzene is a minor product (1%) of the chlorination of benzene and has limited use as a fumigant, when it is released directly into the air. The half-life is estimated to be 14 days.

1,4-Dichlorobenzene is used for the control of moths, in deodorant blocks and for the synthesis of poly(phenylene sulphide) resin (PPS). This resin (Ryton) has a high temperature resistance and in 1987 the US demand was 47×10^3 t.

The EPA estimated (1980) that 2 million workers were exposed to 1,2-dichlorobenzene and 1 million to 1,4-dichlorobenzene. Some 10% of all the 1,2-isomer is released to air and 70% of the 1,4-isomer from mothproofing.

Sweden has banned the use of 1,2-dichlorobenzene whenever direct human contact is possible on the grounds of its persistent mutagenic effect on experimental animals.

The TLVs (TWA) are 1,2-dichlorobenzene 25 ppm ($150\,\mathrm{mg\,m^{-3}}$) and 1,4- dichlorobenzene 10 ppm ($60\,\mathrm{mg\,m^{-3}}$); the latter isomer is rated a group 3 carcinogen, carcinogenic in animals but unlikely to induce human cancers.

Further reading

IARC (1982) *Some Industrial Chemicals and Dyestuffs*, IARC, Lyon, Vol. 29, pp. 213–238.

Kirk–Othmer ECT, 4th edn, Wiley, New York, 1993, Vol.6, pp.88–100.

USEPA (1993) *Manual of Chemical Methods*, 2nd edn, AOAC International, Arlington, VA.

1,2-DICHLOROETHANE CAS No. 107-06-2
LD_{50} oral rat $680\,\mathrm{mg\,kg^{-1}}$; LC_{50} fish $185–550\,\mathrm{mg\,l^{-1}}$

$ClCH_2CH_2Cl$ $C_2H_4Cl_2$ $M = 99.0$

Ethylene dichloride; v.p. 8.50 kPa; odour threshold $25–450\,\mathrm{mg\,m^{-3}}$; $\log P_{ow}$ 1.48. Flammable (flash point 13°) and decomposes in flames to HCl, phosgene, etc. Trade names include Dutch liquid, Brocide and Dichlor-emulsion.

GC determination in air following charcoal absorption gives a sensitivity of 1.2 $\mathrm{\mu g\,m^{-3}}$ with FID and $0.02\,\mathrm{\mu g\,m^{-3}}$ by GC–MS. In water it may be determined by direct injection to a sensitivity of $0.5\,\mathrm{\mu g\,l^{-1}}$.

Ethylene dichloride was 15th in volume production in the USA in 1983 at 5740 $\times 10^3$ t year^{-1}. It is manufactured by the addition of chlorine to ethylene in the liquid or the vapour phase also by the oxidative addition of HCl in the presence of a Cu (II) catalyst. (See also **vinyl chloride**.)

Commercial purity is high at 99%; 0.1% of an alkylamine is added to inhibit the photolytic decomposition to give HCl and chlorine. Over 80% of ethylene dichloride goes to the production of vinyl chloride, other products include 1,1,1-trichloroethane, trichloroethylene and ethyleneamines. Its application as a scavenger for lead in exhaust fumes is declining; it was once used as a fumigant, but this use is now barred in Canada, the UK and the USA. In the troposphere it reacts to give chloroformate, chloroacetyl chloride and decomposition products:

$$O=C\overset{Cl}{\underset{H}{<}} \qquad O=C\overset{CH_2Cl}{\underset{Cl}{<}} \quad \text{with HCl, CO and CO}_2$$

chloroformate   chloroacetyl chloride

The high-boiling residues from manufacturing units were formerly dumped in the North Sea; the present practice is incineration. Ethylene dichloride is removed from air through its reaction with the hydroxyl radical and is rapidly lost from water bodies owing to its volatility.

Levels in air over the remote Pacific Ocean were $0.17\,\mu g\,m^{-3}$, in the rural UK 0.08 $\mu g\,m^{-3}$ and in the urban UK–0.5–$2.1\,\mu g\,m^{-3}$. Near production plants in the USA air levels reached $736\,\mu g\,m^{-3}$ and indoors in Canadian homes ranged from 0.1 to $6.8\,\mu g\,m^{-3}$. Levels in sea water at the mouth of the Mississipi were 0.05–$0.21\,\mu g\,l^{-1}$ and the average in 14 river basins in the USA was $5.6\,\mu g\,l^{-1}$. It is generally below the detection limit of $0.5\,\mu g\,l^{-3}$ in drinking water.

The principal target in humans is the liver but the population exposure in industrial countries is low with a daily intake of 8–$140\,\mu g\,day^{-1}$. Occupational levels on manufacturing sites may reach $6\,\mu g\,m^{-3}$; occupational exposure has been reported in workers in oil refineries and those in an aircraft factory inhaling fumes from gums dissolved in dichloroethane. Exposed workers suffered a significant increase in pancreatic cancer, but this may have been due to contact with other chemicals on the same site. In humans inhalation affects the CNS, cause headaches and dizziness followed by cyanosis and vomiting. Oral exposure causes similar reactions; a lethal dose is 20–50 ml. Metabolism follows the sequence

(cf carbon tetrachloride)

The IARC has concluded that 1,2-dichloroethane is carcinogenic in mice and rats.

Further reading
Environ. Health Criteria, World Health Organization, Geneva, 1995, Vol. 176.

1,1-DICHLOROETHYLENE CAS No. 75-35-4
LD_{50} rat $1500\,mg\,kg^{-1}$; LD_{50} mouse $200\,mg\,kg^{-1}$

$CH_2 = CCl_2$ $C_2H_2Cl_2$ $M = 97.0$

Vinylidene dichloride; b.p. 32°; v.p. 591 Torr (25°); dens. 1.219 (20°); $\log P_{ow}$ 2.02. It has a sweet odour like chloroform. Solubility in water 2.5 $g\,l^{-1}$ (25°), soluble in acetone, benzene, ethanol, etc.

First prepared by Regnault in 1838 and its first copolymers were first made in 1939. Synthesized by the dehydrochlorination of 1,1,2-trichloroethane by alkali; dichloroacetylene is included as an impurity. European production in 1983 was 352×10^3 t; US production in 1992 was 79×10^3 t.

It is used as a copolymer for the packaging of food and copolymers are also extruded to make pipes and tubes.

Vinylidene dichloride may be determined in air by sorption in charcoal followed by GC–MS at a typical level of $1 \mu g l^{-1}$. Water is sampled by purging with nitrogen followed by sorption from the gas stream and GC at a typical level of $2.8 \mu g l^{-1}$. Hexane extraction is employed for soil samples, when GC with EC detection is effective at levels of 1 ppm. It is found in foodstuffs, e.g. in cheese at $0.005 \, mg \, kg^{-1}$.

Airborne levels in the USA ranged between 0.02 and $24.0 \mu g \, m^{-3}$ and it was estimated that 6500 workers were exposed at $90–100 \mu g \, m^{-3}$ ($1 mg \, m^{-3} = 2.96 \, ppm$). Vinylidene dichloride has been detected in the breath of residents of New Jersey at levels up to $14 \mu g \, m^{-3}$. It has been found in industrial wastewater from plastics production at $200 \mu g l^{-1}$ and in the river Rhine at $0.3 - 80 \mu g l^{-1}$.

Vinylidene dichloride is a group 3 carcinogen. The US TLV (TWA) is 5 ppm ($20 mg \, m^{-3}$) and the UK long term exposure limit is 10 ppm ($40 mg \, m^{-3}$).

Further reading

Howard, P.H. (ed) (1989) *Handbook of Environmental Fate and Exposure Data for Organic Chemicals*, Lewis, Chelsea, MI, Vol.1, p. 561.

IARC (1986), *Some Chemicals Used in Plastics and Elastomers*, IARC, Lyon, Vol.39, pp. 195ff.

1,2-DICHLOROETHYLENE CAS No. (mixture) 540-59-0

LD_{50} rat (mixture) > $2 g \, kg^{-1}$

$C_2H_2Cl_2$ $M = 97.0$

cis (Z)   trans (E)

1,2-Dichloroethene, acetylene dichloride. The commercial product is a mixture of the *cis*-isomer (60%) and the *trans*-isomer (40%). The pure substances have the following properties:

	cis (Z)	trans (E)
CAS No.	156-59-2	156-60-5
B.p. (°)	48	60
Dens. (20°) (g ml^{-1})	1.446	1.45
V.p. (20°) (kPa)	35	24
Water solubility (g/100 g)	0.63	0.35

Both compounds are soluble in ethanol, diethyl ether, etc. The commercial mixture is used for degreasing and as a solvent for perfumes and lacquers. It is highly flammable and with alkali gives chloroacetylene, which is also very flammable.

The mixture is synthesized by the direct addition of chlorine to acetylene or from the initial step in the manufacture of vinyl chloride by the oxychlorination of ethylene. It is commonly transported in 30 or 50 gallon steel drums.

Dichloroethylenes are less potent as narcotics than chloroform and they are largely excreted unchanged through the lungs. They are often found on polluted sites, possibly from the bioreduction of the more common tri- and tetrachloroethylene.

The occupational long-term exposure limit is 200 ppm ($733\,\mathrm{mg\,m^{-3}}$) and the short-term limit is 250 ppm ($916\,\mathrm{mg\,m^{-3}}$).

Further reading

Luxon, S.G. (1992) *Hazards in the Chemical Laboratory*, 5th edn, Royal Society of Chemistry, Cambridge.

Richardson, M.L. (1988) *Risk Assessment of Chemicals in the Environment*, Royal Society of Chemistry, Cambridge.

DICHLOROPHENOLS

LD_{50} rat oral $580\,\mathrm{mg\,kg^{-1}}$ (2,4-); LD_{50} rat i.p. $390\,\mathrm{mg\,kg^{-1}}$ (2,6-)

$C_6H_4Cl_2O$ $M = 163.0$

Compound	M.p. (°)	B.p. (°)	Dens. ($g\,ml^{-1}$)	V.p. (25°)	$\log P_{ow}$	Water solubility (25°) ($g\,l^{-1}$)
2,3-Dichlorophenol	60	206	–	0.18	2.39	–
2,4-Dichlorophenol	43	210	1.383	1 mm (50°)	2.92	4.5
2,6-Dichlorophenol	68	220	–	–	2.64	–

Soluble in ethanol, diethyl ether.

All three are severe skin and eye irritants and are moderately toxic by ingestion. 2,4-Dichlorophenol is an experimental teratogen and there is evidence of its inducing cancer in animals; this and the 2,6- isomer bind reversibly to human serum albumen.

A suitable procedure for their separation and determination in air by HPLC is outlined under **Chlorophenols**.

In the USA 2300 kg of 2,4-dichlorophenol were produced in 1983 for synthesis of the weedkiller 2,4-dichlorophenoxyacetic acid (2,4-D).

Further reading

Environmental Health Criteria 93, Chlorophenols Other Than Pentachlorophenol, WHO/IPCS, Geneva, 1989.

1,2-DICHLOROPROPANE CAS No. 78-87-5

LD_{50} oral rat $1.9\,\mathrm{g\,kg^{-1}}$; LC_{50} fish $61 - 320\,\mathrm{mg\,l^{-1}}$. Highly phytotoxic

$$CH_3 - CH - CH_2Cl$$
$$\mid$$
$$Cl$$

$C_3H_6Cl_2$ $M = 113.0$

Propylene dichloride; b.p. 97°; v.p. 42 Torr, 27.9 kPa (20°); dens. 1.159 g ml^{-1}; logP_{ow} 2.28. Water solubility 2.7 g kg^{-1} (20°); soluble in ethanol, diethyl ether.

Manufactured by the addition of chlorine to propylene. It is used in mixtures with dichloropropenes for soil fumigation, also in processing gum and oil, making rubber and wax, in furniture finishes, dry cleaning, degreasing and paint removers. It is an intermediate for tetrachloroethylene.

1,2-Dichloropropane is detected by the methods used for the dichloropropenes; it was found in 11 out of 36 air samples taken in Japanese cities in the range 6.5×10^{-3} – 1.4 µg m^{-3}. In use as a soil nematocide it forms a 50% mixture with dichloropropene but has been condemned because of risk from leaching to groundwater. It is degraded completely by methanogenic bacteria.

In potato-growing regions of the Netherlands dichloropropane has been found in well water at depths of 13 m at average levels of 1.1 µg l^{-1}; in wells in the USA levels in the range 0.4–16 µg l^{-1} have been recorded.

1,2-Dichloropropane can cause dermatitis in groups of exposed workers, notably those who were painters or metal workers. It has not been classified as a human carcinogen.

The WHO guideline for drinking water is 20 µg l^{-1} and the EPA limit is 5 µg l^{-1}. The ACGIH has set the TLV (TWA) at 69 ppm (347 mg m^{-3}).

Further reading
Environmental Health Criteria 146, 1,3-Dichloropropane, 1,2-Dichloropropane and Mixtures, WHO/IPCS, Geneva, 1993.

DICHLOROPROPENES
LD$_{50}$ oral rat 85–94 mg kg^{-1}; LC$_{50}$ fish (96 h) 0.5–9 mg l^{-1}

cis (Z)   trans (E)

	cis (Z)	trans (E)
CAS No.	10061-01-5	10061-02-6
B.p (°)	105	112
V.p.(25°)(Pa)	4850	3560
Dens. (g ml^{-1})	1.22	1.21
Log P_{ow}	1.82 (20°)	2.22 (25°)
Water solubility (g l^{-1})	2.5	2.5

The detection limit in air is 0.005 mg m^{-3} by GLC with FID and 0.001 mg kg^{-1} in water with ECD.

The mixed isomers are manufactured either by the high-temperature chlorination of propylene or by dehydration of 1,3-dichloropropan-2-ol with POCl$_3$ or P$_2$O$_5$ – benzene.

The dichloropropenes have a short half-life in water owing to their volatility. They decompose in air through reaction with OH radicals with a half-life between 7 and 12 days; they also react with ozone with $t_{1/2}$ between 12 and 55 days.

The commercial mixture of geometrical isomers is injected directly into soil as a pre-planting fumigant to control nematodes in vegetables and tobacco. In 1981 over 7.2 $\times 10^6$ kg of mixed 1,3-dichloropropane and dichloropropenes was used in California, while European production was 6.7×10^3 t year^{-1}. Dichloropropane is now restricted because of the risk of groundwater contamination. The dichloropropenes are transformed in soil into 3-chloroallyl alcohol and then by oxidation into 3-chloroacrolein and 3-chloroacrylic acid.

1,3-Dichloropropene induces bladder and lung tumours in mice and rats. It is held that human exposure is unlikely in the general population, occupational exposure is normally below 4.5 mg m^{-3} (1 ppm). Indications are irritation of the eyes and respiratory tract; symptoms of poisoning are nausea and vomiting at levels over 1500 ppm.

In mammals including man the dichloropropenes are excreted following conjugation with GSH:

$$ClCH{=}CH{-}CH_2Cl \xrightarrow{\text{GSH}} ClCH{=}CH{-}CH_2SG \longrightarrow ClCH{=}CH{-}CH_2{-}S{-}CH_2$$

$$CH_3{-}\underset{\underset{O}{\|}}{C}{-}\underset{H}{N}{-}CH{-}CO_2H$$

N-acetyl-S-(3-chloroprop-2-enyl) cysteine

Further reading
Environmental Health Criteria 146, 1,3-Dichloropropane, 1,2-Dichloropropane and Mixtures, WHO/IPCS, Geneva, 1993.

DICHLORVOS CAS No. 67-73-7
(Vapona)
Insecticide, fumigant, anthelmintic. Toxicity class Ib. LD$_{50}$ rat 60–80 mg kg^{-1}; LD$_{50}$ mouse 140 mg kg^{-1}; LD$_{50}$ fish 0.9–2.7 mg l^{-1}

$$MeO{-}\underset{\underset{OMe}{|}}{\overset{\overset{O}{\|}}{P}}{-}O{-}CH{=}CCl_2 \qquad C_4H_7O_4\,PCl_2 \quad M = 221.0$$

Dimethyl 2,2-dichlorovinyl phosphate; v.p. 2.1×10^{-3} mmHg (20°); water solubility 1% at 20°.

Dangerous to fish and bees; rapidly hydrolysed in human blood ($t_{1/2} = 11$ min) to phosphoric acid and non-persistent products. The half-life in mammals is only 25 min but has caused human fatality when inhaled. A Shell product it is listed by the Food and Agriculture Organization under Prior Informed Consent (PIC), whereby governments may register a refusal to import on the grounds of health and/or environmental impact.

Formulated for indoor domestic use as a slow-release fly killer, employed in glasshouses on fruit and vegetables and to control animal flies and fleas.

Codex Alimentarius acceptable daily intake 0.004 mg kg^{-1}(1977). Maximum residue limits cereal grains 2.0, fruits 0.1 mg kg^{-1}

DIELDRIN CAS No. 60-57-1
(Killgerm)
Wide-spectrum insecticide, wood preservative. Toxicity class I. LD$_{Lo}$ oral humans 65 mg kg^{-1}; LD$_{50}$ oral rat 38 mg kg^{-1}; LC$_{50}$ rainbow trout 19 µg l^{-1}

$C_{12}H_8Cl_6O \quad M = 381$

1,2,3,4,10,10-Hexachloro-6,7-epoxy-1,4,4a,5,6,7,8,8a-octahydro-1,4,5,8-dimethano-naphthalene; m.p.177°; v.p. 1.8×10^{-7} Torr (25°); Water solubility $0.1\,\mathrm{mg\,l^{-1}}$.

A human carcinogen and extremely persistent. A 4-year-old boy in South Africa (1989) ingested the solid in mistake for powdered milk, he suffered stomach cramps, vomited and survived only after hospitalization.

Dieldrin is produced from aldrin in soil, water and living organisms. It is very toxic to fish and its use in sheep dips and as a seed dressing led to a decline in raptorial birds. It was consequently banned in sheep dip as early as 1966 in the UK and in the USA and in Canada it is largely restricted to termite control. Dieldrin has been used in timber treatment and is effective against disease vectors such as locusts and the tsetse fly. It is degraded in mammals to water-soluble metabolites including 6,7-*trans*-dihydroxydihydroaldrin.

Codex Alimentarius acceptable daily intake and maximum residue levels are the same as for aldrin. Despite restrictions on its use in developed countries, it is found with endrin and lindane at levels above the ADI in Cairo residents, where bread is the major source. Dieldrin has also been found at a level of 0.13 ppb in the breast milk of a sample of 25 women permanently resident in Delhi.

It can be determined by trace enrichment of water samples, followed by HPLC using an ODS column and UV detection at 220 nm (Braithwaite and Smith, 1990).

In the UK its importation requires Prior Informed Consent.

Reference
Braithwaite, A. and Smith, F.J. (1990) *Chromatographic Methods*, 4th edn, Chapman & Hall, London.

DIETHYLENE GLYCOL CAS No. 111-46-6
LD_{50} rat $20.7\,\mathrm{g\,kg^{-1}}$; LD_{50} guinea pig $7.8\ \mathrm{g\ kg^{-1}}$

$HOCH_2CH_2-O-CH_2CH_2OH \qquad\qquad C_4H_{10}O_3 \quad M = 106.1$

Bis(2-hydroxyethyl) ether, DIGOL, DEG; m.p.$-6.5°$ b.p. 245°; v.p. 1 Torr (92°); dens 1.118. Miscible with water, soluble in ethanol, acetone, diethyl ether.

Used as an anti-freeze, a dyestuff solvent, in glues and pharmaceuticals, as a lubricating agent and in the finishing of wool. The bis(allyl carbonate) is obtained by reaction with phosgene, followed by allyl alcohol:

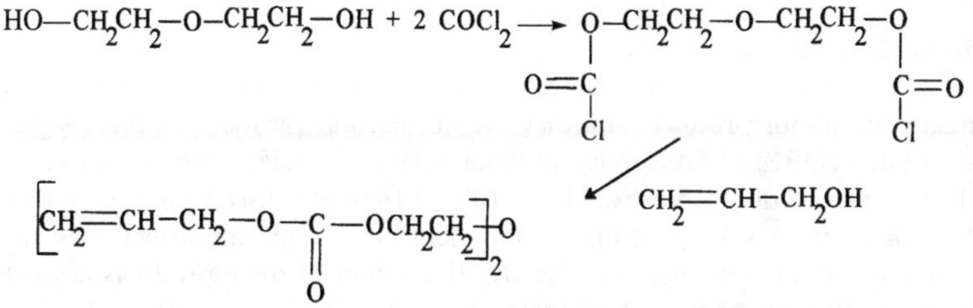

diethylene glycol bis-(allyl carbonate)

Polymerization of this product provides the material for eyeglass lenses.

Diethylene glycol occurs as a contaminant in food and beverages. It is rapidly degraded by oxidizing bacteria and is metabolized by the rat to 2-hydroxyethoxyacetic acid ($HOCH_2CH_2OCH_2CO_2H$), which is excreted.

Diethylene glycol was the agent responsible for the notorious 'elixir of sulphanilamide' incident of 1937 and caused the death of at least 73 residents of Tulsa in the USA. Sulphanilamide was dissolved in this glycol in ignorance of its toxicology, although it is now known that doses of 60 ml taken sequentially can prove fatal. The issuing company in this incident was fined $26 000, but more importantly the case led to a requirement for toxicity testing before a new drug goes on to the market. Ironically, the case was brought in the USA by the Food and Drugs Agency on the grounds of 'misbranding,' since the term 'elixir' implies that the product contains alcohol – had the preparation been called 'solution of sulphanilamide' no prosecution would have been possible.

In the UK the long-term occupational exposure limit is 23 ppm (100 mg m^{-3}). The USSR set a limit in water of 1.0 mg l^{-1}.

Further reading

Chengelis, C.P., Holsen, J.F. and Gad, S.C. (1995) *Regulatory Toxicology*, Raven Press, New York.

DIETHYLHEXYL PHTHALATE CAS No. 117-81-7
(DEHP)

LD_{50} oral rat 25 g kg^{-1}; LD_{50} rainbow trout 540 mg l^{-1}. A group 2B carcinogen

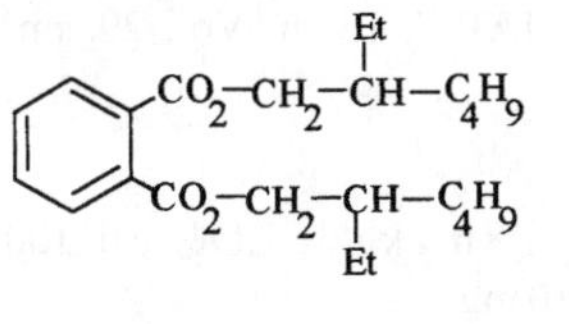

$C_{24}H_{38}O_4$ $M = 390.6$

Di(2-ethylhexyl) phthalate, di-*sec*-octyl phthalate; b.p. 370°, 236° / 1.33 kPa; dens. 0.98 g ml^{-1}; v.p. 8.6×10^{-4} kPa. Water solubility 45 μg l^{-1} (20°), but values up to 350 μg l^{-1} in colloidal suspensions. Log $P_{ow} = 5$, consequently it is highly lipophilic and bioaccumulative with a bioconcentration factor in mosquito larvae of 1320. Miscible with most organic solvents.

Many trade names, e.g. Bisoflex, Eviplast, Octoil. The commercial grades contain traces of phthalic acid.

Some phthalate esters occur naturally in coal and crude oil. Diethyl and dimethyl phthalate are minor products and they are not used as plasticizers; more significant are di-n-butyl (DBP) and diisobutyl phthalate (DiBP). DEHP accounts for half of the total world production, which was 1.1×10^8 t in 1984. It is used to plasticize PVC and finds applications in tubing, medical components, footwear, upholstery, etc., and also as a dielectric fluid in capacitors. The DEHP content of these products lies between 20 and 40%, but can be as high as 55%.

Incineration leads to complete destruction but only at high temperatures – inefficiency leads to air pollution. Levels in urban air can reach 300 ng m^{-3} but are usually < 100 ng m^{-3}.

Analysis is complicated by the risk of contamination from plastic in contact with the sample. It may be detected in air by GC-ECD at 0.5 ng m^{-3}; 2 ng in samples from river water are detected by UV (224 nm) following HPLC.

DEHP is the most widespread of all organic pollutants. Some occurences were as follows: in marine water, Gulf of Mexico (1978) 0.006–0.316 µg l^{-1} and UK estuaries (1981) 0.058–0.078 µg l^{-1}; in freshwater, urban Japan (1974) 0.1–2.2 µg l^{-1} and the Rhine in the Netherlands (1982) up to 4.0 µg l^{-1}. Levels in Rhine sediments were 6500–70 500 µg kg^{-1} (dry weight, 1977) and in the river Usk (UK, 1974) 30 000 µg l^{-1}.

DEHP is known to migrate during 7 days from PVC wrapping to give levels of 4 – 16 mg kg^{-1} in cheese, sausage, flour and rice; it is also found in milk. Some other occurrences (1972) were commercial fishfood 2000–7000 µg kg^{-1}, catfish from Mississipi river 3200 µg kg^{-1}, seal blubber 10 600 µg kg^{-1}.

It is estimated that human intake in the UK is 20 µg/person/day. In the USA occupational exposure of workers in DEHP production was at worst an 8 h TWA of 0.02–4.1 mg m^{-3}. In Sweden 2 h exposure to phthalate esters gave levels in 54 workers in the range of 0.01–2.0 mg m^{-3}.

In the early 1980s DEHP was shown to be teratogenic in mice and rats. At high doses it produced liver tumours in these animals, although the action was not directly with DNA but by the induction of intracellular organelles known as peroxisomes. DEHP is not now used to plasticize food-contact plastics in the USA, although there exists a 3000-fold margin between these doses and the estimated dietary intake of the alternative plasticizers.

Further reading
Environ. Health Criteria, World Health Organization, Geneva, 1992, Vol. 131.
IARC (1982) *Some Industrial Chemicals and Dyestuffs*, IARC, Lyon, Vol. 29, pp. 257–295.

DIFENACOUM CAS No. 56073-07-5
Rodenticide, anticoagulant. Toxicity class I. LD$_{50}$ rat 1.8 mg kg^{-1}; LD$_{50}$ cat 100 mg kg^{-1}; LD$_{50}$ dog >50mg kg^{-1}; LC$_{50}$ rainbow trout 0.10 mg l^{-1}

$C_{31}H_{24}O_3$ $M = 444.5$

3-(3-Biphenyl-4-yl-1,2,3,4-tetrahydro-1-naphthyl)-4-hydroxy coumarin; m.p. 217°; v. p. 0.16 mPa (45°); water solubility $< 10\,\mathrm{mg}\,\mathrm{l}^{-1}$.

Manufactured by Sorex (London). It inhibits the synthesis of blood-clotting factors and is effective against rats and mice which are resistant to other anticoagulants. It is effective for longer than warfarin; vitamin K_1 is an antidote.

DIMETHOATE CAS No. 60-51-5
(Cygon, Dimeton)
Systemic wide-spectrum insecticide, acaricide. Toxicity class II. TD_{Lo} oral humans $256\,\mathrm{mg}\,\mathrm{kg}^{-1}$; LD_{50} rat $300\,\mathrm{mg}\,\mathrm{kg}^{-1}$; LD_{50} quail $84\,\mathrm{mg}\,\mathrm{kg}^{-1}$; LC_{50} rainbow trout $6\,\mathrm{mg}\,\mathrm{l}^{-1}$; LD_{50} bee $0.1\,\mu\mathrm{g/bee}$

$$\begin{array}{ccc} \overset{S}{\underset{|}{\overset{||}{}}} & \overset{O}{\overset{||}{}} & \\ \text{MeO}-\text{P}-\text{S}-\text{CH}_2-\text{C}-\text{NHMe} & & C_5H_{12}NO_3PS_2 \quad M = 229.3 \\ \underset{\text{OMe}}{|} & & \end{array}$$

O,O-Dimethyl-S-methylcarbamoylmethyl phosphorodithioate; m.p. 49°; v.p.1.1 mPa(25°); water solubility 24 g l^{-1} (20°).

Introduced by American Cyanamide. It is metabolized to the phosporothioate and oxidation gives the dimethoxon $MeNHCOCH_2SP(O)(OMe)_2$, which strongly inhibits cholinesterase and is more toxic to birds than it is to mammals.

A group of 35 Malaysian workers spraying dimethoate on six plantations all suffered headaches, sore eyes and nausea and over half the group had blurred or darkened vision.

Dimethoate is used for the control of pests in cereals, citrus fruits, coffee, cotton, tea, tobacco and vegetables. It was associated in Egypt with the poisoning and possible death arising from the use of subsidised cotton insecticides on edible fruits in plastic housing and tunnels.

There are no ambient air standards. The EPA limit in drinking water is $7\,\mu\mathrm{g}\,\mathrm{l}^{-1}$.

Codex Alimentarius acceptable daily intake 0.01 mg kg^{-1} (1987). Maximum residue levels olive oil 0.05, potatoes 0.05, banana. 1.0, sugar beet 0.05 mg kg^{-1}

DIMETHYLFORMAMIDE CAS No. 68-12-2
LD_{50} rat $2.8\,\mathrm{g}\,\mathrm{kg}^{-1}$; LC_{50} bluegill sunfish (96 h) $7.1\,\mathrm{g}\,\mathrm{l}^{-1}$

$$\begin{array}{cc} \overset{O}{\overset{||}{}} & \\ \text{HCNMe}_2 & C_3H_7NO \quad M = 73 \end{array}$$

Formdimethylamide, DMF; m.p. −60°; b.p. 153°; dens. 0.944 (25°); v.p. 0.35 kPa (2.65 Torr at 20°); dielectric constant 36.7. Completely miscible with water and most organic solvents. Widely used as an industrial solvent, e.g. for the manufacture of acrylic fibres and polyurethanes. In Taiwan in 1991 consumption for the production of fibrous polyacrylonitrile was > 30 000 t.

Dimethylformamide does not occur naturally and is produced industrially by:

$$CO + Me_2NH \longrightarrow (Me)_2NCHO$$

It is shipped in tank containers and also 200 kg drums; world production (1980) was 225×10^3 t year^{-1}.

Dimethylformamide has an ammonia-like odour with a detection threshold of $0.15\,\mathrm{mg\,m^{-3}}$. GLC is the method of choice for determination and is applicable to blood levels down to $30\,\mu\mathrm{g\,l^{-1}}$.

Contact with strong oxidizing agents, halogens and organochlorine metal compounds can lead to explosions. In acidic solution (pH 4.0) reaction with sodium nitrate gives a low yield of N-nitrosodimethylamine.

Industrial plant may vent DMF from polyurethane and synthetic leather manufacturing at concentrations of 300–1200 ppm at a rate of 0.3 normalized $\mathrm{m^3\,s^{-1}}$ at 70°; control is exercised by carbon adsorption or by catalytic incineration over 1.0% platinum / alumina (Tsai *et al.*, 1994). Dimethylformamide is of low toxicity but high levels can cause nausea, abdominal pain and vomiting; blood pressure changes have also been reported, although recovery is usually complete. Levels in air close to industries manufacturing polyurethane synthetic leather can reach $0.12\,\mathrm{mg\,m^{-3}}$ and liver damage has been noted in several instances amongst groups of workers after long term exposure; for this reason DMF has been excluded from pharmaceutical and cosmetic products.

The ACGIH has set a TLV (TWA) skin at 10 ppm $(30\,\mathrm{mg\,m^{-3}})$.

Reference

Tsai, W.-T., Chang, Y.-H. and Chang, C.-Y. (1994) Catalytic incineration of organic vapours containing N,N- dimethylformamide, *J. Hazardous Mater.*, **37**, 241–244.

Further reading

Environ.Health Criteria World Health Organization, Geneva, 1991, Vol. 114.

N,N'-DIMETHYL-1,4-PHENYLENEDIAMINE CAS No. 99- 98-9

$\mathrm{LD_{50}}$ rat $50\,\mathrm{mg\,kg^{-1}}$

$\mathrm{C_8H_{12}N_2}$ $M = 136.2$

Dimethyl-*p*- phenylenediamine, *p*-aminodimethylaniline; m.p. 41°; b.p. 262°; 136°/14 Torr. Insoluble in water; soluble in dilute HCl, EtOH, benzene, acetone.

Used as its diazonium salt as a test for phenols and aromatic amines (see **Azo dyestuffs**), also for the photometric determination of sulphate and in the titrimetric determination of vanadium.

Listed by the USEPA as 'extremely hazardous' – poisonous by ingestion, inhalation and by skin contact. It is a dermatitic agent.

Further reading

Lewis, R.J. (ed.) (1996) *Sax's Dangerous Properties of Industrial Materials*, 9th edn, Van Nostrand Reinhold, New York, p.1353.

DIMETHYL SULPHATE CAS No. 77-78-1

$\mathrm{LD_{50}}$ oral rat $440\,\mathrm{mg\,kg^{-1}}$; $\mathrm{LC_{50}}$ (96 h) fish 7.5–15 ppm

$$\text{MeO}-\overset{\displaystyle O}{\underset{\displaystyle O}{S}}\diagup$$

$$C_2H_6O_4S \quad M = 126.1$$

Sulphuric acid dimethyl ester; b.p. 188° (dec.); v. dens. 4.35 (air = 1); v.p. 0.106 kPa (25°), 0.8 Torr; water solubility 28 g l^{-1} with hydrolysis.

Produced by the reaction between dimethyl ether and SO_3:

$$2 \text{ MeO}-\text{Me} + 2 \text{ SO}_3 \longrightarrow \left[\begin{array}{c} \text{Me } O \\ | \quad \| \\ \text{MeO}-S=O \\ + \quad | \\ O^- \end{array} \right] \longrightarrow \text{MeO}-\overset{O}{\underset{O}{S}}\diagup + \text{MeO}-\overset{O}{\underset{O}{S}}\diagup\text{HO} + H_2SO_4$$

with separation by vacuum distillation.

Principal uses are as a methylating agent for phenols and amines and as an intermediate for dyestuffs and pharmaceuticals. Production in the USA (1977) was 45×10^3 t and in Western Europe (1983) 31×10^3 t year^{-1}.

Dimethyl and diethyl sulphate (CAS No. 64-67-5) are acutely toxic to mammals through oral and skin contact, leading to necrosis of the stomach, skin burns, cyanosis and convulsions. In humans, irritation of the eyes occurs at 1 ppm and inhalation at 97 ppm (500 mg m^{-3}) for 10 min can be fatal; there is no characteristic odour or other warning. Following short-term inhalation of the vapour a death followed 11 h after exposure, despite only mild initial symptoms of nausea and irritation of the eyes.

The ACGIH has set the TLV (TWA) at 0.1 ppm (0.52 mg m^{-3}). Dimethyl sulphate has been placed in the IARC group 2A – carcinogenic in animals.

Further reading
Environ. Health Criteria World Health Organization, Geneva, 1984, Vol. 48.

DIMETHYL SULPHIDE CAS No.75-18-3
LD$_{50}$ rat 535 mg kg^{-1}

MeSMe $C_2H_6S \quad M = 62.1$

Methylthiomethane; b.p. 38°; v.p. 420 Torr(20°); v.dens. 2.1; solubility in water 6.3 g l^{-1}. Soluble in most organic solvents. The vapour forms explosive mixtures with air. It has an odour of decayed vegetables, threshold 2×10^{-2} mg m^{-3} (7.8 ppb, 20°).

Dimethyl sulphide is formed as a by-product of the Kraft paper pulp process or by the action of hydrogen sulphide on methanol with a dehydration catalyst; it can be oxidized to the important solvent dimethyl sulphoxide:

$$3 \text{ CH}_3-\text{OH} + 2 \text{ H}_2\text{S} \longrightarrow (\text{CH}_3)_2\text{S} + \text{CH}_3-\text{SH} + 3 \text{ H}_2\text{O}$$

$$\downarrow O_2 + NO_2$$
$$\uparrow [O]$$
$$\text{CH}_3-\overset{\displaystyle}{\underset{\displaystyle O}{S}}-\text{CH}_3 + \text{NO}$$

Dimethyl sulphide is formed at sea by the action of algae upon dimethyl sulphoniopropionate (DMSP):

$$(CH_3)_2 \overset{+}{S} - CH_2 - \overset{\overset{H}{|}}{\underset{\underset{H}{|}}{C}} - CO_2H \longrightarrow (CH_3)_2S + CH_2 = CH - CO_2H$$

Base⁻

In waters around the UK dimethyl sulphide is found at a winter mean of $0.1\,nmol\,l^{-1}$ and at a summer mean of $9.4\,nmol\,l^{-1}$; the concentration of DMSP is 14 times that of dimethyl sulphide. It is the predominant volatile sulphur compound in the oceans and accounts for most of the sulphur transferred from them to the atmosphere. It was found at a concentration of $180-1500\,ng\,m^{-3}$ at a height of 5 m above the sea, falling to $35-135\,ng\,m^{-3}$ at 47 m (Ayres and Gras, 1991). Concentrations above the Great Lakes were in the range $5.2-27\,ng\,l^{-1}$ at the surface and the sulphur transfer to air was estimated at $107\,t\,year^{-1}$ at maximum. The rate of ocean emission is in the range $0.7-13\,mmol\,S\,m^{-2}\,year^{-1}$ and the total global flux is $1.1 \pm 0.5\,Tmol\,year^{-1}$ (Saltzman and Cooper, 1989).

Dimethyl sulphide is used as a gas odorant and as a solvent for inorganic compounds. It occurs with methanethiol in sewage, also in beer at a level of about $20\,\mu g\,l^{-1}$ and as an odorant in garlic and truffles (see also **Methanethiol**).

Neither OSHA nor ACGIH has set a TLV for dimethyl sulphide but in Russia the STEL is 50 ppm ($129\,mg\,m^{-3}$).

References

Ayres, G.P. and Gras, J.L. (1991) Seasonal relationship between cloud condensation nuclei and aerosol methanesulphonate in marine air, *Nature*, **353**, 834–835.

Saltzman, E.S. and Cooper, W.J. (1989) *Biogenic Sulphur in the Environment*, ACS Symposium Series, No.393, American Chemical Society, Washington, DC.

DYSPROSIUM CAS No. 7429-91-6

Dy

Dysprosium, one of the rare-earth metals, has an atomic number of 66, an atomic weight of 162.50, a specific gravity of 8.54 at 20°, a melting point of 1411° and a boiling point of 2562°. It is a bright, lustrous, silvery metal and forms compounds in the +3 oxidation state. There are seven natural isotopes: ^{156}Dy (0.06%), ^{158}Dy (0.1%), ^{160}Dy (2.34%), ^{161}Dy (18.9%), ^{162}Dy (25.5%), ^{163}Dy (24.9%) and ^{164}Dy (28.2%). It is found along with other rare-earth elements in the minerals *monazite, bastnasite, xenotime, fergusonite* and *gadolinite*. The average concentration in the earth's crust is 6 ppm.

Uses.   Laser material. Permanent magnets.

There is a lack of environmental data, therefore all dysprosium compounds should be regarded as toxic. The trichloride, $DyCl_3$, is poisonous to small laboratory animals by injection but is only mildly toxic to humans. Dysprosium salts may exhibit anticoagulant properties.

Further reading
Gschneider, K.A., Jr, and LeRoy, K.S. (eds) (1988) *Handbook on the Physics and Chemistry of the Rare Earths*, North-Holland, Amsterdam.
Kirk–Othmer ECT 4th edn, Wiley, New York, 1995, Vol. 14, pp. 1091–1115.

Dysprosium (III) oxide CAS No. 1308-87-8

Dy_2O_3

Uses. With nickel, in cermets used as nuclear reactor control rods that do not require water cooling.

Toxicity. LD_{50} oral rat $> 5\,g\,k\,g^{-1}$.

Dysprosium (III) nitrate pentahydrate CAS No.10031–49–9

$Dy(NO_3)_3 . 5H_2O$

Uses. Preparation of Dy_2O_3.

Toxicity. LD_{50} oral rat $2386\,mg\,kg^{-1}$; intraperitoneal rat $227\,mg\,kg^{-1}$. Strong oxidizing agent.

ENDOSULFAN CAS No. 115-29-7 (α-form shown) 33213-65-9 (β-form)
(Thiodan, Benzoepin)
Non-systemic broad-spectrum insecticide. Toxicity class I. LD_{50} rat $70\,mg\,kg^{-1}$; LD_{50} rabbit dermal $359\,mg\,kg^{-1}$; LC_{50} fish $200\,\mu g\,l^{-1}$

$C_9H_6Cl_6O_3S \quad M = 406.9$

1,4,5,6,7,7-Hexachloro-8,9,10-trinorborn-5-en-2,3-ylenebismethylene sulphite; the technical grade is a mixture of the α-isomer (66%) and the β-isomer (34%); m.p. α-isomer 109°, β-isomer 213°; v.p. 0.83 mPa (20°); water solubility $0.3\,mg\,l^{-1}$.

Introduced by Hoechst, it is hydrolysed to the diol and sulphur dioxide and is rapidly detoxified in mammals.

Unless applied under proper control, penetration of the skin is dangerous and farmers tend to use it above the recommended levels. It was used in the Philippines against a plague of snails in rice paddies but as a result the fish–rice culture was wiped out; women planting seedlings suffer ill effects as they spend 7 h day^{-1} in the paddies. The government sought to control imports through Prior Informed Consent legislation, but this was frustrated by the non-appearance of their lawyers when the manufacturer won a court order for continuance.

Farmers in the Sudan, who were aware of the risks to humans and animals, used excess of endosulfan to treat maize; 16 lb of the crop were fed in part to poultry, which died. The remainder was sold on to make bread for a funeral party; 31 of the guests died and several hundred were affected, notably by memory loss.

In addition to its use on rice and cereals, endosulfan is also applied to fruit, leaf and root vegetables.

The EPA considers it to be a teratogen. Maximum levels in workplace air are set in the UK and the USA at a (TLV) TWA of $0.1 \, mg \, m^{-3}$.

Codex Alimentarius acceptable daily intake $0.006 \, mg \, kg^{-1}$. Maximum residue levels rice 0.1, milk 0.004, potatoes 0.2, fruit 2.0 mg kg^{-1}

Further reading
Pestic. News 1992, No. 16, 3.

ENDRIN CAS No. 72-20-8
(Endrix, Nendrin)
Insecticide. Toxicity class I. LD_{50} rat $1.3 \, mg \, kg^{-1}$; LD_{50} bee $0.65 \, \mu g/bee$; LC_{50} fish 0.5-$0.7 \, \mu g \, l^{-1}$

$C_{12}H_8Cl_6O \quad M = 380.9$

M.p. $230°$; v.p. 2.7×10^{-7} Torr $(25°)$; water solubility $0.25 \, \mu g \, l^{-1}$ $(25°)$.

Endrin has proved fatal to humans and severe poisoning leads to fits and limb convulsions; it is carcinogenic in rats.Its use has been banned in developed countries and residue levels have declined as a result. In Casco Bay, Maine, the level in sediments in 1990 was in the range 0.06-0.85 ppb. In some Third World countries it continues in use.

Codex Alimentarius acceptable daily intake $0.0002 \, mg \, kg^{-1}$. Maximum residue levels wheat 0.02, milk 0.0008, rice 0.02, meat 0.1 mg kg^{-1}.

Endrin, dieldrin and lindane are all found at levels above the **ADI** in Cairo residents, the principal source being bread. These compounds and **DDT** are also widespread in the Governates of Egypt.

In the USA and UK the threshold limit value is a TWA of $0.1 \, mg \, m^{-3}$.

Further reading
Environ. Health Criteria, World Health Organization, Geneva, 1992, Vol. 130.

EPICHLOROHYDRIN CAS No. 106-89-8
LD_{50} rat $240 \, mg \, kg^{-1}$; LC_{50} bluegill sunfish (96 h) $18 \, mg \, l^{-1}$

$C_3H_5ClO \quad M = 92.5$

1-Chloro-2,3-epoxypropane; b.p. $117°$; v.p. 16 Torr $(25°)$; dens. 1.18; log P_{ow} 0.45. Water solubility $60 \, g \, l^{-1}$ $(20°)$; miscible with acetone, $CHCl_3$, diethyl ether, CCl_4.

Synthesized from propylene via allyl chloride:

The major outlet is by reaction with bisphenol A to give epoxy resins and also for glycerol synthesis and as a fumigant.

Epichlorohydrin is hydrolysed to 1-chloropropane-2,3-diol in water with a half-life of 8 days; it reacts with the OH radical in air with a half-life of 4 days. The bioconcentration factor is 4.6 and accumulation is therefore unlikely.

The threshold odour level is $0.3\,mg\,m^{-3}$ and exposure results in irritation of the eyes and nose. Epichlorohydrin is poisonous by ingestion.

The NIOSH analytical procedure will detect the range $2–60\,mg\,m^{-3}$ for a 20 l air sample, by adsorption on charcoal and GC with Chromosorb 101. During production of epoxy resins and glycerol, levels in air have been found in the range 3-15 ppm ($11–57\,mg\,m^{-3}$).

The IARC has placed epichlorohydrin in category 2A – carcinogenic in animals, inadequate evidence in humans but a suspected human carcinogen. The ACGIH have proposed a TLV (TWA) of 0.5 ppm ($1.9\,mg\,m^{-3}$).

Further reading
Environ. Health Criteria, World Health Organization, Geneva, 1984, Vol. 33.
Witcoff, H.A. and Reuben, B.G. (1996) *Industrial Organic Chemicals*, Wiley, New York, pp. 187–188.

ERBIUM CAS No. 7440-52-0

Er

Erbium, a member of the rare-earth elements of the lanthanide series, has an atomic number of 68, an atomic weight of 167.26, a specific gravity of 9.066 at 25°, a melting point of 1529°, and a boiling point of 2863°. The pure metal is soft and malleable and has a bright, silvery, metallic lustre. It forms compounds in the +3 oxidation state. Naturally occurring erbium is a mixture of six stable isotopes: ^{162}Er (0.14%), ^{164}Er (1.61%), ^{166}Er (33.60%), ^{167}Er (22.95%), ^{168}Er (26.8%), and ^{170}Er (14.9%). Nine radioactive isotopes have also been characterized.

Found in the minerals *gadolinite, monazite and bastnasite*. Average concentration in the earth's crust is ~ 3.5 ppm.

Uses of erbium and erbium salts. Nuclear and metallurgical industries. Alloys. Lasers. Dopants in semiconductors.

No biological role. Erbium salts are poisonous to small animals by injection routes. Levels in human organs are < 1 ppm. Few data are available. Like all lanthanide elements and their compounds they should be handled with care, avoiding the risk of inhaling. Erbium dust is flammable.

Further reading
Gschneider, K.A., Jr and LeRoy, K.S. (eds) (1988) *Handbook on the Physics and Chemistry of the Rare Earths*, North-Holland, Amsterdam.
Kirk–Othmer ECT, 4th edn, Wiley, New York, 1995, Vol. 14, pp. 1091–1115.

Erbium chloride hexahydrate CAS No. 10025-75-9

$ErCl_3.6H_2O$

Uses. Preparation of erbium salts and complexes.

Toxicity. LD_{50} oral mouse $4417\,mg\,kg^{-1}$; intraperitoneal guinea pig $128\,mg\,kg^{-1}$.

Erbium oxide CAS No. 12061-16-4

Er_2O_3

Uses. Catalyst. Ceramics. Semiconductor dopant. As a pink powder, pigment for glass. Sintering aid. Additive in Y–Fe and Y–Al garnets.

Toxicity. N.a.

ETHANETHIOL CAS No. 75-08-1
LD_{50} rat 1.96 g kg^{-1}

CH_3CH_2SH C_2H_6S $M = 62.1$

Ethyl mercaptan; b.p. 36°; v.p. 442 Torr (20°); dens. 0.839; v. dens. 2.14. Very flammable. Solubility in water 6.7 g l^{-1} (20°), soluble in acetone, ethanol, diethyl ether. Powerful leek-like odour, threshold 1 ppb; it is added to natural gas to give warning of leakage; rapidly detoxified by hypochlorite solutions.

Ethanethiol is used as an intermediate in the manufacture of plastics, insecticides and antioxidants. It is formed in vinous fermentation, petroleum and sewers and occurs in the atmosphere of the planet Jupiter.

Ethanethiol is a common air pollutant. Human volunteers inhaled air containing 10 mg m^{-3} (4 ppm) for 3 h over 5–10 days and experienced periodic nausea and irritation of the mucous membrane, lips, mouth and nose. There were no symptoms at 0.4 ppm (1 mg m^{-3}), but at high levels it affects the CNS.

It can be identified by HPLC–MS of a derivative:

formed by reaction with a primary amine and phthaldehyde (Simpson *et al.*, 1983).

The ACGIH has set the TLV (TWA) at 0.5 ppm (1.3 mg m^{-3}). In the UK the STEL is 2 ppm (3 mg m^{-3}).

Reference
Simpson, R.C., Spriggle, J. E. and Veening, H. (1983) Off-line chromatographic MS studies of *o*-phthalaldehyde–primary amine derivatives, *J. Chromatogr*, **261**, 407–414.

Further reading
Cremlyn, R.J. (1996) *An Introduction to Organosulphur Chemistry*, Wiley, Chichester.

ETHANOL CAS No. 64-17-5
LD_{50} rat 3.4 g kg^{-1}; LC_{50} (7 day) guppy 11 g l^{-1}

CH_3CH_2OH C_2H_6O $M = 46.1$

Ethyl alcohol; m.p. − 114°; b.p. 78.5°; v.p. 43 Torr (20°) = 6.5% in saturated air; dens. 0.79; log P_{ow}−0.30.). Highly flammable. Solubility in water > 100 g l^{-1} (23°); soluble in diethyl ether, acetone, benzene, etc.

First produced from ethylene by Faraday in 1828. The older process, still used in Russia, is the indirect hydration of ethylene:

$$2\,CH_2{=}CH_2 \;+\; H_2SO_4 \longrightarrow CH_3{-}CH_2{-}O{-}\overset{\displaystyle O}{\underset{\displaystyle EtO}{\overset{\|}{S}}}{=}O \xrightarrow{H_2O} 2\,EtOH + H_2SO_4$$

More recently production depends upon direct hydration of ethylene by passing it mixed with steam over a catalyst, such as phosphoric acid, suspended on Celite or another clay-based medium.

Ethanol is widely used as a solvent, in organic synthesis, as an antiseptic and in perfumery. The worldwide demand in 1991 was 19×10^9 l and a large proportion of this was added to motor fuel; Brazil consumed 12×10^9 l and the USA 3.6×10^9 l.

The acute toxicity is low in animals. In humans signs of over-exposure include drowsiness and lack of coordination.

Beverages.   Beer was produced by the ancient Egyptians 5000 years ago. Starch is the basis of industrial production by fermentation:

$$\text{starch}(C_6H_{10}O_5)_n \xoverset{nH_2O}{\longrightarrow} nC_6H_{12}O_6 \xoverset{nH_2O}{\longrightarrow} nH_2O + 2nCO_2$$

A national brewery produces about 1 million barrels (1 barrel = 31 gallons = 117.3 l). The usual feed is barley and owing to the high molecular weight of the component starch a preliminary treatment is required to produce the brewer's malt:

1. The water take-up is increased from ca 12% to over 40%.
2. Drain and allow the action of natural hydrolytic enzymes for several days. This product may be dried to a few per cent water and stored.
3. In the 'mashing' process the grain is mixed with water and maintained at 65–68° for 1–2 h when the α- and β-amylases solubilize it completely to give a 'wort,' separable from the insoluble mash and with 94% of its specific gravity due to carbohydrates.
4. The wort is boiled with hops to denature the malt enzymes and so halt the degradation of large carbohydrates.
5. The hopped wort is cooled, clarified, oxygenated and fermented with yeast.

A 100 lb batch of sugar produces 48 lb of ethanol and 46 lb of CO_2, and also 3 lb of glycerol. The flame-drying of barley malts was a source of nitrosamines in beer, but changes in the process ensure that the level is less than 5 ppb. In the USA individual exposure due to beer drinking was estimated at $0.34\,\mu g\,day^{-1}$.

A traditional measure contains 7–9 g of ethanol and this will be contained in 24 ml of whisky, gin or vodka, 60 ml of sherry, 120 ml of wine or 280 ml of beer. Should a 70 kg person drink 1 pint of beer (560 ml; 16 g EtOH) then the peak blood level will be 30 mg/100 ml after 1 hour, declining by metabolism in the liver over 4–6 h. In the UK the legal limit for drivers is 80 mg/100 ml.

The breathalyser depends upon the visual indication when orange potassium dichromate, acidified with sulphuric acid and suspended on silica gel, is reduced by ethanol to green chromium sulphate. For an accurate determination gas chromatography is the method of choice. Use of a capillary GC column coated with Carbowax

400 allows the quantification of ethanol and other components, e.g. acetaldehyde, ethyl acetate, methanol and higher alcohols.

Physicians advise that 3 pints of beer or one bottle of wine per day constitute the upper safe level; this must vary with individuals. In the USA 12% of the population are held to be 'heavy drinkers' and in the UK 0.5 million are thought to have a 'drinking problem.' Chronic consumption of alcohol leads to loss of weight and malnutrition (floridity). Damage to the brain has been proved after death by a comparison with the brains of older deceased persons – typically the membrane around the brain is thickened, surface furrows are widened and ventricles enlarged. These changes are found in drinkers many years before death; brain damage also occurs as a result of a deficiency of thiamine (vitamin B_1). Liver enlargement results from the accumulation of triglycerides and proteins – prolongation will lead to cirrhosis with accumulation of serous fluid and liver disease. Women are more susceptible than men and there is a risk of foetal damage in the early weeks of pregnancy, with consequential growth retardation and neurological dysfunction in the child.

In the workplace the ACGIH has set a TLV (TWA) of 1000 ppm (1880 mg m^{-3}).

Further reading

Birch, G.C. and Lindley, M.G. (eds) (1985) *Alcoholic Beverages*, Elsevier, London, various chapters.

Denney, R.C. (1997) None for the Road: understanding drink-driving, Shaw, Crayford.

Kirk–Othmer ECT, 4th edn, Wiley, New York, 1994, Vol. 9, pp. 812–849.

ETHYLENE CAS No. 74-85-1

$CH_2 = CH_2$ C_2H_4 $M = 28.0$

Ethene; m.p. $-169°$; b.p. $-104°$; v.p. 10 Torr ($-132°$); log P_{ow} (calc.) 1.13. Very flammable. Solubility in water at $0°$ and 101 kPa $= 226$ ml l^{-1}. Very soluble in EtOH.

Ethylene is produced naturally by various fruits, including apples, pears and peaches. Produced by the cracking of petroleum fractions, typically from naphtha, b.p. 20–200° (C_4–C_{12}) at temperatures in the range 750–950°. The reaction is controlled by dilution in steam, to prevent the continued reaction of the desired products, and also by rapid quenching. At a reactor pressure of 172 kPa the principal products from naphtha are methane 18%, ethylene 34%, propylene 14%, benzene 6.4% and 1,3-butadiene 4.5%. These proportions may be changed by varying the severity of the reaction conditions.

Levels in air are principally derived from leakage at production plants, from the exhaust of vehicles and from wood burning. Some typical levels are urban air 12–250 ppb, petrol engine 15% (vol.) of emissions, wood burning 2 kg t^{-1} and emissions from cigarettes 1200 µg/cigarette.

Ethylene reacts in air with OH radicals with a half-life of 1 day; the reaction with ozone has a half-life of 4 days.

In the USA in 1994 ethylene production was 48 $\times 10^9$ lb, a level exceeded only by sulphuric acid, nitrogen and oxygen. The principal products obtained worldwide from ethylene in 1990 were polyethylene 34 $\times 10^6$, ethylene dichloride 10 $\times 10^6$, Ethylene oxide–ethylene glycol 8.2 $\times 10^6$, ethyl benzene→styrene 4.5 $\times 10^6$ and vinyl acetate 1.2 $\times 10^6$ t.

Polyethylene was first made in 1898 by Von Pechmann by reaction of diazomethane (CH_2N_2) and was obtained as a semi-crystalline solid, m.p. ca 120°, and in 1935 by Perrin of ICI. The ICI product was low-density polyethylene (LDPE, 0.880–0.915 g cm^{-3}), now produced in the molecular weight range 500–60 000 at temperatures of 130–300° at pressures of about 800 atm (Carraher, 1996). LDPE may be obtained with branched chains, which increases the volume and reduces the density as does copolymerization with propylene. Antioxidants such as BHT and UV stabilizers such as benzophenones are also added and the product will also include residues from the peroxide used as the initiator. LDPE is used to make blown film, non-toxic food packaging, bags, bin-liners, coatings, mouldings and adhesives.

High-density polyethylene (HDPE, > 0.940 g cm^{-3}) is produced at temperatures of 50–100° and atmospheric pressure through the action of Ziegler catalysts (cf. propylene) or of Phillips catalysts, based on chromium oxides. HDPE has a molecular weight ranging from 500 up to several million and is used as a replacement for paper and glass, in waxes and as pipes for water, sewage and gas.

Ethylene is not an irritant but at levels in air >35% it can cause memory disturbance; at 50% it causes unconsciousness. It is not carcinogenic, but has been reported to form the carcinogen ethylene oxide in experimental animals through the action of cytochrome P450 oxidase.

The ACGIH have clasified ethylene as a simple asphyxiant, slightly stronger than nitrous oxide, with no other significant effect. No controls on workplace levels have been defined.

Reference
Carraher, C.E.(1996) *Polymer Chemistry*, 4th edn, Marcel Dekker, New York.

Further reading
Chemical Safety Data Sheets (1993) Royal Society of Chemistry, Cambridge, Vol. 5, pp. 143–147.
Kirk–Othmer ECT, 4th edn, Wiley, New York, 1994, vol.9, pp. 877–915; 1996, vol. 17, pp. 702–784.

ETHYLENE DIBROMIDE CAS No. 106-93-4
(Dibrome).
Nematicide, insecticide. LD$_{50}$ rat 200 mg kg^{-1}; LC$_{50}$ (48 h) sunfish 18 mg l^{-1}.

$BrCH_2CH_2Br$ $\qquad$ $C_2H_4Br_2$ $\quad$ $M = 187.8$

1,2-Dibromoethane; m.p. 9°; b.p. 131° v.p. 1.5 kPa(25°); water solubility 4.3 g kg^{-1} (30°)

Introduced by Dow Chemical, who no longer make it. Used for the control of nematodes and soil pests, for the fumigation of buildings including homes, as a scavenger for lead in petrol and as a general industrial solvent.

Ethylene dibromide is a suspected human carcinogen and prolonged inhalation produces liver damage. Limits in the workplace in the UK TWA 4 mg m^{-3}, NIOSH 0.4 mg m^{-3}.

ETHYLENE GLYCOL CAS No. 107-21-1
LD$_{50}$ rat 6.1 g kg^{-1}; LC$_{50}$ (24 h) goldfish > 5 g l^{-1}

HOCH$_2$CH$_2$OH $\qquad\qquad$ C$_2$H$_6$O$_2$ $\quad M = 62.1$

Ethane-1,2-diol; m.p. − 13°; b.p. 198°; v.p. 0.06 Torr (20°); at 20° the saturation air level is 79 ppm; threshold odour concentration 90 mg m^{-3}. Ethylene glycol is completely miscible with water and markedly depresses its freezing point.

First prepared by Wurtz in 1859. Over 50% of the ethylene oxide produced is now hydrolysed to produce ethylene glycol, under neutral conditions a 90% yield is obtained together with diethylene glycol (HOCH$_2$CH$_2$OCH$_2$CH$_2$OH) and higher glycols. In 1993 the production capacity of the USA was 3.5 × 10^6 t year^{-1}.

It is used as a raw material for polyethylene terephthalate (PET):

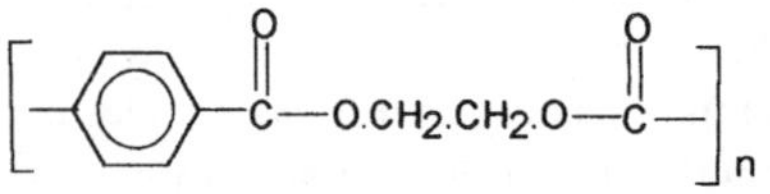

About 65% of this production is used to make the fibres known under the trade names Terylene and Dacron; another use is in the manufacture of soft drinks bottles. Ethylene glycol is also used as an antifreeze for engines, runways and in paints, also in heat transfer fluids, the dehydration of natural gas and to make the esters of lauric and stearic acids. Ethylene dinitrate is a liquid explosive.

Some human volunteers when exposed to 12 ppm (30 mg m^{-3}) for 4 weeks experienced throat irritation and headache; complaints only became significant at a level of 140 ng m^{-3}. At 25° its saturation level in air exceeds the TLV for mist and vapour of 50 ppm (127 mg m^{-3}).

Further reading

ACGIH (1991) *Documentation of TLVs and Biological Exposure Indices*, 5th edn, ACGIH, Cincinnati, OH.
Kirk–Othmer ECT, 4th edn, Wiley, New York, 1994, Vol.12, pp. 695–714.

ETHYLENE GLYCOL DINITRATE CAS No. 628-96-6
LD$_{50}$ rat 616 mg kg^{-1}; LD$_{Lo}$ (96 h) perch 5 mg l^{-1}

O$_2$NO—CH$_2$CH$_2$—ONO$_2$ $\qquad\qquad$ C$_2$H$_4$N$_2$O$_6$ $\quad M = 152.1$

Ethylene dinitrate, EGDN; f.p. −22°; v.p. 0.05 Torr (20°); dens. 1.49 (20°); v.dens. 5.25. Explodes on heating. Insoluble in water but soluble in most organic solvents.

It is used as a fuel additive but importantly as a constituent of dynamite, which is composed of EGDN (60–80%) and nitroglycerine (40–20%). The mixture has a lower melting point than pure nitroglycerine and its use avoids the hazard of working with frozen dynamite. The physiological effects of the two compounds are very similar.

Industries using EGDN at levels of 1–2 mg m^{-3} (0.5 ppm) experience no difficulties. Exposure to levels of 0.3 mg m^{-3} affect the heart through dilation of the coronary arteries and cause headache. It is easily adsorbed through the skin and only a 25 min exposure at a level of 0.7 mg m^{-3} reduces the blood pressure. A Swedish study in 1977 attributed increased deaths from heart disease to ethylene dinitrate.

The WHO recommends, and the UK has imposed, a level for all nitrates in drinking water of 50 mg l^{-1}. The ACGIH has set a TLV (TWA) of 0.05 ppm (0.31 mg m^{-3}) for EGDN and the UK has set the STEL at 0.2 ppm (1.2 mg m^{-3}).

Further reading
ACGIH (1991) *Documentation of TLVs and Biological Exposure Indices*, 5th edn,
    ACGIH, Cincinnati, OH.

ETHYLENE OXIDE CAS No. 75-21-8
(Anprolene, Melgas, Sterigas, Oxyfume).
LD_{50} oral rat $72\,mg\,kg^{-1}$; LC_{50} fish ca $90\,mg\,l^{-1}$

C_2H_4O $M = 44.1$

Dimethylene oxide; b.p. 10.4°, v.p. 146 kPa (1095 Torr, 20°), dens. 0.87 g ml^{-1};
log P_{ow} 0.30. Very soluble in water. Highly flammable. May polymerize explosively in
contact with metal chlorides and oxides, with acids and bases or on heating.

Ethylene oxide has an ethereal odour, perception level $470\,mg\,m^{-3}$. Detection level
in air by GLC $0.27\,mg\,m^{-3}$ (FID) and by IR spectrometry $1.8\,mg\,m^{-3}$; detection level
in water $2\,mg\,l^{-1}$ (FID).

It is manufactured by the oxygenation of ethylene catalysed by silver oxide. In 1981
production in Western Europe was 1370×10^3 t; in the USA production in 1983 was
2540×10^3 t. It is used as an intermediate for glycol, ethers and ethanolamines and in
the manufacture of polyethylene terephthalate for fibres, films and bottles. 1% is lost
during production and in use as a sterilant.

Ethylene oxide has a residence time in air of 5–6 days; it is removed by reaction
with OH radicals. It is metabolized by either hydrolysis:

1, 2-ethanediol $\longrightarrow$ glyoxylic acid $\longrightarrow$ formic acid $+ CO_2$

or the action of $GSH(GSCH_2CH_2OH)$ and excretion in urine as *N*-acetyl-*S*-
(2-hydroxyethyl)cysteine :

$$HO_2C—CH—CH_2—S—CH_2CH_2OH$$
$$\mid$$
$$NHCOCH_3$$

Human occupational exposure is highest in hospitals and during the manufacture
of health instruments. Short term levels (5 min–2 h) range between 100 and
$1800\,mg\,m^{-3}$. It is very soluble in blood and causes headaches, nausea, vomiting
and skin injury, including dermatitis. It irritates the eye and can damage the
cornea.

The IARC has concluded that ethylene oxide is a proven carcinogen in animals. It
is teratogenic in mice, rats, rabbits and monkeys. Sax regards it as a confirmed
human carcinogen, moderately toxic by inhalation when high concentrations cause
pulmonary oedema.

Further reading
ECETOX (1984) *Technical Report 11, Ethylene Oxide Toxicology*, European Chemi-
    cal Industry Ecology and Toxiology Centre, Brussels.
Environ Health Criteria, World Health Organization, Geneva, 1985, Vol. 55.
Kirk–Othmer ECT, 4th edn, Wiley, New York, 1994, Vol. 9, pp. 915–959.

ETHYLENETHIOUREA CAS No. 96-45-7
LD_{50} rat $1800\,mg\,kg^{-1}$; LC_{50} fish $7\,g\,l^{-1}$

$C_3H_6N_2S$ $M = 102.2$

Imidazolidine-2-thione; m.p. 200°; water solubility $2\,g\,l^{-1}$ (30°).

This compound may be associated with the dithiocarbamate pesticides (see **Mancozeb**) either as a residue from their preparation or from cooking produce tainted with the pesticide. Studies have shown that substantial decomposition occurs on boiling in water for 15 min.

Ethylenethiourea has a low acute toxicity, with LD_{50} $545\,mg\,kg^{-1}$ for female rats, but was maternally toxic at $80\,mg\,kg^{-1}$; there is evidence of carcinogenicity in animals.

Adverse effects on the thyroid have been reported but in a study involving 2000 exposed workers no case of thyroid cancer was found.

Ethylenethiourea is also used as a cross-linking catalyst in the rubber industry.

No control limits have been set in the UK or by the USEPA. However, some US States specify drinking water levels in the range $0.25-3\,\mu g\,l^{-1}$.

EUROPIUM CAS No. 7440-53-1

Eu

Europium, a member of the lanthanide group of elements, has an atomic number of 63, an atomic weight of 151.96, a specific gravity of 5.24 at 20°, a melting point of 822° and a boiling point of 1597°. A silvery white metal, it forms compounds in the +2, and +3 oxidation states. Although many isotopes of europium are known, only two occur naturally, ^{151}Eu (47.8%) and ^{153}Eu (52.2%). The average concentration in the earths crust is ~ 1.1 ppm. *Bastnasite* and *monazite* are the principal ores containing europium.

Uses.   Laser material. Europium-activated yttrium vanadate as a red phosphor in colour TV tubes. Eu^{2+} salts are used to give blue phosphors in TV tubes. Europium complexes are used as shift reagents in nuclear magnetic resonance spectrometry.

No biological role. Europium compounds are rarely encountered in the environment. Levels in humans very low ppm concentrations. Europium salts are toxic when injected into laboratory animals. They may act as anticoagulants. The metal dust is a fire and explosion hazard.

Further reading
Gschneider, K.A., Jr and LeRoy, K.S. (eds) (1988) *Handbook on the Physics and Chemistry of the Rare Earths*, North-Holland, Amsterdam.
Kirk–Othmer ECT, 4th edn., Wiley, New York, 1995, Vol. 14, pp. 1091–1115.

Europium(III) chloride hexahydrate CAS No. 13759-92-7

$EuCl_3.6H_2O$

Uses. Preparation of other europium salts.

Toxicity. Poison by intravenous route. Skin irritant. LD_{50} oral mouse 3527 mg kg^{-1}; i.p.r guinea pig 156 mg kg^{-1}.

Europium(III) oxide CAS No. 1308-96-9

Eu_2O_3

Uses. Phosphor activator in X-ray tubes. Catalyst.

Toxicity. LD_{50} oral rat > 5 g kg^{-1}.

FENITROTHION CAS No. 122-14-5
(Sumithion, Dicofen). Insecticide. Toxicity class II. LD_{50} rat 1700 mg kg^{-1}; LC_{50} (48 h) carp 4 mg l^{-1}

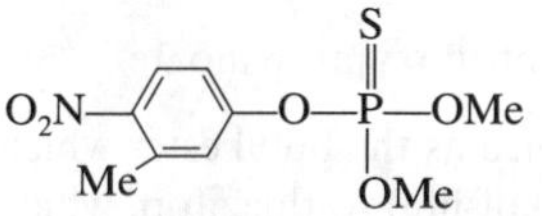

$C_9H_{12}NO_5PS$ $M = 277$

O,O-Dimethyl-*O*-4-nitro-*m*-tolyl phosphorothioate; b.p.145°/0.1 mmHg; v.p. 18 mPa (20°); water solubility 30 mg l^{-1}. Introduced by Sumitomo Chemical. The principal metabolites are fenitrooxon (P=O for P=S), 3-methyl-4-nitrophenol and the *S*-Me isomer.

The level in soil decays sharply with $t_{1/2} = 13$ h. It is used for control of insects in cereals, soft fruit, vines, rice, sugar and vegetables, also in households for control of flies,cockroaches etc.
No standards have been set in air, soil or water.
Codex Alimentarius acceptable daily intake 0.005 mg kg^{-1} (1988). Maximum residue levels wheat flour 2.0, white bread 0.2, polished rice 1.0, potatoes 0.05, fruit 0.5–2.0 mg kg^{-1}.

FENVALERATE CAS No. 51630-58-1
(Sumicidin, Belmark)
Non-systemic broad-spectrum insecticide, acaricide. Toxicity class II. LD_{50} rat 451 mg kg^{-1}; LC_{50} rainbow trout 3.6 μg l^{-1}; Toxic to bees. LD_{50} 0.2 μg/bee

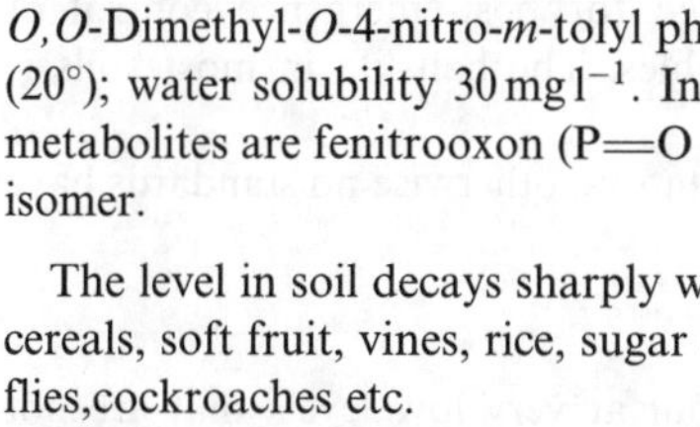

$C_{25}H_{22}ClNO_3$ $M = 419.9$

(*RS*)-α-Cyano-3-phenoxybenzyl (*RS*)-2-(4-chlorophenyl)-3-methylbutyrate as a racemic mixture; v.p. 1.1×10^{-8} Torr (25°) or 19.2 μPa; water solubility < 1 mg l^{-1}; viscous yellow liquid.

Introduced by Sumitomo Chemical. It is rapidly detoxified by mammals and is used against pests which are resistant to other classes of insecticide on fruit, olives, hops, cereals, potatoes, sugar, tobacco, etc. With monocrotophos(Azodrin) it is the most widely used pesticide for cotton crops.

No standards for air, soil or water have been set in the UK or USA.

Codex Alimentarius acceptable daily intake $0.02 \, \text{mg kg}^{-1}$ (1986). Maximum residue levels wheat flour 0.2, milk 0.1, root vegetables 0.05, citrus fruit $2.0 \, \text{mg kg}^{-1}$. It is widely used and the EC has a lower limit of $0.05 \, \text{mg kg}^{-1}$ for many commodities.

FLUAZIFOP CAS No. 69335-91-7; 69806-50-4(butyl ester)
(Fusilade)
Selective systemic herbicide. Toxicity class II/III. LD_{50} rat $2000 \, \text{mg kg}^{-1}$; LC_{50} fish $0.5{-}1.4 \, \text{mg l}^{-1}$

$C_{15}H_{12}F_3NO_4$ $M = 327.3$;
$C_{19}H_{20}F_3NO_4$ $M = 383.4$

butyl ester Butyl(*RS*)-2-[(4,5-trifluoromethyl-2-pyridoxy)phenoxy]propanoate.

Introduced by ICI Plant Protection Division. It is applied as the butyl ester which is taken up by leaves and then hydrolysed in one week (half-life) to fluazifop, which is translocated in plants and grasses. It is used for post-emergence control of weeds in sugar beet, potatoes, fruit and vegetables. The half-life in most soils is 3 weeks.

The UK has set an MRL of $0.1 \, \text{mg kg}^{-1}$ for potatoes, otherwise no standards have been set; it is of low toxicity to birds and bees.

FLUORIDES CAS No. 16984-48-8
Fluorides are widespread in the environment, but at very low levels that are not considered to be harmful.

Fluorides (from the reaction of fluorine with metals/non-metals) are used in making steel (as fluxes), chemicals, ceramics, lubricants, dyes, plastics and pesticides. Toothpaste and mouthwashes have fluorides added to prevent dental caries.

Inorganic fluorides are highly irritating and toxic. Acute effects can arise from their reaction with acid or water to give hydrogen fluoride. Chronic fluorine poisoning or fluorosis occurs mainly among miners of cryolite due to the formation of calcium fluoride in bones. The estimated lethal dose for humans is in the range 2.5–5.9 g of fluoride. Sub-lethal doses can cause very severe nausea, vomiting, diarrhoea, abdominal burning and cramp-like pains. They also cause attacks of asthma and severe bone changes, making normal movements painful. Irritants to the eyes, skin and mucous membranes; impairment of growth in young workers. Organic fluorides are generally less toxic than other halogenated hydrocarbons owing to the strength of the C—F bond.

Metal fluorides are used as fluxes because of their corrosive nature and thus require careful handling. Brick kilns are a localized source of F compounds since they emit mixtures of HF, SiF_4 and H_2SiF_6 together with other pollutants such as SO_2. This fluorine-containing material and associated particulates settling on surrounding grazing areas result in ingestion of F^- ion by livestock. Continuous exposure can lead to a toxic condition called fluorosis, which affects bones and teeth. Plants including some cereals and trees are also affected by F^- toxicity. Some metal fluorides are less toxic than others owing to their insoluble nature.

Fluoride ion ($\sim 1 \text{mg} \text{l}^{-1}$) added to many public water supplies has been shown to provide a dramatic reduction in dental decay. High concentrations in water can give mottled teeth and long term exposure to concentrations of 3–$6 \text{mg} \text{F}^- \text{l}^{-1}$ can cause skeletal fluorosis with denser bones, joint pain and limited joint movement. Some skin medicines and cancer treatment drugs also contain fluorides.

Covalent fluorides produced in commercial quantities include SbF_3, SbF_5, BF_3, BrF_3, BrF_5, ClF_5, SiF_4 and SF_6. Most of these compounds are extremely hazardous substances because they are violently hydrolysed by water to produce HF. Silicon tetrafluoride is a hazard in the steel industry from the use of *fluorspar* (CaF_2) as a flux. Freons, chlorofluorohydrocarbons (CFCs), used as refrigerants, on escaping to the atmosphere, cause a reduction of the ozone layer (see **Ozone**).

Occupational exposure limits.   OSHA TLV (TWA) $2.5 \text{mg} \text{m}^{-3}$. OSHA TLV (TWA) $0.2 \text{mg} \text{F}^- \text{m}^{-3}$ in air.

Further reading
Thompson, R. (ed.) (1995) *Industrial Inorganic Chemicals. Production and Uses*, Royal Society of Chemistry, Cambridge, Ch. 8.

FLUORINE CAS No. 7782-41-4

F (stable as F_2)

Fluorine, a pale, yellow–green gas, is a member of Group 17 of the Periodic Table, and has an atomic number of 9, an atomic weight of 18.998, a specific gravity of 1.696 at STP, a melting point of $-219.6°$ and a boiling point of $-188.1°$. It has only one stable isotope, ^{19}F, in nature and forms only one stable ion , F^-, with an oxidation number of -1.

The earth's crust contains 200–1000 ppm of fluorine (as F^-), placing it 13th in order of terrestrial abundance. Important minerals include *fluorospar*, CaF_2, *fluor-oapatite* $Ca_5(PO_4)_3F$, *and cryolite*, Na_3AlF_6. Fluorine is manufactured by the electrolysis of potassium fluoride dissolved in hydrogen fluoride. Typical environmental concentrations are: sea water 1.5, river water 0.1–2, air 1–6 and human bones 2000–12 000 ppm. Total body content (adult, 70 kg) 800 mg.

Uses.   Fluorine is used in rocket fuels, glass and enamels. Fluorinating agent. Manufacture of Teflon (PTFE). Manufacture of uranium (as UF_6).

Fluorine, the most reactive element known, is a dangerous material but may be handled safely using proper precautions. Fluorine is extremely corrosive and irritating to the skin. Inhalation at even low concentrations irritates the respiratory tract. At higher fluorine concentrations, inhalation may result in severe lung congestion. The manifestations of overexposure include irritation of the eyes and burns to the eyes, skin and respiratory tract. Suggested emergency exposure limits (EC) for humans are 15 ppm for 10 min, 10 ppm for 30 min and 7.5 ppm for 60 min. Because of the corrosive effect and discomfort associated with the inhalation of F_2, and because fluorine is not converted in the body to fluoride ion, toxicity does not occur.

Occupational exposure limit.   US PEL (TWA) 1 ppm ($1.6 \text{mg} \text{m}^{-3}$). UK short-term limit 1 ppm ($1.5 \text{mg} \text{m}^{-3}$).

Further reading
ECETOX (1991) *Technical Report 30(4)*, European Chemical Industry Technology and Toxicology Centre, Brussels.
Environmental Health Criteria 36. Fluorine and Fluorides, WHO/IPCS Geneva, 1984.
IARC (1982) *Monograph 27*, pp. 237–303.
IARC (1987) *Monograph*, Suppl. 7, pp. 208–210.
King, R.B. (ed.) (1994) *Encyclopedia of Inorganic Chemistry*, Wiley, Chichester, Vol. 3, pp. 1223–1245.
Kirk–Othmer ECT, 4th edn, Wiley, New York, 1994, Vol. 11, pp. 241–466.
Thompson, R. (ed.) (1995) *Industrial Inorganic Chemicals. Production and Uses*, Royal Society of Chemistry, Cambridge, Ch. 8.
S.I. 1991 No. 472 The Environmental Protection (Prescribed Processes and Substances) Regulations 1991, HMSO, London, 1991.
ToxFAQs (1993) *Fluorine, Hydrogen Fluoride, and Fluorine*, ATSDR, Atlanta, GA.

FOOD CHAINS

An illustration is provided by mirex, which undergoes bioconcentration from a level in water of $25\,pg\,l^{-1}$ via plankton–particulates–smelt–salmon–gulls' eggs–beluga whale to 10^8 times initial concentration. (See also **Polychlorobiphenyls**.)

FORMALDEHYDE CAS No. 50-00-0

LC_{50} (30 min) inhalation rat $0.82\,mg\,l^{-1}$; LC_{50} alage $0.3\,mg\,l^{-1}$; LC_{50} fish 62–$500\,mg\,l^{-1}$; LC_{50} *E. coli* $1\,mg\,l^{-1}$

$$H_2C{=}O \qquad\qquad CH_2O \quad M = 30.0$$

Methanal, methylaldehyde; m.p. $-90°$; b.p $-20°$; v.p. $25\,hPa$ $(20°)$; v. dens 1.03 $(20°)$; $\log P_{ow}$ 0.00. Henry constant $0.02\,Pa\,m^3\,mol^{-1}$, but less volatile than water. It is flammable and forms explosive mixtures with air; with HCl it reacts to form the highly carcinogenic chloromethyl ether, $ClCH_2-O-CH_2Cl$.

Commonly supplied as formalin, a 30–50% solution in water with methanol added to reduce polymerization. Trioxane is the trimer $(HCHO)_3$ and paraformaldehyde is another polymer of between 8 and 100 units.

Formed naturally in large amounts by oxidation of hydrocarbons. Tropospheric methane generates $4 \times 10^{11}\,kg\,year^{-1}$ as compared with US industrial production (1985) of $9.4 \times 10^8\,kg\,year^{-1}$ and worldwide production (1984) of 5.7×10^9 $kg\,year^{-1}$. US production (1990): $3.2 \times 10^9\,kg\,year^{-1}$.

Formaldehyde is manufactured from methanol by the dehydrogenation of methanol, which is mixed with air and passed over a bed of catalyst silver at atmospheric pressure and 700°.

It is used as a sterilant, in food preservation, household cleaners, for the synthesis of phenolic plastics, melamine–formaldehyde resins and urea–formaldehyde plastics for thermal insulation of buildings. Its use by gardeners to control slugs and snails presents a risk to humans and animals.

A typical level in rainout is $150\,\mu g\,l^{-1}$ but this can rise to over $1000\,\mu g\,l^{-1}$. Some levels $(\mu g\,l^{-1})$ found in air were as follows: Baltic coast 0.7–2.7, HCHO; Amazon 1.7–7.4, all aldehydes; Central Pacific 0.1–0.8, HCHO.

Formaldehyde is produced from wood burning – in Germany particle board factories had levels up to $40\,mg\,m^{-3}$ in exhausted air. Wood-fired furnaces may initially emit levels of $1000 \, \text{mg m}^{-3}$, falling to $100 \, \text{mg m}^{-3}$. Particle board within modern houses emits formaldehyde and concentrations indoors may reach $0.35 \, \text{mg m}^{-3}$, while mainstream cigarette smoke contains $60–130 \, \text{mg m}^{-3}$. Formaldehyde enters water bodies by elution from melamine and urea resins; water at $100°$ extracts $18 \, \text{mg l}^{-1}$ from urea–formaldehyde in 30 min. It is also detected in foodstuffs: pear 60, pig meat 20, apple 17, freshwater fish 9, tomatoes $6 \, \text{mg kg}^{-1}$.

Formaldehyde is carcinogenic in animals but there is limited evidence of a human effect. Humans can detect it at 1ppm $(1.2 \, \text{mg m}^{-3})$ but a 5 h exposure at lower levels $(0.3 \, \text{mg m}^{-3})$ causes irritation of the eyes, nose and throat. Severe exposure leads to weakness, headaches, abdominal pain, CNS depression, coma and death from respiratory failure or circulatory collapse.

The WHO recommends a maximum intake of $0.2 \, \text{mg day}^{-1}$ in drinking water. The ACGIH has set a STEL of 0.3 ppm $(0.37 \, \text{ng m}^{-3})$ and has classed formaldehyde as an A2 carcinogen.

Further reading
Environ. Health Criteria, World Health Organization, Geneva, 1989, Vol. 89.

FRANCIUM CAS No. 7440-73-5

Fr

Francium, a radioactive member of Group 1 (the alkali metals) of the Periodic Table, has an atomic number of 87, an atomic weight of 223, a melting point of $27°$ and a boiling point of $677°$. It has an oxidation state of $+1$. Over 20 isotopes of francium are known. The longest lived, ^{223}Fr, obtained by α-disintegration of ^{227}Ac (4.95 MeV), has a half-life of 22 min. This is the only isotope of francium that occurs naturally (in uranium minerals). Since there is less than $30 \, \text{g}$ of francium at any time in the total crust of the earth, this element is not considered harmful or dangerous.

Further reading
King, R.B. (ed.), *Encyclopedia of Inorganic Chemistry,* Wiley, Chichester, 1994, Vol. 1, pp. 35–54.

GADOLINIUM CAS No. 7440-54-2

Gd

Gadolinium, a member of the rare earth elements, has an atomic number of 64, an atomic weight of 157.25, a specific gravity of 7.90 at $20°$, a melting point of $1314°$ and a boiling point of $3264°$. It is a silvery white metal forming compounds in the $+3$ oxidation state. Seven isotopes that occur naturally are ^{152}Gd (0.20%), ^{154}Gd (2.18%), ^{155}Gd (14.80%), ^{156}Gd (20.47%), ^{157}Gd (15.65%), ^{158}Gd (24.84%) and ^{160}Gd (21.86%).

Found in several minerals including *gadolinite, monazite* and *bastnasite*. Average concentration in earth's crust is $\sim$7 ppm. Levels of gadolinium in the human body are very low.

Uses. Manufacture of gadolinium yttrium garnets for microwave applications. Superconductivity. Addition to alloys to improve high-temperature working and chemical inertness. In the nuclear industry as components of fuel or control rods, where they trap neutrons. Contrast agents in nuclear magnetic imaging.

Some gadolinium compounds may be carcinogenic and highly toxic. Tumorigenic data are reported for experimental animals. Gadolinium compounds are skin and eye irritants. The metal dust is a fire and explosion hazard.

Further reading

Gschneider K.A., Jr, and LeRoy, K.S. (eds) (1988) *Handbook on the Physics and Chemistry of the Rare Earths*, North-Holland, Amsterdam.

Kirk–Othmer ECT, 4th edn, Wiley, New York, 1995, Vol. 14, pp. 1091–1115.

Gadolinium chloride hexahydrate CAS No. 13450-84-5

$GdCl_3.6H_2O$

Uses.   Preparation of gadolinium compounds.

Toxicity.   A skin and eye irritant. Poison by intraperitoneal injection. LD_{50} oral mouse $2000\,mg\,kg^{-1}$.

Gadolinium(III) nitrate hexahydrate CAS No. 19598-90-4

$Gd(NO_3)_3.6H_2O$

Uses.   Preparation of gadolinium compounds.

Toxicity.   Moderately toxic when swallowed.

Gadolinium(III) oxide CAS No. 12064-62-9

Gd_2O_3

Uses.   Phosphor additive. Single-crystal material. Bubble domain. Special glasses.

Toxicity. LD_{50} oral rat $> 5\,g\,kg^{-1}$.

GALLIUM CAS No. 7440-55-3

Ga

Gallium, a member of Group 13 of the Periodic Table, has an atomic number of 31, an atomic weight of 69.72, a specific gravity of 5.904 at 29.6°, a melting point of 29.76°, a boiling point of 2204°, and exists as two stable isotopes, ^{69}Ga (60.11%) and ^{71}Ga (39.89%). Gallium is a beautiful, lustrous silvery liquid or a grey solid. Main oxidation states of +3 and +1. Uniformly distributed, constituting 19 ppm of the earth's crust, 32nd in order of relative abundance. Closely associated with aluminium minerals. Emitted into the atmosphere from metallurgical plants and from coal burning. Gallium is a by-product of the aluminium industry. Typical concentrations: raw water $0.09\,\mu g\,l^{-1}$; soils 1–$70\,mg\,kg^{-1}$ (as Ga^{3+}); plants 3–$30\,mg\,kg^{-1}$ (Ga, dry ash weight); sea water ~0.03 ppb, the principal species being $Ga(OH)_4^-$.

Uses.   Light-emitting diodes, semiconductor technology (GaAs, GaP, GaN) and in optical apparatus. $^{67}Gallium$ citrate is important as a clinical diagnostic tool in nuclear medicine for the localization of bronchogenic carcinomas, Hodgkin's disease and non-Hodgkin's lymphomas.

The toxicity of gallium and its compounds is usually very low. In healthy human tissue gallium concentrations are in the range <1–8 ppb. The element is a potent inhibitor of protein synthesis and the haem pathway enzyme aminolevulinic acid dehydratase. Recent studies using rats have shown that gallium causes several alterations in gene expression in renal proximal tubule cells *in vitro*, including the induction of haem oxygenase. In industry, potential health hazards come from ingestion of, or from skin contact with, gallium salts and the manufacture of semiconductors containing gallium. Inhalation by a worker of a gallium salt aerosol with a concentration of 50 mg m^{-3} caused renal damage. Ingestion or inhalation of GaAs particulate may cause cancer, especially in lung tissue.

Further reading

Chemical Safety Data Sheets (1989) Royal Society of Chemistry, London, Vol.2, pp. 160–162.

ECETOCX (1991) *Technical Report 30(4)*, Euopean Chemical Industry Ecology and Toxicology Centre, Brussels.

Fowler, B.A., Yamauchi, H., Conner, E.A. and Ankkerman, M. (1993) Cancer risks for humans from exposure to the semiconductor metals, *Scand. J. Work Environ. Health,* **19**, 101–103.

King, R.B. (ed.) (1994) *Encyclopedia of Inorganic Chemistry*, Wiley, Chichester, Vol. 3, pp. 1249–1281.

Kirk–Othmer ECT, 4th edn, Wiley, New York, 1994, Vol. 12, pp. 299–317.

Gallium arsenide CAS No. 1303-00-0

GaAs

Uses. Semiconductor in high-temperature rectifiers. High-speed microcircuits.

Toxicity. LD$_{50}$ oral rat, mouse >10 g kg^{-1}; LD$_{50}$ intraperitoneal mouse 4.7 g kg^{-1}; systemic arsenic poisoning.

Gallium chloride CAS No. 13450-90-3

GaCl$_3$

Uses. Chemical agent.

Toxicity. Poisonous when injected. LD$_{50}$ oral rat 4700 mg kg^{-1}; LD$_{50}$ intraperitoneal mouse 37–93 mg kg^{-1}

Gallium nitrate CAS No. 13494-90-1

Ga(NO$_3$)$_3$.9H$_2$O

Uses. Cytotoxic/anticancer drug. It can reverse bone degeneration, osteoporosis. Catalyst. Antineoplastic agent. Its clinical use is limited owing to nephrotoxicity and hepatotoxicity.

Toxicity. LD$_{50}$ oral rat, mouse 2.7–4.36 g kg^{-1}; LD$_{50}$ intravenous mouse 55 mg kg^{-1}.

Gallium Oxide CAS No. 12024-21-4

Ga$_2$O$_3$

Uses. Catalyst. Manufacture of ceramics.

Toxicity. LD$_{50}$ oral mouse 10 g kg^{-1}.

GERMANIUM CAS No. 7440-56-4

Ge

Germanium, a member of Group 14 of the Periodic Table, has an atomic number of 32, an atomic weight of 72.61, a specific gravity of 5.32 at 20°, a melting point of 938.25°, a boiling point of 2833°, and five stable iotopes: ^{70}Ge (20.23%), ^{72}Ge (27.66%), ^{73}Ge (7.73%), ^{74}Ge (35.94%) and ^{76}Ge (7.44%). Germanium is a brittle, grey-white lustrous crystalline metalloid. Principal oxidation states are +2 and +4.

Rare element, 1.5 ppm in the earth's crust, 53rd in order of relative abundance. Germanium minerals are extremely rare but the element is widely distributed in trace amounts. Obtained as a by-product from the flue dusts of smelters processing zinc ores. The average germanium concentrations in various media are water $0.03\,\mu g\,l^{-1}$; soil $\sim 1\,mg\,l^{-1}$; foods $\sim 2\,\mu g\,g^{-1}$ wet weight; natural biological samples $0.1–1\,\mu g\,g^{-1}$; healthy human tissues (heart, liver, brain, etc.) $<1\,\mu g\,g^{-1}$ wet weight. Daily human diet, $\sim 1.5\,mg$ of germanium.

Uses. Semiconductors. Infrared windows and detector. Special alloys. For example, silver–copper–germanium alloys are used in dental castings. Organogermanium compounds have been used as antitumour agents

Germanium compounds are considered to be of a low order of toxicity, rare instances of poisoning have been reported. Experimental LD_{50} $500–5000\,mg\,kg^{-1}$. The experimental animals suffer from hypothermia, diarrhoea and respiratory and cardiac failure. No sign of poisoning from germanium metal up to 1.4 g. Studies using germanium-based antitumour agents have shown that this element is capable of inhibiting DNA, RNA and protein in *in vitro* systems with impairment of cell production and cell maturity. Germanium compounds are also known to be immunosuppresive and may also interfere with drug metabolizing systems.

Futher reading

Fowler, B. A., Yamamauchi, H., Conner, E.A. and Akkerman M. (1993) Cancer risks for humans from exposure to the semiconductor metals, *Scand. J. Work Environ. Health*, **19**, No. S1, 101–103.

Gerber G.B. (1988) Germanium, in Seiler, H.G., Sigel, H. and Sigel, A. (eds), *Handbook of the Toxicology of Inorganic Compounds*, Marcel Dekker, New York, Ch. 24.

King, R.B. (ed.) (1994) *Encyclopedia of Inorganic Chemistry*, Wiley, Chichester, Vol. 3, pp. 1282–1316.

Kirk–Othmer ECT, 4th edn, Wiley, New York, 1994, Vol. 12, pp. 540–554.

Koffmaier, P. (1994) Complexes of metals other than platinum as antitumour agents, *Eur. J. Clin. Pharmacol*, **47**, 1–16.

Thompson, R. (ed) (1995) *Industrial Inorganic Chemicals. Production and Uses*, Royal Society of Chemistry, Cambridge.

Germanium(IV) bromide CAS No. 13450-92-5

GeBr$_4$

Uses. Chemical agent.

Toxicity. Poison by intravenous route. LD_{50} intravenous mouse $56\,mg\,kg^{-1}$.

Germanium(IV) chloride CAS No. 10038-98-9

GeCl$_4$

Uses. Optical fibres. Preparation of organogermanium compounds.

Toxicity. Poison by intraperitoneal route. Mildly toxic by inhalation. Inhalation of large amounts of $GeCl_4$ can result in bronchitis and pneumonia. Mutagen. A skin, eye and mucous membrane irritant. LC_{50} (2 h) inhalation mouse $44 \, mg \, m^{-3}$; LD_{50} intravenous mouse $56 \, mg \, kg^{-1}$.

Germanium hydride CAS No. 7782-65-2
(Germane)

GeH_4

Uses. Chemical vapour deposition (CVD) of silicon and germanium.

Occupational exposure level. USA PEL (TWA) 0.2 ppm ($0.63 \, mg \, m^{-3}$).

Toxicity. Toxic at 100 ppm to humans, death at 150 ppm.

Germanium(IV) oxide CAS No. 1310-53-8

GeO_2

Uses. Catalyst. Alloys and glasses.

Toxicity. Poisonous by intraperitoneal route. Chronic ingestion of 710 ppm GeO_2 in water inhibits growth in chickens. The dioxide stimulates the generation of red blood cells. LD_{50} oral mouse $1250 \, mg \, kg^{-1}$; intraperitoneal rat $750 \, mg \, kg^{-1}$.

Alkylgermanium compounds are much less toxic than the corresponding tin or lead compounds. LD_{50} oral mouse/rat of alkylgermanium compounds 300–$2870 \, mg \, kg^{-1}$.

GLYPHOSATE CAS No. 1071-83-6
(Roundup, Rodeo, Stirrup)
Non-selective broad-spectrum herbicide, chemical intermediate. Toxicity class III.
LD_{50} rat $5600 \, mg \, kg^{-1}$; LC_{50} trout $86 \, mg \, l^{-1}$; LD_{50} bee $0.1 \, mg/bee$

$$\begin{array}{c} CO_2H \quad\quad O \\ | \quad\quad\quad || \\ CH_2-N-CH_2-P(OH)_2 \\ H \end{array} \quad\quad C_3H_8NO_5P \quad M = 169.1$$

N-(Phosphonylmethyl)glycine; m.p.230° (dec.); v.p. 3.0×10^{-7} Torr; water solubility $12 \, g \, l^{-1}$.

Introduced by ICI Agrochemicals. It is absorbed on leaves and then acts to block enzymic synthesis of amino acids; it is used in orchards, vineyards and for plantation crops–tea, coffee, rubber, bananas; also for aquatic weed control and to regulate the growth of sugar cane.

Glyphosate is not carcinogenic and is excreted unchanged by mammals. It is immobile in soils, with a half-life of <60 days and is approved in the UK for use on weeds in or near water courses; the maximum permitted level in water is $0.2 \, mg \, l^{-1}$. It is not metabolized in plants.

Codex Alimentarius acceptable daily intake $0.3 \, mg \, kg^{-1}$ (1986). Maximum residue levels wheat 5.0, wheat flour 0.5, milk 0.1, meat $0.1 \, mg \, kg^{-1}$.

GOLD CAS No. 7440-57-5

Au

Gold, a member of Group 11 of the Periodic Table, has an atomic number of 79, an atomic weight of 196.97, a specific gravity of 19.3 at 20°, a melting point of 1064°, a boiling point of 2856° and exists as one stable isotope, ^{197}Au (100%). Gold is a lustrous, yellow, malleable and ductile metal. Principal oxidation states are 0, +1 and +3.

Widely dispersed, but at a very low concentration, about 0.004 ppm of the earth's crust. Found as native gold and as tellurides, $AuTe_2$ and $(Ag,Au)Te_2$. It is also found in sea water as the gold complex $[AuCl_4]^-$.

Uses. Jewellery. Coinage. Dental fillings. Gold plating.

The most important biological effect of gold is in complexes such as thio salts in their anti-arthritic activity (anti-inflammatory property). Examples of gold complexes are trisodium gold(I) bis(thiosulphate), $Na_3[O_3S_2 - Au - S_2O_3]$, and disodium gold(I) thiomalate:

$$Au—S—CH—CH_2—CO_2^-\ 2Na^+$$
$$\qquad\quad |$$
$$\qquad\quad CO_2^-$$

These complexes often give rise to many side effects, which include changes in blood, teeth and kidneys and jaundice.

Gold can become nephrotoxic to some arthritic patients. Other side effects include aplastic anaemia and dermatitis. Gold (radioactive ^{198}Au in colloidal form) and a number of gold complexes possess antitumour activity (cf. **Platinum**). At various levels they become toxic to plants and animals, including humans.

Since concentrations in the environment are very low (~ng quantities), gold is not known to cause any environmental or occupational health problems.

Toxicity. LD_{Lo} intravenous rat 58 mg kg^{-1}.

Further reading
Brown, D.H. and Smith, W.E. (1990) The chemistry of gold drugs used in the treatment of rheumatoid arthritis, *Chem. Soc. Rev.*, **9**, 217–240.
England, M.W. (1989) Toxicity of gold in a variety of different species, *Health Phys.*, **57** (Suppl. 1), 115–119.
King, R.B. (ed.) (1994) *Encyclopedia of Inorganic Chemistry*, Wiley, Chichester, Vol. 3, pp. 1320–1340.
Kirk–Othmer ECT, 4th edn., Wiley, New York, 1994, Vol. 12, pp. 738–767.
Martindale. The Extra Pharmacopoeia, 30th edn, Pharmaceutical Press, London, 1993, p. 1375.
Rapson, W.S. (1984) Skin contact with gold and gold alloys, *Gold Bull.*, **17**, 102–108.

Gold(III) chloride CAS No. 13453-07-1

$AuCl_3$

Uses. Preparation of gold complexes. Analytical chemistry.

Toxicity. LD_{50} subcutaneous rat 1500 mg kg^{-1}.

Sodium tetrachloroaurate(III) dihydrate CAS No. 13874-02-7

$NaAuCl_4.2H_2O$

Uses. Preparation of other Au(III) compounds and complexes.

Toxicity. LD_{50} intravenous mouse 41 mg Au kg^{-1}; intraperitoneal mouse 72 mg kg^{-1}.

HAFNIUM CAS No. 7440-58-6

Hf

Hafnium, a member of Group 4 of the Periodic Table, has an atomic number of 72, an atomic weight of 178.49, a specific gravity of 13.31 at 20°, a melting point of 2227° and a boiling point of 4602°. It is a silvery, ductile, lustrous metal forming compounds in the +4 oxidation state. Natural isotopes are ^{174}Hf (0.16%), ^{176}Hf (5.20%), ^{177}Hf (18.60%), ^{178}Hf (27.31%), ^{179}Hf (13.63%) and ^{180}Hf (35.10%). Most zirconium minerals such as *zircon* contain 1–5% of hafnium. The relative abundance in the earth's crust is ~2.8 ppm. The metal is obtained by high-temperature reduction of $HfCl_4$ with magnesium or by thermal decomposition of HfI_4. About 50 t are used annually.

Uses. Nuclear reaction control rods. Alloys with iron, niobium, tantalum and titanium and other metals. Gas-filled and incandescent lamps. In nickel-based superalloys.

Many hafnium compounds are poisonous. Intravenous and intraperitoneal injection into rats caused liver damage. Hafnium dust is flammable.

Occupational Exposure Limits. US TLV (TWA) 0.5 mg m^{-3}.

Toxicity. LD_{50} oral rabbit 8000 mg kg^{-1}.

Further reading
ECETOX (1991) *Technical Report 30(4)*, European Chemical Industry Technology and Toxicology Centre, Brussels.
Filov, V.A. *et al.* (eds) (1993) *Harmful Chemical Substances*, Ellis Horwood, New York, Vol. 1, pp. 614–619.
King, R.B. (ed.) (1994) *Encyclopedia of Inorganic Chemistry*, Wiley, Chichester, Vol. 8, pp. 4475–4488.
Kirk–Othmer ECT, 4th edn, Wiley, New York, 1994, Vol. 12, pp. 863–880.
SI 1991 No. 472 The Environmental Protection (Prescribed Processes and Substances) Regulations 1991, HMSO, London, 1991.

Hafnium(IV) chloride CAS No. 13499-05-3

$HfCl_4$

Uses. Preparation of hafnium metallocene, a Ziegler–Natta-type catalyst. Preparation of metal by Kroll process (reduction by magnesium).

Toxicity. Irritating to the eyes, skin and respiratory system. LD_{50} oral rat 362 mg kg^{-1}; intraperitoneal mouse 135 mg kg^{-1}.

HELIUM CAS No. 7440-59-7

He

Helium is a member of Group 18 (noble gases) of the Periodic Table, has an atomic number of two, an atomic mass of 4.0026, a specific gravity of 0.1785 at 0° and 1 atm, a melting point of −272.2° and a boiling point of −268.93°. Natural helium consists almost entirely of one isotope, ^{4}He, with a very small amount of ^{3}He. It is a colourless, odourless, tasteless, inert gas and the oxidation state is usually zero. Helium is obtained from natural gas wells. Apart from hydrogen, helium is the most abundant element in the universe. It is present in the earth's atmosphere 1 in 200 000 parts.

Uses. Inert gas in arc welding. Inert atmosphere in growing silicon and germanium crystals. Artificial atmosphere for deep sea divers (80% He:20% O_2). Pressurised liquid fuel for rockets. Filling balloons. Carrier gas in chromatography.
   Helium is an asphyxiant.

Further reading
Kirk–Othmer ECT, 4th edn, Wiley, New York, 1995, Vol. 13, pp. 1–53.
King, R.B. (ed.) (1994) *Encyclopedia of Inorganic Chemistry*, Wiley, Chichester, Vol. 5, pp. 2660–2680.

HENRY'S LAW
Henry's law states that: at equilibrium the solubility of a gas is directly proportional to its partial pressure:

$$p_i = Kx_i$$

This is illustrated by the solubility of oxygen in water. From air at STP, 1 kg of water dissolves 17 mg of oxygen, but at high altitude under a presure of 0.5 atm, 1 kg of water dissolves 8.5 mg of oxygen.
   The relationship may be written as

$$H'_c = C_{sg}/C_{sl}$$

where H'_c = Henry's constant, C_{sg} = concentration in the gas phase and C_{sl} = concentration in the liquid phase, or

$$H_c = P_{vp}/S$$

where P_{vp} = vapour pressure in Pa and S = aqueous solubility in $mol\,m^{-3}$. The units of H_c are $Pa\,m^3\,mol^{-1}$.
   Substances with a large Henry's constant (H_c) are more likely to volatilize from aqueous solution and so be more readily distributed in the environment. Benzene with a Henry's constant of $550\,Pa\,m^3\,mol^{-1}$ readily volatilizes. The constant is sometimes quoted in $atm\,m^3\,mol^{-1}$ and the conversion between units depends on the relation $1\,Pa = 9.869 \times 10^{-6}$ atm. Therefore, for benzene, $H_c = 550\,Pa\,m^3\,mol^{-1} = 550 \times 9.869 \times 10^{-6} = 5.4 \times 10^{-3}\,atm\,m^3\,mol^{-1}$.
   Values of Henry's constant ($atm\,m^3\,mol^{-1}$) range widely for different structural types:

Compound	H_c	Compound	H_c
Cyclohexane	0.19	Ethyl chloride	8.5×10^{-3}
Acetone	3.7×10^{-5}	Dieldrin	5.8×10^{-5}
Methyl ethyl ketone	1.05×10^{-5}	2,3,7,8-TCDD	3.25×10^{-5}
Methylamine	1.11×10^{-5}	Pentachlorophenol	2.75×10^{-6}
Ethylene glycol	6×10^{-8}	Methyl parathion	1.0×10^{-7}

For an extensive compilation of physical data, including vapour pressures, aqueous solubilities and Henry's law constants for alkanes, alkenes, halo-organics, PAHs and some pesticides, see Mackay and Shiu (1981).

Reference
Mackay, D. and Shiu, W.Y. (1981) *J. Phys. Chem. Ref. Data*, **10**, 1175–1199.

HEPTACHLOR CAS No. 76-44-8
(Velsicol)
Insecticide. Toxicity class II. LD_{50} rat $200 \, mg \, kg^{-1}$; LC_{50} rainbow trout $7 \, \mu g \, l^{-1}$

$C_{10}H_5Cl_7$ $M = 373.3$

1, 4, 5, 6, 7, 8, 8-Heptachloro-3a, 4, 7, 7a-tetrahydro-4,7-methanoindene; m.p.(pure) 95–96°, (technical grade) 46–74°; v.p. 53 mPa (25°); water solubility $0.06 \, mg \, l^{-1}$ (25°).

Introduced by Velsicol Chemical. The technical material contains 72% heptachlor with 22% *trans*-chlordane (see **Chlordane**).

Used for the control of ants and termites in soil and for household insects. Heptachlor is carcinogenic in mice and rats and is metabolized to the epoxide which accumulates in body fat. The half-life in soil is almost 1 year.

Regulatory limits: in workplace air in the UK, TWA $0.5 \, mg \, m^{-3}$; in the USA (ACGIH) $0.05 \, mg \, m^{-3}$; in drinking water EPA $0.4 \, \mu g \, l^{-1}$. *Codex Alimentarius* acceptable daily intake $1 \, \mu g \, kg^{-1}$. Maximum residue levels cereal grains 0.02, milk 0.006, poultry meat 0.2, vegetables $0.05 \, mg \, kg^{-1}$

HEXACHLOROBENZENE CAS No. 118-74-1
LD_{50} rat $10 \, g \, kg^{-1}$; LC_{50} (96 h) for five freshwater species 0.05–$0.2 \, mg \, l^{-1}$

C_6Cl_6 $M = 284.8$

Perchlorobenzene, Amatim, HCB; m.p. 230°; b.p. 326°; dens. 2.04; v.p. 1.1×10^{-5} Torr; v. dens. 9.8; log P_{ow} 6.44. Solubility in water $6.2 \, \mu g \, l^{-1}$, soluble in benzene, chloroform, diethyl ether; slightly soluble in cold ethanol.

Hexachlorobenzene was first prepared by Lorentz in 1893. It is obtained industrially by the action of chlorine–$FeCl_3$ on benzene at 150–200°, or by recovery from the waste tar from the production of tetrachloroethylene, which also contains hexachlorobutadiene. HCB is also found in residues from the production of trichloroethylene and vinyl chloride.

HCB is used as a plasticizer for PVC and in synthesis; for example, it is hydrolysed under forcing conditions by alkali to pentachlorophenol. HCB is used as a fungicidal fumigant to control bunt on wheat seeds and for treatment of other cereals. Production in Japan (1977) was 300×10^3 kg. It is a metabolite of Lindane and occurs in ground-water in Europe and the USA at levels between $2.5 \, \mathrm{ng \, l^{-1}}$ and $2 \, \mathrm{\mu g \, l^{-1}}$.

HCB has been detected in foods at low levels, e.g in beef at $12 \, \mathrm{\mu g \, kg^{-1}}$, in pork at $7 \, \mathrm{\mu g \, kg^{-1}}$, in milk at $4.2 \, \mathrm{\mu g \, l^{-1}}$ and in butter at $0.15 \, \mathrm{mg \, kg^{-1}}$, also in tinned fruit and vegetables. It was detected in human milk in France and Ghana at $1.0 \, \mathrm{mg \, l^{-1}}$.

Hexachlorobenzene is sampled from air by trapping on Chromosorb 101, followed by hexane extraction and GC–ECD with a sensitivity of $28 \, \mathrm{ng \, m^{-3}}$. Samples are obtained from milk by extraction with petrol ether and GC–ECD with a sensitivity of $5 \, \mathrm{\mu g \, l^{-1}}$. HCB has been declining in milk in the UK since a high point in 1979–80 when it was at $60 \, \mathrm{\mu g \, l^{-1}}$.

In 1959, grain treated with HCB in Turkey was sent for human consumption, which lead to 4000 cases of porphyria following an estimated daily intake of 50–$200 \, \mathrm{mg \, day^{-1}}$. The death rate from porphyria was 14% of cases and active symptoms persisted in some individuals for 20 years after ingestion.

At a daily dose of $12 \, \mathrm{mg \, kg^{-1}}$ HCB induced liver cell tumours in mice and a positive indication was also found in hamsters. It has been rated as a group 2B carcinogen by the IARC with sufficient evidence in animals and as a possible risk to humans. The ACGIH give a TLV (TWA) of $0.025 \, \mathrm{mg \, m^{-3}}$ and class it as a group A3 carcinogen, i.e. unlikely to produce human cancers.

Further reading

Coultate, P. (1996) *Food, the Chemistry of its Components*, 3rd edn, Royal Society of Chemistry, Cambridge, p. 285.

IARC (1982) *Some Industrial Chemicals and Dyestuffs*, IARC, Lyon, Vol. 29, pp. 93–148.

HEXACHLOROBUTADIENE CAS No. 87-68-3

LD_{50} rat $65 \, \mathrm{mg \, kg^{-1}}$; LC_{50} (96 h) goldfish $0.09 \, \mathrm{mg \, l^{-1}}$

$C_4Cl_6 \quad M = 260.7$

Perchlorobutadiene, HCBD; b.p. 210–220°; v.p. 22 Torr (100°); dens. 1.665; v. dens. 9.0. Water solubility $2 \, \mathrm{mg \, l^{-1}}$; soluble in ethanol, diethyl ether.

HCBD was first prepared by Krafft in 1877 and is now obtained industrially from the high-boiling residues from the production of tri- and tetrachloroethylene and carbon tetrachloride. It has been detected in air in the vicinity of these plants at levels up to $460 \, \mathrm{\mu g \, m^{-3}}$.

It is used for the recovery of chlorine from waste gas streams and as a fumigant in soil and on vines. The USSR used 800×10^3 kg in 1975 and US production in that year was 3.6×10^6 kg.

HCBD may be determined by GLC in a similar manner to hexachlorobenzene. It has been found in fresh milk at $0.08 \, \mathrm{\mu g \, kg^{-1}}$ and in wine using TLC at $5 \, \mathrm{\mu g \, l^{-1}}$. The most likely source of human exposure is in drinking water, where it has been detected at 0.7 ppb. The bioconcentration factor in fish was 5800.

Hepatic disorders are induced by HCBD in mice. Vineyard workers showed symptoms of exposure – hypotension, bronchitis, heart disease and hepatitis.

The ACGIH has set a TLV (TWA) of 0.02 ppm (0.21 mg m^{-3}) and rates hexachlorobutadiene as an A2 carcinogen, carcinogenic in animals at a level relevant to human experience.

Further reading

Selinger, B. (1986) *Chemistry in the Market Place*, Australian National University Press, Canberra.

HEXACHLOROCYCLOPENTADIENE CAS No. 77-47-4
LD$_{50}$ rat 510 mg kg^{-1}; LC$_{50}$ fish 78–130 mg l^{-1}

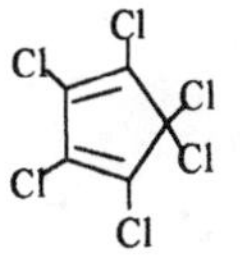

C_5Cl_6 $M = 272.8$

1, 2, 3, 4, 5, 5′-Hexachloro-1,3-cyclopentadiene (HEX); b.p. 239°/753 Torr; m.p. $-10°$; dens. 1.71 (20°); v.p. (25°) 10.7 Pa (0.08 mmHg); log P_{ow}5.04 (28°); 1 ppm = 11.3 mg m^{-3}

It has a pungent musty odour and when pure is light yellow in colour, λ_{max} 323 nm (log $I_o/I = 3.2$). It is non-flammable, is strongly adsorbed on soil and rapidly volatilizes from water. In the mass spectrum the M − 35 ion is the most abundant.

The estimated world production (1988) was 15×10^3 t. When manufactured by the action of chlorine on cyclopentadiene it is only 75% pure; the product includes lower chlorinated cyclopentadienes with hexachlorobenzene and octachlorocyclopentene; the latter gives a good quality product when it is thermally dechlorinated at 470°.

Hexachlorocyclopentadiene is a key intermediate in the synthesis of chlorinated pesticides, e.g. endrin, endosulfan and mirex. It is also used to make the fire retardants Dechlorane Plus and chlorendic anhydride and was once used as a general biocide.

Hexachlorocyclopentadiene is rapidly degraded in the environment – the residence time in air is about 5 h where the detection level is 0.4 µg m^{-3}. On photolysis in water the principal degradation products are 2,3,4,4,5-pentachloro-2-cyclopentenone (shown), hexachloro-2-cyclopentenone and hexachloro-3-cyclopentenone; subsequent oxidation gives pentachloro-2, 4-pentadienoic acid (shown) and thence pentachlorobutadienes by decarboxylation.

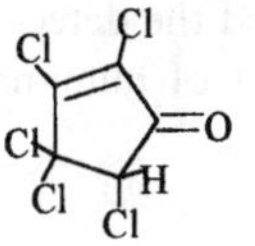

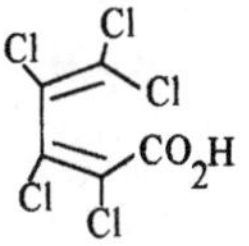

HEX is acutely toxic to marine organisms. Human exposure risk relates to an incident in March 1977 at a wastewater treatment plant at Louisville, Kentucky. This was the result of the illegal dumping of 6 t of the impure compound into a sewer line – levels in sewage reached $1000\,mg\,l^{-1}$ with airborne samples reaching 400 ppb ($4.5\,mg\,m^{-3}$). Symptoms of exposure included irritation of the eyes and throat, headaches; mild abnormalities of liver function were seen but no long term effects are on record.

The risk of human exposure is limited as only two major plants are in operation – Shell International in the Netherlands and Velsicol in the USA.

Further reading
Environ. Health Criteria World Health Organization, Geneva, 1991, Vol. 120.

HEXACHLOROPHENE CAS No. 70-30-4
LD_{50} rat $56\,mg\,kg^{-1}$

$C_{13}H_6Cl_6O_2$ $M - 406.9$

2, 2'-Methylenebis(3,4,6-trichlorophenol), HCP, Dermadex; m.p. 165°; log P_{ow} 7.54. Soluble in acetone, $CHCl_3$, benzene, EtOH.

Antiseptic, constituent of germicidal soaps. The monosodium salt is a soil fungicide and 45×10^3 kg were produced in the USA in 1975.

Prepared by the condensation of 2,4,5-trichlorophenol with formaldehyde; its use was questioned because of the contamination of this phenol with 2,3,7,8-TCDD – the level of the dioxin in 2,4,5-T is now restricted to $< 15\,\mu g\,kg^{-1}$.

Ingestion of hexachlorophene leads to nausea, vomiting, diarrhoea and damage to heart muscles; it is harmful in contact with the skin. Brain lesions formed in rats and monkeys at levels only slightly above those experienced by people using HCP-containing soaps, toothpaste and shampoos.

Hospital operating theatre staff had blood levels in the range $0.07–0.22\,mg\,l^{-1}$ but these rapidly returned to normal when out of contact. Burns patients treated liberally with HCP experienced abnormal body temperatures, headaches and convulsions. Its use is now restricted in the USA.

HCP can be determined by TLC and HPLC. Use of GC–ECD enabled the detection of levels of $1\,\mu g\,kg^{-1}$ in food and $10\,\mu g\,kg^{-1}$ in the hexane extracts of human adipose tissue. The estimated bioaccumulation factor is 317 000.

Further reading
IARC (1979) *Some Halogenated Hydrocarbons*, IARC, Lyon, Vol. 20, pp. 241–257.

n-**HEXANE** CAS No. 110-54-3

$CH_3(CH_2)_4CH_3$ C_6H_{14} $M = 86.2$

B.p. 69°; dens. 0.66; v.p. 150 mmHg (25°); v. dens 2.97; log P_{ow} (25°) 3.6.

Commercial hexane is a mixture including n-hexane (20–80%), 2- and 3-methylpentane, 2,3-dimethylbutane, cyclohexane and isomers of pentane. Other solvents including acetone and trichloroethylene occur with minor amounts of phthalate esters.

Hexane is highly flammable and soluble in ethanol and most other solvents. Its petrolic odour is detectable at a level of 210 mg m^{-3} (60 ppm). It is determined by GC with FID or MS detection.

It is isolated from natural gas and crude oil for use in the extraction of vegetable oils such as soybean and peanuts, also as a general solvent and in polymerization of rubber.

The atmospheric half-life is estimated as < 2 days – occupational exposure is largely by inhalation and through the skin, the eyes may be affected.

A former textileworks site at Carrbrook, UK Derbyshire, was used after 1970 by Chemstar for the recovery of solvents from industrial residues and these were stored in metal drums close to homes. In September 1981, an escape of hexane was ignited by the burners of a steam boiler; the resultant fire killed an employee and destroyed most of the buildings. A survey 2 years later revealed extensive and persistent soil pollution on the site (Craig and Grzonka, 1994).

Oxidative metabolites include 2,5-hexanedione, $CH_3COCH_2CH_2COCH_3$, which is held responsible for neurotoxic effects in humans, these have been noted at levels in the range 70–352 mg m^{-3} (20–100 ppm) and include vertigo and giddiness. Longer term exposure can lead to peripheral neuropathy with weakness of the extremities and diminished reflexes.

The ACGIH has set the TLV (TWA) at 50 ppm (176 mg m^{-3}) for n-hexane and for other isomers at 500 ppm (1760 mg m^{-3}).

Reference
Craig, T. and Grzonka, R. (1994) *Land Contam. Reclam.*, **2**, 19–25.

Further reading
Health and Safety Guide, World Health Organization, Geneva, 1991, Vol. 122.

HOLMIUM CAS No. 7440-60-0

Ho

Holmium, a member of the lanthanide or rare earth elements, has an atomic number of 67, an atomic weight of 164.93, a specific gravity of 8.85 at 20°, a melting point of 1474° and a boiling point of 2695°. Pure holmium has a metallic to bright silver lustre and forms compounds with an oxidation state of +3. Natural holmium consists of only one isotope, ^{165}Ho

A rare element, 1.3 ppm in the earth's crust. Holmium occurs in *gadolinite, monazite* and other rare earth minerals.Very low concentrations in human organs.

Uses. Additive to electro-light sources and garnets. Holmium blue in glasses and ceramics. Electronics.

The element and its compounds seem to be of low toxicity.

Further reading
Gschneider, K.A., Jr, and LeRoy, K.S. (eds) (1988) *Handbook on the Physics and Chemistry of the Rare Earths*, North-Holland, Amsterdam.

Kirk–Othmer ECT, 4th edn, Wiley, New York, 1994, Vol. 14, pp. 1091–1115.

Holmium(III) chloride CAS No. 10138-62-2

$HoCl_3$

Uses. Preparation of holmium compounds.

Toxicity. LD_{50} oral mouse 5165 mg kg^{-1}, intraperitoneal mouse 312 mg kg^{-1}.

HYDRAZINE CAS No. 302-01-2

NH_2NH_2

Uses. Reducing agent. Rocket fuel.

Hydrazine and its methyl derivatives are used in large amounts, $> 20\,000$ t year^{-1}, as rocket fuel. Since they are flammable there is occupational risk of fire and explosion. Accidental ingestion of hydrazine is rare; 20 ml of a 6% aqueous solution taken orally by a technician resulted in vomiting, weakness, sleepiness and arrhythmia. A high incidence of heart attacks and increased risk of cancer have been reported in hydrazine manufacturing workers.

Occupational exposure limits. US TLV (TWA) 0.01 ppm (0.013 mg m^{-3}).

Toxicity. Flammable liquid. Toxic and corrosive substance. Skin and eye irritant to humans. The liquid causes severe eye and skin burns. Probable carcinogen to experimental animals, possible carcinogen to humans. Teratogenic and reproductive effects on mice and rats. LD_{50} oral rat, mouse 60 mg kg^{-1}; LC_{50} (4 h) inhalation mouse 252 ppm.

Further reading
ECETOX (1991) *Technical Report 30(4)*, European Chemical Industry Technology and Toxicology Centre, Brussels.
Environmental Health Criteria 68 : Hydrazine, World Health Organization, Geneva, 1987.
Kirk–Othmer ECT, 4th edn, Wiley, New York, 1995, Vol. 13, pp. 560–605.

Hydrazine hydrate CAS No. 10217-52-4

$NH_2NH_2.H_2O$

Uses. Reducing agent. Preparation of silver and copper mirrors. Removal of oxygen in hot water boilers. Corrosion inhibitor. Rocket propellant.

Toxicity. May cause cancer. Eye irritant. LD_{50} oral rat 129 mg kg^{-1} ; oral rabbit 55 mg kg^{-1}.

Further reading
See **Hydrazine**.

Hydrazine sulfate CAS No. 10034-93-2

$NH_2NH_2.H_2SO_4$

Uses. In gravimetric analysis of nickel, etc. Antioxidant in soldering flux. Reducing agent. Destroying fungi and moulds.

Toxicity. Carcinogenic and tumorigenic to laboratory animals. LD_{50} oral rat, mouse 434–601 mg kg^{-1}.

Further reading
See **Hydrazine**.

HYDROGEN CAS No. 1333-74-0

H, stable form H_2

Hydrogen, the first element in the Periodic Table, has an atomic number of one, an atomic weight of 1.0078, a specific gravity of 0.08988 g l^{-1} at $-253°$, a melting point of $-259.14°$, and a boiling point of $-252.87°$. There are three isotopes of hydrogen : ordinary hydrogen or protium, ^1H, deuterium (CAS No. 7782-39-0) with an atomic weight of 2, ^2H or D, and the unstable isotope tritium, ^3H, with an atomic weight of three. One part of deuterium is found to 6000 ordinary hydrogen atoms. Tritium atoms are found in even smaller quantities. Tritium is produced in nuclear reactors and is used in the production of the hydrogen bomb.

Hydrogen is the most abundant of all the elements in the universe. On earth, hydrogen occurs principally in combination with oxygen in water, but is also present in organic matter. It is present as the free element in the atmosphere, ca 1 ppm by volume. Total body content (adult, 70 kg) 7000 g.

Uses. Vast quantities of hydrogen are required in the Haber ammonia process. It is also used in large quanties in methanol production, hydrogenation of oils and fats, hydrocracking, hydrodesulfurization, etc., and as a fuel. It is also used in the manufacture of hydrochloric acid, for the reduction of metallic ores and for filling balloons. Liquid hydrogen is important in cryogenics and in the study of superconductors. Deuterium is commonly available as deuterium oxide (heavy water), which is used as a moderator to slow neutrons. It is also used as a solvent in nuclear magnetic resonance spectrometry. Tritium is used as a radioactive agent in luminous paints and as a tracer.

Hydrogen combines with other elements, sometimes explosively, to form hydrides. It can act as a simple asphyxiant. Ordinary hydrogen readily exchanges with deuterium, e.g. in a mixture of water and heavy water. Replacement of more than 30% of hydrogen in the body water of mammals has disastrous effects including central nervous system disturbance, muscle tremors, kidney dysfunction, anaemia, disturbed carbohydrate metabolism, etc.

Further reading
ECETOX (1991) *Technical Report 30(4)*, European Chemical Industry Technology and Toxicology Centre, Brussels.
King, R.B. (ed.) (1994) *Encyclopedia of Inorganic Chemistry*, Wiley, Chichester, Vol. 3, pp. 1444 1471.
Kirk–Othmer ECT, 4th edn, Wiley, New York, 1993, Vol. 8, pp. 1–29; 1995, Vol. 13, pp. 838–894.
Thompson, R. (ed) (1995) *Industrial Inorganic Chemicals. Production and Uses*, Royal Society of Chemistry, Cambridge.
Thomson, J.F. (1964) *Biological Effects of Deuterium*, Pergamon Press, Oxford.

HYDROGEN AZIDE CAS No. 14343-69-2
(Hydrazoic acid)

HN$_3$

Uses.   Preparation of azides.

Occupational exposure limits.   N.a.
   Danger of explosion risk. Shock/heat sensitive. Strong irritant to eyes and mucous membranes.

Toxicity.   Toxic.

HYDROGEN BROMIDE CAS No. 10035-10-6

HBr

Uses.   Preparation of organic and inorganic bromides. Hydrobromic acid.

Occupational exposure limits.   OSHA PEL (TWA) 3 ppm ($10 \, mg \, m^{-3}$).

Toxicity.   Poisonous gas. A corrosive irritant to eyes, skin and mucous membranes. LC_{50} inhalation rat 2858 ppm h^{-1}; LD_{50} intraperitoneal rat 76 mg kg^{-1}.

Further reading
See **Bromine**.

HYDROGEN CHLORIDE CAS No. 7647-01-00

HCl

A human poison. Hydrogen chloride is a highly corrosive gas, irritating to the eyes, skin and mucous membranes at concentrations around 35 ppm. Mildly toxic by inhalation. Soluble in water forming hydrochloric acid, a strong corrosive acid.

Occupational exposure limit.   OSHA PEL (TWA) 5 ppm Cl.

Toxicity. LC_{Lo} inhalation human (30 min) 1300 ppm.

Further reading
See **Chlorine**

HYDROGEN CYANIDE CAS No. 420-05-3

HCN

Uses.   Organic synthesis. Production of monomers of synthetic fibres and plastics.

Occupational exposure limits.   OSHA TLV (TWA) 5 mg CN m^{-3}.

Toxicity.   Lethal concentrations 200 mg m^{-3} for 10 min.
   Man-made sources of HCN arise from the industrial manufacture of aromatics, coke plants, decomposition of nitrogen-containing plastics by heat, etc. Smoke from one cigarette produces 150–300 µg of cyanide. Hydrogen cyanide is a very poisonous gas, rapidly causes asphyxia by blocking respiratory enzymes and impairing tissue respiration, leading to respiratory paralysis. Following poisoning, symptoms of salivation, nausea, lower jaw stiffness, convulsions, cardiac arrhythmia, respiratory stimulation (hyperpnea). Exposure to concentrations of 100–200 ppm for periods of 30–60 min can cause death. The fatal dose of HCN for humans is around 50 mg.

Further reading
See **Cyanides**.

HYDROGEN FLUORIDE CAS No. 7664-39-3

HF

Hydrogen fluoride is produced on a commercial scale as a by-product from the manufacture of phosphate fertilizer from rock phosphate apatite, $Ca_3(PO_4)_5$:

$$Ca_5F(PO_4)_3 + H_2SO_4 + 2H_2O \longrightarrow CaSO_4.2H_2O + HF + H_3PO_4$$

Uses. Hydrogen fluoride is used mainly to make cryolite (see aluminium) and chlorofluorocarbons (CFCs). Etching of light bulbs.

Anhydrous hydrogen fluoride or hydrofluoric acid (aqueous HF) is extremely corrosive to the skin, eyes, mucous membranes and lungs. A corrosive irritant to the eyes and skin at 1.5 ppm. HF produces severe skin burns that are slow in healing. Concentrations of 50–250 ppm are very dangerous. A human poison by all routes to the body. Prolonged exposure causes permanent damage to heart and lungs and possibly death.

Occupational exposure limits. OSHA PEL (TWA) $0.2 \, mg \, F^- \, m^{-3}$. ACGIH TLV (TWA) 3 ppm F^-. MAC drinking water $4 \, mg \, F^- \, l^{-1}$.

Toxicity. LC_{50} (30 min) human 30 ppm.

Further reading
ECETOX (1991) *Technical Report 30(4)*, European Chemical Industry Technology and Toxicology Centre, Brussels.
Environmental Health Criteria 36 . Fluorine and Fluorides, WHO/IPCS, Geneva, 1984.
IARC (1982) *Monograph* 27, pp. 237–303.
IARC (1987) *Monograph*, Suppl. 7, pp. 208–210.
King, R.B. (ed.) (1994) *Encyclopedia of Inorganic Chemistry*, Wiley, Chichester, Vol. 3, pp. 1223–1245.
Kirk–Othmer ECT, 4th edn., Wiley, New York, 1994, Vol. 11, pp. 241–466.
Thompson, R. (ed.) (1995) *Industrial Inorganic Chemicals. Production and Uses*, Royal Society of Chemistry, Cambridge, Ch. 8.
S.I. 1991 No. 472 The Environmental Protection (Prescribed Processes and Substances) Regulations 1991, HMSO, London, 1991.
ToxFAQs (1993) *Fluorine, Hydrogen Fluoride, and Fluorine*, ATSDR, Atlanta, GA.

HYDROGEN IODIDE CAS No. 10034-85-2

HI

Uses. Manufacture of hydriodic acid.

Toxicity. Poison by inhalation and ingestion. A corrosive irritant to skin, eyes and mucous membranes.

Further reading
See **Iodine**.

HYDROGEN PEROXIDE CAS No. 7722-84-1

H_2O_2

Uses.   Bleach for textiles, paper pulp, leather, oils and fats, Manufacture of sodium perborate and percarbonate for detergents. Manufacture of organic peroxy compounds.

Occupational exposure limit.   OSHA PEL (TWA) 1 ppm.

Toxicity.   Poison by inhalation, ingestion and skin contact. A corrosive irritant to skin, eyes and mucous membranes. Possible carcinogen. Tumorigenic and mutation data. A very powerful oxidant. It can easily cause blistering of the skin.

   Inorganic peroxides are used as strong oxidizing agents and are also used as bleaches. Organic peroxides tend to be unstable (explosive) and are often used as polymerization catalysts.

HYDROGEN SELENIDE CAS No. 7783-07-5

H_2Se

Uses.   Preparation of zinc selenide and other Selenides.

Occupational exposure limit.   OSHA PEL (TWA) 0.05 ppm Se.

Toxicity.   A deadly poison by inhalation. Very poisonous irritant to skin, eyes and mucous membranes.Can cause damage to the lungs and liver as well as conjunctivitis. Repeated exposure to 0.3 ppm proved fatal to experimental animals by causing pneumonitis.

Further reading
See **Selenium**.

HYDROGEN SULPHIDE CAS No. 7783-06-4

H_2S

A human poison by inhalation. Concentrations of 20–150 ppm cause irritation of the eyes. At higher concentrations may cause irritation of the mucous membranes and the upper respiratory tract. Prolonged exposure may result in pulmonary oedema. A 30 min exposure to 500 ppm H_2S results in headaches, dizziness and excitement and may lead to bronchitis and paralysis of the respiratory centre and asphyxation. It is a common air pollutant.

Uses.   Preparation of sulphides.

Occupational exposure limits.   OSHA PEL (TWA) 10 ppm; STEL 15 ppm.

Toxicity.   Very toxic. LC_{Lo} inhalation human 600 ppm (30 min); LC_{50} inhalation rat 444 ppm.

Further reading
See **Sulphur**.

HYDROQUINONE CAS No. 123-31-9
LD_{50} animals 300–1300 mg kg^{-1}; LD_{50} cat 42–86 mg kg^{-1}; LC_{50} (96 h) fish 0.04–0.1 mg l^{-1}

$C_6H_6O_2$ $M = 110.1$

1,4-Dihydroxybenzene, 1,4-benzenediol; Tecquinol; white, odourless crystals, m.p. 174°; b.p. 287°; dens. 1.33(15°); v.p. 2.4×10^{-3} Pa (1.8×10^{-5} Torr) at 25°; v. dens. 3.8; log P_{ow} 0.59. Solubility in water 57 g/100 g (25°); soluble in acetone and other polar organic solvents, almost insoluble in benzene and CCl$_4$.

Hydroquinone occurs naturally in plants and animals. The plant sources include coffee beans and the glucoside is found in blueberries and cranberries. It is the precursor of the quinone ejected defensively by the bombardier beetle and is formed from phenol by the action of cytochrome P450.

Hydroquinone is a reducing agent and reacts rapidly with oxygen in alkaline solution; products include p-benzoquinone, hydroxy-p-benzoquinone and the oligomers. It may be determined by oxidimetric titration with ceric sulphate and also by colorimetry, e.g. via reaction with phloroglucinol.

The trimethylsilyl ether may be quantified by GC–FID, while UV absorption at 290 nm leads to detection in HPLC.

It is produced commercially by the oxidation of aniline, by the hydroxylation of phenol or by the hydroperoxidation of diisopropylbenzene (see **Cumene**).

Hydroquinone is widely used as a photographic developer, in lithography and for industrial X-ray film. It is used to make antioxidants for protection of rubber and foodstuffs. It is an ingredient of many hair dyes and skin care products, including skin lighteners (CIR Expert Panel, 1986). In 1992 world production was 35×10^3 t (USA 16 000, Europe 11 000, Japan 6000 t).

In humans deaths have been reported through the accidental ingestion of hydroquinone at levels in the range 80–200 mg kg^{-1}. At lower levels of 500 mg day^{-1} over 5 months no changes were observed in blood or urine and inhalation of the dust gave no systemic effects.

There is current concern about the action of skin lighteners, which are applied to reduce ethnic prejudice, as these contain 1.5–2% of hydroquinone. This may result in patchy pigmentation disorders, dermatitis and ochronosis.

The ACGIH has set the TLV (TWA) at 0.4 ppm (2 mg m^{-3}).

Reference
CIR Expert Panel (1986) *J. Am. Coll. Toxicol.*, **5**, 123–165.

Further reading
Environ. Health Criteria, World Health Organization, Geneva, 1994, Vol. 157.

HYPOCHLOROUS ACID CAS No. 7790-92-3

HOCl

Uses. The acid and its salts are used as bleaching agents (see **Calcium hypochlorite**).

Toxicity. Toxic by ingestion and inhalation. Powerful irritant to skin, eyes and mucous membranes. Teratogenic to laboratory animals.

Further reading
Kirk–Othmer ECT, 4th edn, Wiley, New York, 1993, Vol. 5, pp. 932–967.

IMAZALIL CAS No. 35554-44-0; 60534-80-7(sulphate)
(Fungaflor)
Broad-spectrum systemic fungicide. Toxicity class II. LD_{50} rat 220–340 mg kg^{-1}; LC_{50} rainbow trout 1.5 mg l^{-1}. Not dangerous to bees

$C_{14}H_{14}Cl_2N_2O$ $M = 297.2$

(±)-1-(β-Allyloxy-2,4-dichlorophenylethyl)imidazole; m.p. 53°; v.p. 0.16 mPa (20°); water solubility 0.18 g l^{-1}.

First reported by Laville in 1973 and introduced by Janssen Pharmaceutical. It has protective and curative action and is used for control of fungal diseases on fruit, vegetables and ornamental plants.

Imazalil is rapidly metabolized by mammals with loss of the allyl group; no workplace standards have been set.

Codex Alimentarius acceptable daily intake 0.03 mg kg^{-1} (1991). Maximum residue levels wheat 0.01, potatoes 5.0, citrus fruits, 5.0 mg kg^{-1}.

INDIUM CAS No. 7440-74-6

In

Indium, a member of Group 13 of the Periodic Table, has an an atomic number of 49, an atomic weight of 114.82, a density of 7.31 at 20°, a melting point of 156.61°, a boiling point of 2072° and two isotopes, ^{113}In (4.29%) and the radioactive ^{115}In (95.71%, $t_{1/2} = 4.4 \times 10^{14}$ years). Indium is a soft, silver-white metal with oxidation states of +1, +2 and +3.

An uncommon element, 0.1 ppm in the earth's crust, 63rd in relative abundance. Found with zinc ores and ores of tin, manganese, tungsten, copper, iron, lead, cobalt and bismuth. Indium enters the environment through smelting and refining of lead/zinc ores and burning of fossil fuels. Coal can contain up to 4.2 mg kg^{-1}. In sea water, with typical concentrations of $< 1 \times 10^{-7}$ mg l^{-1}, it is present in the ionic form, $In(OH)_2^+$, in soil 0.2 mg kg^{-1} and in the air 20–1000 pg In m^{-3}. World production is $\sim$ 80 t year^{-1}

Uses. Surface coatings, bearing and dental alloys. Low-power sodium lamps. Batteries. Semiconductors (InSb, InAs, InP) and electronics. Manufacture of yellow glass. The radionuclide ^{111}In($t_{1/2} = 67$ h) has been used as a diagnostic tool for malignant lesions.

Indium compounds are harmful by inhalation and in contact with skin exposures. The potentially greatest hazard is likely to arise from the use of indium (e.g. as InAs and InP) together with antimony and germanium in the electronics industry due to fumes. Studies have shown that indium interferes with haem synthesis and is a potent inhibitor of aminolevulinic acid dehydratase in several tissues. Indium exposure is also associated with a general suppression of protein synthesis. Overall data suggest that the carcinogenic properties of indium are probably linked to alterations in the synthesis and maintenance of enzyme systems which metabolize organic carcinogens. Indium can affect the heart, liver, kidneys and the blood. Workers engaged in indium production complained of pains in joints/bones, decay in teeth, nervous/stomach upsets, cardiac pain and general weakness.

Occupational exposure limits. OSHA PEL (TWA) $0.1 \, \text{mg m}^{-3}$ (as In). UK short term limit $0.3 \, \text{mg m}^{-3}$ (as In)

Further reading
ECETOX (1991) *Technical Report 30(4)*, European Chemical Industry Ecology and Toxicology Centre, Brussels.
Fowler, B.A., Yamauchi, H., Conner, E.A. and Akkerman, M. (1993) Cancer risks for humans from exposure to the semiconductor metals, *Scand. J. Work Environ. Health*, **19**, 101–103.
Hayes, R.L. (1988) Indium, in Seiler, H.G., Sigel, H. and Sigel, A. (eds), *Handbook on Toxicity of Inorganic Compounds*, Marcel Dekker, New York, Ch. 27, pp. 323–336.
Kirk–Othmer ECT, 4th edn, Wiley, New York, 1995, Vol. 14, pp. 155–160.
SI No. 472 The Environment Protection (Release into Air of Prescribed Substances) 1991, HMSO, London. 1991.

Indium(III) chloride CAS No. 10025-82-8

$InCl_3$

Uses. In electroplating.

Toxicity. LD_{50} subcutaneous rabbit $2350 \, \mu\text{g kg}^{-1}$; LD_{50} intraperitoneal mouse $9500 \, \mu\text{g kg}^{-1}$. Poison by most routes. When inhaled can cause damage to the respiratory system and lungs.

Indium(III) nitrate CAS No. 13465-14-0

$In(NO_3)_3 . xH_2O$

Uses. Chemical agent.

Toxicity. Experimental teratogenic and reproductive effects. A severe skin irritant.

Indium(III) sulphate hydrate CAS No. 13464-82-9

$In_2(SO_4)_3 . xH_2O$

Uses. Chemical agent.

Toxicity. Moderately toxic by ingestion.

IODATES CAS No. 15454-31-6

IO_3-

Variable toxicity. General eye, skin and mucous membrane irritants. Powerful oxidants. Similar to bromates and chlorates.

IODIDES CAS No. 20641-54-5

I^-

Similar to bromides. Prolonged exposure to iodides may produce 'iodism', which is manifested by skin rash, running nose, headaches, etc. In severe cases, weakness, anaemia, loss of weight and general depression may occur.

IODINE CAS No. 7553-56-2

I (stable as I_2)

Iodine, a member of the halogen group, Group 17 of the Periodic Table, has an atomic number of 53, an atomic weight of 126.90, a specific gravity of 4.93 at 20°, a melting point of 113.5°, a boiling point of 184.4° and exists as one stable isotope, ^{127}I. Although a non-metal, iodine exists in the form of rhombic, violet–black crystals with a metallic lustre. Iodine occurs in various oxidation states from −1 to +5 and +7.

Found in igneous rocks, brine lakes, sea water and seaweed; 62nd element in order of abundance in the earth's crust, to the extent of 0.46 ppm. At one time iodine was extracted from seaweed. Sodium iodate ($NaIO_3$) and sodium periodate ($NaIO_4$), which occur as impurities in Chile saltpetre ($NaNO_3$) deposits, are now the main source of iodine. Typical environmental concentrations are sea water 0.05 ppm; river water 2 ppb; human body (adult, 70 kg) 30 mg.

Uses. Manufacture of iodine compounds. In germicides, antiseptics. Catalyst. Analytical chemical reagent. In lubricants, in dyestuffs. Antihyperthyroidism.

Iodine is an essential nutrient that is required for the functioning of thyroid hormones. Although iodine is bioaccumulated by a range of microorganisms, ranging from soil bacteria to marine algae, it is a toxic element to most species, e.g. LC_{50} (24 h) channel catfish 0.44 mg l^{-1}; LD_{50} oral rabbit, rat, mouse 10–22 g kg^{-1}; LD_{Lo} oral dog 800 mg kg^{-1}. A human poison by ingestion, inhalation and skin absorption. Doses of 2–3 g have been fatal. Depending on the level of dosage, symptoms may include diarrhoea, thyroid hyperfunction, reproductive and gene mutation. Serious exposures are rare in industry owing to the low volatility of the solid. Initial signs are irrritation and burning of the eyes, lacrimation, coughing and irritation of the nose and throat. Ingestion of large quantities will cause abdominal pain, nausea, vomiting and diarrhoea. Chronic ingestion of ca 200 mg day^{-1} can result in thyroid disease.

Although several radioactive species of iodine are known (iodine-129/131/132/133/134/135), only iodine-129/131 have a global significance. Iodine-129 ($t_{1/2} = 1.7 \times 10^7$ years) is naturally produced in the upper atmosphere from interaction of high-energy particles with xenon, from spontaneous fission of uranium-235 and as a fission product from nuclear explosions. Not considered to be a problem but may become so with increased use of power reactors. Iodine-131 ($t_{1/2} = 8$ days) can be a major soil contaminant after nuclear accidents (Marples, 1986). Also, because of the rapid transfer from grass → cow → cow's milk → humans, concentrations of this isotope can rise to dangerous levels with the banning of the sale of milk and the slaughter of animals over a wide area for several months. Escape of a small amount of iodine-131 from a nuclear reactor (experimental Stationary Low Power Reactor) occurred in Idaho, USA and large amounts from the Chernobyl Reactor Accident in the Ukraine (April 1986).

Occupational exposure limits. US TLV (TWA) and UK short-term limit 0.1 ppm $(1 \, mg \, m^{-3})$.

Reference
Marples, D.R. (1986) *Chernobyl and Nuclear Power in the USSR*, Macmillan, Basingstoke.

Further reading
Occupational Exposure Limits: Criteria Document Summaries 1993, HMSO, London, 1993.
ECETOX (1991) *Technical Report 30(4)*, European Chemical Industry Ecology and Toxicology Centre, Brussels.
Kirk–Othmer ECT, 4th edn, Wiley, New York, 1995, Vol. 14, pp. 709–737.
Martindale. The Extra Pharmacopoeia, 30th edn, Pharmaceutical Press, London, 1993, p. 970.
Smerdely, P. *et al.* (1967) Hypothyroidism, *Lancet*, **ii**, 661–664.
Ullmann's Encyclopedia of Industrial Chemicals, Radionuclides, VCH, Weinheim, 1989, Vol. A14, pp. 381–391.

Iodine chloride CAS No. 7790-99-0

ICl

Uses. In analytical chemistry as Wijs' solution used to determine the iodine value of fats and oils.

Toxicity. Poison. LD_{Lo} oral rat $50 \, mg \, kg^{-1}$.

Iodine(V) fluoride CAS No. 7783-66-6

IF_5

Uses. Fluorinating agent.

Toxicity. Harmful.

Iodine(V) oxide CAS No. 12029-98-0

I_2O_5

Uses. Preparation of iodates.

Toxicity. Severe irritant when inhaled.

IPRODIONE CAS No. 36734-19-7
(Rovral)
Contact fungicide with preventative and curative action. Toxicity class IV. LD_{50} rat $>2 \, g \, kg^{-1}$; LC_{50} rainbow trout $4 \, mg \, l^{-1}$

$C_{13}H_{13}Cl_2N_3O_3$ $M = 330.2$

3-(3,5-Dichlorophenyl)-N-isopropyl-2,4-dioxoimidazolidine-1-carboxamide;   m.p. 134°, v.p. 5 $\times 10^{-7}$ Pa (25°), water solubility 13 mg l^{-1}.

Introduced by Rhone-Poulenc Agrochimie and of major usage in Brazil and elsewhere for the control of fungal infections in vegetables, fruit, cereals, potatoes, cotton and sunflowers, etc.

Iprodione is rapidly degraded by plants and animals and in the soil, initially by reaction of the cyclic imide and then to 3,5-dichloroaniline and further products.

Codex Alimentarius acceptable daily intake 0.3 mg kg^{-1} (1977). Maximum residue levels husked rice 3.0, fruits 5–10 mg kg^{-1}.

IRIDIUM CAS No. 7439-88-5

Ir

Iridium, a member of Group 9 of the Periodic Table, has an atomic number of 77, an atomic weight of 192.22, a specific gravity of 22.65 at 20°, a melting point of 2446° and a boiling point of 4428°. It is a silver-white, very hard metallic element and forms compounds in the +3 and +4 oxidation states. Although many isotopes of Ir are recognized, only two occur naturally, ^{191}Ir (37.3%) and ^{193}Ir (62.7%).

Iridium occurs uncombined with platinum and other metals of this group in alluvial deposits. Relative abundance in the earth's crust is 1×10^{-3} ppm. It is recovered as a by-product of the nickel mining industry.

Uses. Manufacture of apparatus and crucibles for high-temperature work. Electrical contacts. Hardening agent for platinum. Alloy with osmium for manufacture of pen nibs and compass bearings. Spark plugs.

The pure metal is chemically inert even to aqua regia. The chlorides IrCl$_3$ (CAS No. 10025-83-9) and IrCl$_4$ (CAS No. 16941-92-7) are eye and skin irritants and are moderately toxic to experimental animals by ingestion. There are no reports of health effects from workers handling iridium and its compounds. Iridium dust is flammable. The ^{190}Ir and ^{192}Ir radioisotopes are used in clinical radiography and most reports of the toxicity of iridium refer to these isotopes.

Further reading
Kirk–Othmer ECT, 4th edn, Wiley, New York, 1996, Vol. 19, pp. 347–406.
Ullmann's Encyclopedia of Industrial Chemicals, Radionuclides, VCH, Weinheim, 1992, Vol. A21, pp. 75–131.

IRON CAS No. 7439-89-6

Fe

Iron , a member of Group 8 of the Periodic Table, has an atom number of 26, an atomic weight of 55.85, a specific gravity of 7.90 at 20°, a melting point of 1538° and a boiling point of 2861°. The element is a white, malleable metal and forms compounds principally in the +2 and +3 oxidation states. There are four naturally occurring isotopes: ^{56}Fe (91.75%), ^{54}Fe (5.85%), ^{57}Fe (2.12%) and ^{58}Fe (0.28%). Two artificial isotopes are produced from nuclear weapons testing . Iron-59 ($t_{1/2} = 44.56$ days) is the activation product of the reaction $^{58}Fe(n, \gamma)^{59}Fe$. Iron-55 ($t_{1/2} = 657$ days) has a biological half-life in the human body of 2000 days and is also produced in light water reactors (LWRs) by the reactions $^{54}Fe(n, \gamma)^{55}Fe$ and $^{56}Fe(n, 2n)^{55}Fe$.

One of the most abundant elements on earth, approximately 4.7% (47 000 ppm) of the earth's crust. It is widely distributed as oxides and carbonates, of which the main ones are *haematite* (Fe_2O_3), *magnetite* (Fe_3O_4), *limonite* (approximately $2Fe_2O_3.3H_2O$) and *siderite* ($FeCO_3$). The metal is obtained by the high temperature reduction of the iron(III) oxide by coke. The major source of atmospheric contamination is from the iron and steel industry, electric arc furnaces, welding, incineration and the burning of fossil fuels. Typical concentrations in the environment are : soils 7000–42000 ppm, ground water <0.5–100 mg l^{-1}, drinking water up to 670 μg l^{-1}, MAC 200 μg Fe l^{-1}, sea water 0.01–0.14 mg l^{-1}, air 0.9–1.2 μg m^{-3}, adult male 49 mg kg^{-1}, adult female 39 mg kg^{-1}, adult daily intake 8–18 mg day^{-1} and total body content (adult , 70 kg) 4.2 g.

Uses. Iron and its alloys are used mainly in the construction, transportation, machine and energy industries.

Iron is essential for healthy growth of all living organisms. It is needed for a variety of biochemical reactions, DNA synthesis, formation of haemoglobin and oxygen transport, electron transfer and catalytic effects, e.g. cytochromes, catalases and peroxidases. In the human body there are 3–4 g of iron, most of which is involved in erythropoiesis – formation of erythrocyte (red blood cells). Most of the remaining iron is stored in the liver and spleen as ferritin and hemosiderin. Iron in excess is potentially toxic to all forms of life and by all routes of exposure: toxic to plants at concentrations ca 200 mg Fe l^{-1}; toxic to human at concentrations ca 200 mg Fe day^{-1}. Iron dust, usually in the form of the oxide can cause conjunctivitis, choroiditis, retinitis and siderosis of tissues, pulmonary fibrosis and increased risk of lung cancer. The inhalation of large amounts of iron dust can result in iron pneumoconiosis (arc welder's lung). Acute iron poisoning from ingested iron is only likely with medicinal iron dosage. The average human lethal oral dose is 200–300 mg Fe kg^{-1} body weight. Toxic manifestations include haemorrhagic gastritis and effects on the cardiovascular system and central nervous system.

Excess iron occurs in individuals with very high rates of erythropoiesis (e.g. thalassaemia and sideroblastic anaemia). These large amounts of iron are stored in the liver and can cause damage to liver tisssue. The toxicity of iron to cells is believed to be derived from its ability to catalyse the production of the hydroxyl radical, a potent oxidizing agent. Deficiency in iron is much more common. In plants iron is required for chlorophyll synthesis. When the minimal level of iron is not available, plants develop chlorosis, which is manifested in yellowing of green leaves. For humans the effect is anaemia, and infants and menstruating or pregnant women are most vulnerable. Symptoms include persistent tiredness, paleness, tired and aching muscles, headaches and dizziness. Therapeutic doses of around 250 mg day^{-1} are the usual remedy.

Occupational exposure limit. N.a. Iron dust is an explosion hazard.

Further reading
Elinder, C.-G. (1986) *Iron*, in Friberg, L. Nordberg, G.F. and Voux, V.B. (eds), *Handbook of the Toxicology of Metals*, 2nd edn, Elsevier, Amsterdam, Vol. II, pp. 276–297.
Friberg, L. (1986) *Handbook of the Toxicology of Metals*, 2nd edn, Elsevier, Amsterdam, Vols 1 and 2.
Kirk–Othmer ECT, 4th edn, Wiley, New York, 1995, Vol. 14, pp. 829–902.
SI 1991 No. 472 The Environmental Protection (Prescribed Processes and Substances) Regulations 1991, HMSO, London, 1991.

Iron(II) chloride CAS No. 7758-94-3

$FeCl_2$

Uses.   Reducing agent. Mordant in dyeing.

Toxicity. LD_{50} intraperitoneal mouse 59 mg kg^{-1}; oral rat 450 mg kg^{-1}.

Iron(III) chloride CAS No. 7705-08-0

$FeCl_3$

Uses.   Photography. Ink. Catalyst for organic reactions.

Toxicity.   Carcinogenic. Eye, skin and mucous membrane irritant. LD_{50} oral mouse 1280 mg kg^{-1}; intravenous mouse 142 mg kg^{-1}.

Iron(III) oxide CAS No. 1309-37-1

Fe_2O_3

Uses.   Manufacture of iron metal and steel.

Occupational exposure limits.   OSHA PEL (TWA) dust/fumes 2 mg Fe m^{-3}; respirable fraction 5 mg Fe m^{-3}.

Toxicity. LD_{50} intraperitoneal mouse, rat 5400–5500 mg kg^{-1}.

Iron pentacarbonyl CAS No. 13463-40-6

$Fe(CO)_5$

Uses.   Preparation of carbonyl complexes.

Occupational exposure limits.   OSHA PEL (TWA) 0.1 ppm Fe; STEL 0.2 ppm (Fe).

Toxicity.   Highly toxic by inhalation, skin contact, ingestion, etc. Can be fatal after continuous exposure. Injury to the kidneys, liver and brain. LD_{50} oral rat 25 mg kg^{-1}; oral cat 100 mg kg^{-1}. LC_{50} (30 min) inhalation mouse 2190 mg m^{-3}.

Iron(II) sulphate heptahydrate CAS No. 7782-63-0

$FeSO_4.7H_2O$

Uses.   Iron supplement in medicine. Moss killer.

Toxicity. LD_{50} oral mouse 1520 mg kg^{-1}; intraperitoneal rat 5500 mg kg^{-1}.

KETONES
(See **Acetone, Methyl ethyl ketone**)

The higher molecular weight ketones are used as solvents for cellulose acetate, nitrocellulose lacquers and natural and synthetic gums and resins. The properties of those subject to control by the principal environmental agencies are set out in the following two tables.

Physical properties

Ketone	CAS No.	Mol. wt	b.p. (°)	water solubility (g l^{-1}) (20°)	V.p. (Torr) (°)	Log P_{ow}
Cyclohexanone	108-94-1	98.1	156	23	6.0 (30)	0.81
Diisobutyl	103-83-8	142.2	168	0.4	1.65 (25)	—
Di-*n*-propyl	123-19-3	114.9	145	low	5.2 (20)	1.98
Ethyl *n*-amyl	106-68-3	128.2	168	low	2.0 (25)	—
Methyl isobutyl	108-10-1	100.2	117	17	19 (25)	—
Methyl *n*-butyl	591-78-6	100.2	127	17	11 (25)	1.38
Methyl *n*-propyl	107-87-9	86.1	102	59	35 (25)	0.91

Toxicity data

Ketone	TLV (TWA) (ppm) (mg m^{-3})	LD$_{50}$ rat (g kg^{-1})
Cyclohexanone	25 (100)	1.4
Diisobutyl	25 (145)	5.8
Diethyl	200 (705)	2.1
Di-*n*-propyl	50 (233)	3.0
Ethyl *n*-amyl	25 (131)	0.41[a]
Methyl isobutyl	50 (205)	—[b]
Methyl *n*-butyl	5 (20) skin	2.4
Methyl *n*-propyl	200 (705)	3.7

[a] 0.41 mouse i.p
[b] Rats exposed to 100 ppm showed no symptoms after 2 weeks.

Cyclohexanone

$C_6H_{10}O$

Intermediate in the production of adipic acid and caprolactam. A solvent for nitrocellulose, etc.

Diethyl ketone

$C_5H_{10}O$

Occurs naturally in soft woods and other plants; a flavour component of the shrimp. Forms an explosive peroxide with H_2O_2–HNO_3. It is a solvent for paints; with a half-life in river water of 13 h and bioconcentration factor of 3.3, pollution is unlikely.

Diisobutyl ketone

$C_9H_{18}O$

A solvent for synthetic resins, nitrocellulose, lacquers and rubber. In humans it irritates the eyes and the odour is unpleasant at levels >25 ppm.

Di-*n*-propyl ketone

$C_7H_{14}O$

An aroma component of plants, cooked meat and fish and of the urine of the red fox, *Vulpes vulpes*. A solvent for nitrocellulose, oils, resins and polymers.

Ethyl *n*-amyl ketone

$C_8H_{16}O$

A flavouring agent and an aroma component of plants. Occurs in dairy and meat products.

Methyl iso butyl ketone

$C_6H_{12}O$

A solvent for nitrocellulose, cellulose ethers, fats, oils, gums and resins. It is a propellant for CS gas canisters.

Methyl *n*-butyl ketone

$C_6H_{12}O$

A solvent for lacquers, resins, oils and fats and an ink thinner. Notwithstanding the LD_{50} value for the rat, it is rated as more toxic than other methyl ketones. In 1973 there was an outbreak of toxic distal neuropathy in a cohort of fabric printers. The workforce of 1157 was employed in an atmosphere with up to 36 ppm of this ketone and 86 cases were confirmed – the compound was also found to be neuropathic in rats, cats and dogs; it shares a common metabolic path with *n*-hexane The stricter TLV was set in consequence of these results.

Methyl *n*-propyl ketone

$C_5H_{10}O$

This is emitted from foods including smoked meats.

KRYPTON CAS No. 7439-90-9

Kr

Krypton, a member of Group 18 of the Periodic Table, has an atomic number of 36, an atomic weight of 83.80, a specific gravity of $3.749 \, \text{g} \, \text{l}^{-1}$ at STP, a melting point of $-157.4°$ and a boiling point of $-153.22°$. It is one of the noble gases and the stable oxidation state is 0. Naturally occurring krypton contains six stable isotopes: ^{78}Kr (0.35%), ^{89}Kr (2.25%), ^{82}Kr (11.60%), ^{83}Kr (11.5%), ^{84}Kr (57.0%) and ^{86}Kr (17.30%). Krypton is present in air to the extent of 1 ppm. It is obtained by the liquefaction and separation of air.

Uses. It is used with argon as a low-pressure filling gas for fluorescent lights.

Krypton acts as a simple asphyxiant. The radioisotope ^{85}Kr($t_{1/2} = 10.7$ years) is produced primarily from fission. About 50 000 Ci of this isotope were released as a result of the accident at Three Mile Island, Bethesda, Maryland.

Further reading

Krypton-85 in the Atmosphere – With Specific Reference to the Public Health Significance of the Proposed Controlled Release at Three Mile Island, Bethesda, Maryland, National Council on Radiation Protection and Measurements, Washington, DC, 1980.

Kirk–Othmer ECT, 4th edn, Wiley, New York, 1995, Vol. 13, pp. 1–53.

Ullmann's Encyclopedia of Industrial Chemicals. Radionuclides, VCH, Weinheim, 1993, Vol. A17, pp. 485–539.

LANTHANUM CAS No. 7439-91-0 AND THE LANTHANIDES

The lanthanides consist of a group of 15 inner-transition elements in the Periodic Table with atomic numbers from 57 (lanthanum) to 71 (lutetium). All exhibit an oxidation state of +3, and some also of +2 and +4. The elements scandium and yttrium are often included in this group. The majority of lanthanides are bright, silver-grey metals which are malleable and ductile. The main physicochemical properties are listed in the following table; see separate entries for each element for more information.

Element	Symbol	Atomic No.	Atomic weight	Melting point (°)	Boiling point (°)
Lanthanum	La	57	138.90	920	3454
Cerium	Ce	58	140.12	798	3257
Praseodymium	Pr	59	140.90	931	3212
Neodymium	Nd	60	144.24	1010	3127
Promethium	Pm	61	147	1080	2460
Samarium	Sm	62	150.36	1072	1778
Europium	Eu	63	151.96	822	1597
Gadolinium	Gd	64	157.25	1311	3233
Terbium	Tb	65	158.93	1360	3041
Dysprosium	Dy	66	162.50	1409	2335
Holmium	Ho	67	164.93	1470	2720
Erbium	Er	68	167.26	1522	2510
Thulium	Tm	69	168.93	1545	1727
Ytterbium	Yb	70	173.04	824	1193
Lutetium	Lu	71	174.97	1656	3315

Although the lanthanides are found in many different minerals, the two main sources for commercial production are *monazite* and *bastnasite*. In 1990, world consumption was ~ 35 000 t. China has reserves to meet global demand for the next 1000 years. Cerium is the most abundant rare earth and is ~ 100 times more abundant than cadmium. Coal-fired electric power plants in Europe were projected to release over 900 kg each of cerium, lanthanum and neodymium into the environment in 1985. These quantities exceed the environmental release of mercury and cadmium 50-fold. The average concentrations in the earth's crust vary from lanthanum (30 ppm) to europium (0.5 ppm). Promethium is a man-made element and is not found naturally. Metallic lanthanum exists as two isotopes, ^{138}La (0.09%, $t_{1/2} = 1.06 \times 10^{11}$ years) and ^{139}La (99.91%).

Present in sea water at levels < 0.005 ppm and in soils near lanthanide deposits at 1 ppm. In humans, lanthanides tend to accumulate in the liver and bones. In healthy adults: liver $0.005 \mu g\,g^{-1}$ dry ash, kidney $0.002 \mu g\,g^{-1}$ dry ash, bone $0.2–1.0 \mu g\,g^{-1}$ dry ash.

Uses. Special steels and magnetic alloys, including permanent magnets. Carbon-arc electrodes, fluorescent and halogen lamps. Manufacture and polishing of glass lenses. In optical glass lanthanum improves refractive index and dispersion. Porcelain enamel glazes. Phosphors in colour television sets. Catalysts. Superconductors.

Most lanthanides and their compounds show medium to high toxicity to laboratory animals. There are no reports of carcinogenicity for the lanthanides. In general, the toxicity of lanthanides decreases as the atomic number increases. Studies on animals show that toxicity depends on how the lanthanide ion is introduced. Little appears to be absorbed in oral uptake since lanthanide compounds are not very soluble. Considerably more toxic in intravenous administrations. Inhalation of vapour and dust should be avoided as solid compounds if inhaled tend to remain in the lungs and are eliminated only slowly owing to their insolubility. Lanthanides tend to eliminate in the liver, spleen, kidneys and bone. Symptoms of acute toxicity in rats included writhing, ataxia, laboured respiration and sedation. Some lanthanides have been reported to exhibit antitumour activity. The mechanism involved in the antitumour activity may be because of antagonism of lanthanide ions to the metabolism of Ca^{2+} and Mg^{2+}. A few compounds are phytotoxic.

In humans, chronic exposure to the lanthanides occurs primarily via inhalation in occupational settings and causes granulamatous lesions in the lung. Reports of lanthanide pneumoconiosis occurring in photoengravers, lithogravers and smelter workers. The concentration of lanthanides (La, Ce, Nd, Sm, Eu, Tb, Yb and Lu) found in asmelter's lung was found to be 2–16 times than that in normal lungs. Target organs after ingestion are the liver (hepatic necrosis), kidney, spleen and gastrointestinal tract. Rare earth elements possess pharmacological properties and are capable of reducing blood coagulation.

Lanthanum powder is flammable in air.

Further reading

Bulman, R.A. (1988) Yttrium and lanthanides, in: Seiler, H.G., Sigel, H. and Sigel, A. (eds) (1988) *Handbook on Toxicity of Inorganic Compounds*, Marcel Dekker, New York, pp. 769–785.

Gschneidner, K.A. Jr, and LeRoy, K.A. (eds), (1988) *Handbook on the Physics and Chemistry of the Rare Earths*, North-Holland, Amsterdam.

Haley, P.J. (1991) Health effects of lanthanides, *Health Phys.*, **61**, 809–20.

Hirano, S. and Suzuki, K.T. (1996) Exposure, Metabolism, and Toxicity of Rare Earths and Related Compounds, *Environ. Health Perspect.*, **104**, *Suppl. 1.*

Kirk–Othmer ECT, 4th edn, Wiley, New York, 1995, Vol. 14, pp. 1091–1115.

SI 1991 No. 472 The Environmental Protection (Prescribed Processes and Substances) Regulations 1991, HMSO, London, 1991.

Sulotto, F., Romano, C., Berra, A., Botta, G.C., Rubino, G.R., Sabbinoni, E. and Pietra, R. (1986) Rare-earth pneumoconiosis: a new case, *Am. J. Ind. Med.*, **9**, 567–575.

Venugopal. B. and Luckcy, T.D. (1978–79) Toxicity of Group III Metals, in *Metal Toxicity in Mammals*, Plenum, New York, Vol. 2, pp. 135–157.

Lanthanum chloride heptahydrate CAS No. 10025-84-0

$LaCl_3.7H_2O$

Uses. Chemical agent.

Toxicity. LD_{50} oral rat $4184\,mg\,kg^{-1}$; intraperitoneal rabbit $148\,mg\,kg^{-1}$.

Lanthanum(III) nitrate hexahydrate CAS No. 100587-94-8

$La(NO_3)_3.6H_2O$

Uses. Chemical agent.

Toxicity. Irritant to skin, eyes and mucous membranes. LD_{50} intraperitoneal mouse $410\,mg\,kg^{-1}$.

LEAD CAS No. 7439-92-1

Pb

Lead, a member of Group 14 of the Periodic Table, has an atomic number of 82, an atomic weight of 207.19, a specific gravity of 11.34 at 20°, a melting point of 327.5° and a boiling point of 1749°. Naturally occurring lead consists of four natural isotopes: ^{208}Pb (52.4%), ^{206}Pb (24.1%), ^{207}Pb (22.1%) and ^{204}Pb (1.4%). The longest lived naturally occurring radioactive isotope is $^{210}Pb(t_{1/2} = 23.3\,years)$ from radon decay. Two oxidation states occur, +2 and the less stable +4.

Widely occurring; over 200 minerals of lead are known. The average concentration in the earth's crust is about16 ppm. The most important lead minerals are *galena* (lead sulphide), *cerussite* (lead carbonate) and *anglesite* (lead sulphate). Typical environmental concentrations are as follows: soil 16 ppm, dependent on locality 14–500 ppm; air <0.01–$2\,\mu g\,m^{-3}$, $> 10\,\mu g\,m^{-3}$ when poor air quality; drinking water 1–40 $\mu g\,l^{-1}$; lakes and river water 0.1–10 $\mu g\,l^{-1}$; sea water 0.02 ppb; sewage sludge up to 1000 ppm; city dust up to 5000 ppm ; cigarette 1–4 μg; milk 0.01–0.08 ppm; human hair 1.2–25 $\mu g\,g^{-1}$; human dietary intake 100–500 $\mu g\,day^{-1}$; human bone 0.2–10 ppm. Average total body content (adult, 70 kg) 80 mg.

Uses. Car batteries. Antiknock agents, tetramethyl- and tetraethyllead. Paint. Glass. Ceramics. Lead sheets, cables, solder. Bearing alloys. For X-ray and atomic radiation protection. $^{203}Pb(t_{1/2} = 2\,days)$ is used in diagnostics and $^{212}Pb(t_{1/2} = 11\,h)$ in radio(immuno)therapy.

Lead is a cumulative poison, increasing amounts build up in the body and eventually reach a point at which symptoms and disabilities occur. Poisoning by lead and its compounds is one of the commonest occupational diseases. The problem arises because lead is produced in such large quantities ($>3.5 \times 10^6\,t\,year^{-1}$). Lead is capable of breaking down the blood–brain barrier and is a suspected carcinogen and proven teratogen for laboratory animals.

A major source of lead in the environment is the emission from car exhausts. Tetramethyl- and lead tetraethyllead are used as antiknock agents in petrol although this use is decreasing owing to increased sales of 'lead-free' petrol and diesel. Organolead compounds are much more toxic than inorganic lead compounds. They are rapidly absorbed by the respiratory and gastrointestinal system and through the skin. The toxic species, $(CH_3)_3Pb^+$ and $(C_2H_5)_3Pb^+$, derived from the antiknock agents, are formed in the liver and are highly dangerous since they readily cross the blood–brain barrier and the placenta.

Other important sources of lead in air are lead mining, smelting and refining, discharge of industrial and municipal waste and combustion of fossil fuels. Lead of small particle size ($< 10^{-6}$ cm) is easily transported through air. In major cities the lead concentration in air can be up to $40 \mu g\, m^{-3}$.

The main pathways of lead entry into soil are deposition from the atmosphere, crudely purified waste water and sewage sludge, lead arsenate – used as an agricultural spray – and lead particles from car exhausts. Kerbside concentrations on major roads can exceed 1000 ppm.

Apart from the natural lead in water, the natural concentration is increased by the use of lead pipes, lead in food container tins and pottery and the biomethylation of lead.

The reported lead contents in various items and categories of food are higly variable. Only about 5–10% of ingested lead is absorbed through the gastrointestinal tract but 30–60% of inhaled lead is retained in the lungs, depending on the particle size of airborne lead. The absorbed lead is transported to other organs by complexing with blood erythrocytes. The half-life in blood is about 18 days.

Once lead is inside the body, through either ingestion or inhalation, there are two main areas where lead causes a problem, namely the formation of blood and the central nervous system. Lead affects the nervous system in many ways, e.g. cell metabolism, membranes, nerve transmission–synthesis and release of acetylcholine and myelin formation. The effect of lead on the nervous system varies with the duration and intensity of exposure. Symptoms include altered behaviour, loss of memory, motor dysfunction, loss of appetite, malaise, anaemia. Lead encephalopathy is accompanied by severe cerebral oedema, increase in cerebral spinal fluid pressure and neural degeneration.

Lead effects on blood formation are primarily due to interference in the activities of the enzymes δ-aminolevulinic acid dehydratase (ALAD), which catalyses the formation of porphobilinogen from δ-aminolevulinic acid (ALA), and haem synthetase, which incorporates iron into protoporphyrin IX. The occurrence of unreacted ALA in the urine is a very characteristic indication of lead poisoning.

In adults, the normal level of lead in blood is less than $30–40 \mu g\, dl^{-1}$. Concentrations between 40 and $80 \mu g\, dl^{-1}$ are acceptable but may cause some metabolic effects. Lead levels greater than $120 \mu g\, dl^{-1}$ are dangerous and will cause functional changes. Children are more at risk and their normal levels should be below $10 \mu g\, dl^{-1}$. The unborn are also at risk since lead passes through the placental barrier. There is experimental evidence that blood lead levels around $10 \mu g\, dl^{-1}$ can diminish the IQ scores of children. Lead accumulates in bones, where it replaces calcium.

Lead and its compounds are capable of exerting direct effects on the living cell; most inorganic lead compounds have a similar mode of action. The difference between their toxicity is mainly due their solubility in particular gastric juices. The lethal dose of lead usually ranges from 155 to $455 \, mg\, kg^{-1}$. Cases of acute and chronic poisoning ($> 5 \, mg\, kg^{-1}$) have mainly occurred in the home environment. Drinking water from lead pipes in certain parts of the UK can have levels up to $8500 \mu g\, l^{-1}$. MAC (UK): $50 \mu g\, Pb\, l^{-1}$. Another common source of domestic poisoning by lead is hand-crafted lead-glazed earthenware. The leaching of lead by apple juice stored in a glazed vessel for 3 days produced a toxic concentration of $1300 \mu g\, l^{-1}$. Sixty people (four fatalities) were poisoned by jam stored in glazed pots containing a lead level of $120 \, mg\, l^{-1}$. Lead poisoning from illicit whisky stills which contained lead compounds and lead solder has been reported. In oriental countries, poisoning has occurred from lead-containing facial cosmetics. A worker in a British battery factory developed bilateral wrist drop, neuropathy due to lead poisoning; tests showed that he had plumboporphyria (ALAD deficiency). Burning of car batteries in a Chicago neighbourhood led to 405 children being affected by fumes; only 25% recovered completely, the remainder have permanent neurological damage of varying severity, from palsy to blindness. In Glasgow, UK, 11 people using oxyacetylene burners to cut lead-painted metal girders suffered severe lead poisoning. Lung cancer mortality was shown to be higher than normal for workers employed in a lead smelter. Similarly, a case study of glasswork employees showed an increased risk of death from stomach cancer, lung cancer and cardiovascular disorders.

Occupational exposure limits. OSHA PEL (TWA) 0.05 mg Pb m^{-3}.

Toxicity. LC$_{50}$ (28 days) rainbow trout 0.22 mg kg^{-1}; TD$_{Lo}$ oral woman 450 mg kg^{-1}; TD$_{Lo}$ oral mouse 4800 mg kg^{-1}; LD$_{50}$ oral rat 23 g kg^{-1}.

Further reading
ECETOX (1991) *Technical Report 30(4)*, European Chemical Industry Technology and Toxicology Centre, Brussels.
Environmental Health Criteria 85. Lead: Environmental Aspects, WHO/IPRCS, Geneva, 1989.
IARC Monograph 23, IARC, Geneva, 1980, pp. 325–415.
King, R.B. (ed.) (1994) *Encyclopedia of Inorganic Chemistry*, Wiley, Chichester, Vol. 4, pp. 1944–1964.
Kirk–Othmer ECT, 4th edn, Wiley, New York, 1995, Vol. 15, pp. 69–158.
Needleman, H.L. (1995) Environmental lead and children's intelligence, *Br. Med. J.*, **310**, 408.
Pocock, S.J., Smith, M. and Baghurst, P. (1994) Environmental lead and children's intelligence: a systematic review of the epidemiologic evidence, *Br. Med. J.*, **309**, 1189–1197.
SI1991 No. 472 The Environmental Protection (Prescribed Processes and Substances) Regulations 1991, HMSO, London, 1991.
ToxFAQs (1993) *Lead*, ATSDR, Atlanta, GA.

Lead acetate trihydrate CAS No. 6080-56-4

Pb(CH$_3$COO)$_2$.3H$_2$O

Uses. Dyeing of textiles. Waterproofing. Paint varnishes. Lead drier. Analytical reagent.

Toxicity. Carcinogenic to laboratory animals. Mutation and nerve damage. Toxic by inhalation. LD$_{50}$ oral rat 4665 mg kg^{-1}; intraperitoneal mouse 174 mg kg^{-1}.

Lead arsenate CAS No. 7645-25-2

AsH$_3$O$_4$. xPh

Uses. Insecticide and pesticide.

Toxicity. Poison. Confirmed human carcinogen.

Lead carbonate CAS No. 598-63-0

PbCO$_3$

Uses. As basic lead carbonate, pigment in paints and wood preserver.

Occupational exposure limits. ACGIH TLV (TWA) 0.15 mg Pb m^{-3}.

Toxicity. Poison. LD_{Lo} oral human $571 \, \text{mg kg}^{-1}$.

Lead chromate CAS No. 7758-97-6

$PbCrO_4$

Uses. Yellow pigment.

Occupational exposure limits. ACGIH TLV (TWA) $0.5 \, \text{mg Pb cm}^{-3}$.

Toxicity. IARC Group 1 Carcinogenic. Target organs gastrointestinal tract, central nervous system, kidneys and blood.

Lead cyanide CAS No. 592-05-2

$Pb(CN)_2$

Uses. CI Pigment Yellow 48.

Toxicity. Poison.

Lead nitrate CAS No. 10099-74-8

$Pb(NO_3)_3$

Uses. Match-head ingredient. Mordant in dyeing and printing in textiles. Oxidizing agent.

Occupational exposure limits. ACGIH TLV (TWA) $0.15 \, \text{mg Pb m}^{-3}$.

Toxicity. Poison. LD_{50} intravenous rat $93 \, \text{mg m}^{-3}$.

Lead oxide yellow CAS No. 1317-36-8

PbO

Uses. Yellow pigment. Glass manufacture. Batteries.

Toxicity. LD_{50} intraperitoneal mouse $17\,700 \, \text{mg kg}^{-1}$.

Tetraethylead CAS No. 78-00-2

$(C_2H_5)_4Pb$

Uses. Antiknock reagent in petrol.

Occupational exposure limit. OSHA PEL (TWA) $0.075 \, \text{mg kg}^{-1}$ (skin).

Toxicity. IARC Group 3 (unclassified). Possible mutagen. Target organs blood, CNS and reproductive systems. LD_{50} oral rat $12\,300 \, \mu\text{g m}^{-3}$.

Tetramethyllead CAS No. 75-74-1

$(CH_3)_4Pb$

Uses. Antiknock reagent in petrol.

Occupational exposure limit. N.a. Compound is flammable.

Toxicity. Poison by all routes. Target organ CNS. LD_{50} oral rat $105 \, \text{mg kg}^{-1}$.

LINDANE CAS No. 58-89-9
(BHC, Gammexane)
Insecticide. Toxicity class II. LD_{50} rat $90 \, \text{mg kg}^{-1}$; LC_{50} rainbow trout $0.06 \, \text{mg kg}^{-1}$. Carcinogenic in rats and mice

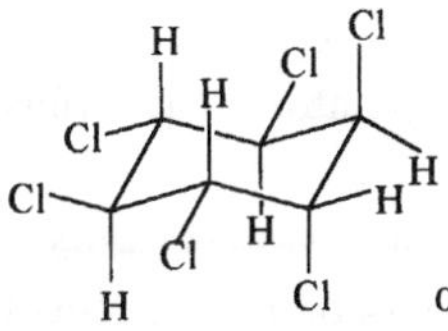

α-form

$C_6H_6Cl_6$ $M = 290.8$

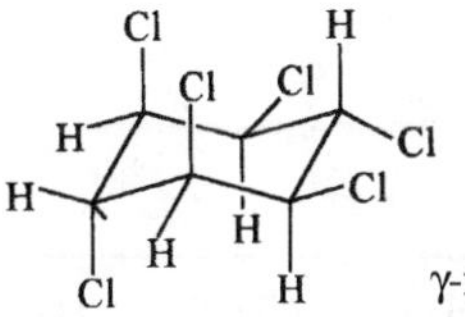

γ-form

Isomers of hexachlorocyclohexane, α and γ shown; m.p.113°(γ); v.p. 1.9×10^{-2} Pa (20°); water solubility 200 mg l^{-1} (20°).

First synthesized by Faraday in 1825 by the addition of three molecules of chlorine to benzene; its description as benzene hexachloride (BHC) is a misnomer.

The technical mixture includes five geometrical isomers of which the α-isomer(70%) predominates; the γ-isomer (15%) or lindane is the only one effective as an insecticide. The mixture was introduced during World War II and it is estimated that 10^7 t were produced up to 1992; it was formerly used in sheep dips. Owing to its persistence and detection in air, soil, water and animal and human tissue, the use of BHC has ceased; it has been replaced by restricted applications of the pure active γ-isomer, lindane.

Lindane is fairly rapidly transformed by animals and is excreted as a chlorophenol. The levels in human adipose tissue declined from a mean of 0.37 ppm in 1970 to 0.1 ppm in 1983.

Sediments in Casco Bay (Maine, 1990) contained only low levels of < 0.5 ppb, but the distribution in ocean air (see table) is evidence of continued major use of the cheaper technical mixture in Asian countries.

Distribution of hexachlorocyclohexanes in ocean air (1990) (mean levels pg m^{-3}) (Iwata *et al.*, 1993)

α-Isomer	Ocean	Lindane
520	North Pacific	76
8600	Bay of Bengal	1100
200	North Atlantic	66
810	South China Sea	500

India produced 20 000 t in 1991. A study of 35 Malaysian workers from six different plantations showed that all had experienced muscular weakness on exposure to lindane; most became dizzy and had stomach pains and nausea.

Control levels have been set in the UK and USA as TWA 0.5 mg m^{-3}; the WHO gives the maximum acceptable level in drinking water as 0.3 μg l^{-1}. At present the minimum detectable level is 0.02 ppm.

Codex Alimentarius: acceptable daily intake $8\,\mu g\,kg^{-1}$ (γ-isomer only, 1989). Maximum residue levels in the UK: meat 1.0, milk 0.008, potatoes 0.05, fruits $0.5-3.0\,mg\,kg^{-1}$.

Reference
Iwata, H.,Tanabe, S.,Sakai, N.and Tatsukawa, R.(1993) Distribution of persistent organochlorines in the oceanic air and surface seawater and the role of oceans on their global transport and fate, *Environ.Sci. Technol.*, **27**, 1080–1083.

LIQUIFIED PETROLEUM GAS

The principal components are:

	$CH_3CH_2CH_2CH_3$	CH_3CHCH_3 $\vert$ CH_3	$CH_3CH_2CH_3$
	n-butane	isobutane	propane
Mol. wt	58.1	58.1	44.1
CAS No.	106-97-8	75-28-5	74-98-6
B.p. (°)	−0.5 (101 kPa)	−12 (101 kPa)	−42
V.p. (kPa)	356 (38°)	498 (38°)	1310 (38°)
Flash point (°)	−138	–	−104
Water solubility ($\mu g\,l^{-1}$)	61	–	$65\,ml\,l^{-1}$ (18°)
Log P_{ow}	2.89	–	2.36

Liquified petroleum gas (LPG) is obtained from natural gas and also from the refining of crude oil. It was first produced in 1914 and was then marketed in 100 lb cylinders as fuel for domestic and industrial use; growth in production has gone hand- in-hand with bulk delivery through pressurized pipelines and transportation by road tanker (10 000 gal) and railway wagons (30 000 gal). Storage is at high pressure above ground in tanks holding some 180 000 gal and also below ground in refrigerated containment.

In the USA in 1991 $66 \times 10^6\,m^3$ were used for petrochemicals, $20 \times 10^6\,m^3$ for domestic and commercial heating and $29 \times 10^6\,m^3$ for blending motor fuels to raise their volatility. The domestic fuel Calorgas is composed largely of the less volatile butane fractions, otherwise the typical composition is 60% butanes –40% propane. Minor components of LPG include ethane, the isomeric butenes and *n*-pentane.

Propane is used in steam reforming to synthesize carbon monoxide and hydrogen:

$$C_3H_8 + 3H_2O \xrightarrow{800°} 3CO + 7H_2$$

and for the synthesis of propylene. The butanes are used principally for the blending of petrol. LPG is a propellant for aerosols.

No adverse effects have been found in animals or humans on exposure to propane, although a dizzy reaction has been observed in humans after a few minutes at the high level of 10^5 ppm. *n*-Butane is regarded as more toxic than propane with an LC_{50} inhalation in the rat of $660\,mg\,m^{-3}$ (277 ppm); it accounts for 30% of all solvent abuse by glue sniffers and is obtained by them from lighter refills.

The greatest risks from LPG are those of fire and explosion – $1\,m^3$ will vaporize to produce $250\,m^3$ of highly flammable gas. A third of all major losses in hydrocarbon processing arise from unconfirmed vapour cloud explosions (UVCEs). A large loss occurred at Romeoville, Illinois, on 23 July 1984 when a tower of 55 × 8.5 ft for processing propane was ruptured – the subsequent explosion broke windows 6 miles away and, with additional property damage by fire, losses were estimated at 10^8 at 1986 prices (Davenport, 1988). An even worse incident occurred in Mexico City on 19 November 1984 at a propane/butane storage facility. This had a capacity of 4×10^6 gal and was originally built in open country, but some 40 000 poorer people made their homes around it. When fire erupted, 500 residents living within 1000 ft of the centre were killed.

The ACGIH has set a TLV (TWA) for butane of 800 ppm (1900 mg m^{-3}). The UK long-term limit is 600 ppm (1430 mg m^{-3}).

Reference
Davenport, J.A. (1988) *J. Hazardous Mater.* **20**, 3–20.

LITHIUM CAS No. 7439-93-2

Li

Lithium, a member of Group 1 of the Periodic Table, has an atomic number of 3, an atomic weight of 6.94, a specific gravity of 0.534 at 20°, a melting point of 180.54°, and a boiling point of 1342°.Lithium is a silvery, very reactive metal forming compounds in the +1 oxidation state. Although several isotopes are known, only two occur naturally: ^{6}Li (7.5%) and ^{7}Li (92.5%).

The earth's crust on average contains about 20 ppm of lithium. Found in the minerals *lepidolite, spodumene* (LiAlSi$_2$O$_6$), *petalite* (LiAlSi$_4$O$_{10}$), and *amblygonite* [(*Li, Na*)*AlPO*$_4$(*F, OH*)] and lithium micas. Lithium is currently recovered from various brine deposits. The pure metal is produced by electrolytic reduction of the fused chloride. Typical concentrations in the environment are : oceans 180 ppb; river water 2.0 µg l^{-1}; soil ~50 mg kg^{-1}; plants and vegetables 0.01–30 mg kg^{-1} dry weight; blood < 5 µg ml^{-1} ; air concentration in manufacturing plant of lithium compounds 0.68–0.80 mg m^{-3}. Average total human body content 2 mg.

Uses. Heat transfer agent. Synthesis of organic compounds. Lithium batteries. Special glass and ceramics. Nuclear weapons.

Lithium is toxic to plants and animals. The toxicity depends on the solubility of lithium salts in water, and the anion. For example , 1.2–4.0 mg l^{-1} of lithium chloride is moderately toxic to plants, but severe toxic effects are observed in the concentration range 4–40 mg kg^{-1}. The uptake is lessened by the presence of calcium.

Lithium enters the human body by ingestion and primarily affects the mobilization of calcium into and out of cells by inhibiting certain enzymes in inisitol phosphate pathways. Lithium poisoning stimulates glycogenolysis, inhibits glycogenesis and affects the citric acid cycle. Symptoms of low stimulation response, diarrhoea, respiratory problems, etc., from poisoned animals also showed that the electrolytic balance in blood, internal organs and the CNS was altered. Cardiac and renal functions were also impaired. Severe lithium poisoning led to hypertension, epileptic seizures, convulsions, mental deterioration, coma and death. Hypothyroidism is one of the long-term side effects of lithium treatment.

Some lithium compounds, especially the carbonate, are used in psychiatry (600–1800 mg day^{-1}). The difference between therapeutic levels of lithium in blood

(1 mmol $Li^+ l^{-1}$) and toxic levels is small. Concentrations of 2 mmol l^{-1} are associated with toxic symptoms, 4 mmol l^{-1} can be fatal. Overdosing in the treatment of manic depression is the main cause of death from ingestion of lithium compounds.

Occupational exposure limit.   N.a. Lithium ignites in air.

Further reading
Domingo, J.L. (1994) Metal-induced developmental toxicity in mammals. A review, *J. Toxicol. Environ Health*, **42**, 123–141.
Kirk–Othmer ECT, 4th ed., Wiley, New York, 1995, Vol. 15, pp. 434–463.
Wennberg, A. (1994) Neurotoxic effects of selected metals, *J. Work. Environ. Health*, **20**, 65–71.

Lithium bromide CAS No. 7550-35-8

LiBr

Uses.   Antidepressant.

Toxicity. LD_{50} oral mouse 1840 mg kg^{-1}; intraperitoneal mouse 1160 mg kg^{-1}.

Lithium carbonate CAS No. 554-13-2

Li_2CO_3

Uses.   Pyrotechnics. Ceramics. Glass. Plastics. Flux. In the treatment of manic depressive psychosis; in the UK about 1 in every 1500 people have been treated.

Toxicity.   Human carcinogen. Acts as a strong base; the solution in water is very caustic. Over-exposure may lead to kidney and/or liver damage. LD_{50} oral rat 525 mg kg^{-1}; intraperitoneal mouse 236 mg kg^{-1}; oral dog 500 mg kg^{-1}.

Lithium chloride CAS No. 74477-41-8

LiCl

Uses.   Dry storage batteries. Flux.

Toxicity.   Human poison. Irritant. Can cause dermatitis. LD_{50} oral mouse 1165 mg kg^{-1}; LD_{Lo} oral human 200 mg kg^{-1} ; intraperitoneal cat 492 mg kg^{-1}.

Lithium hydroxide CAS No. 1310-65-2

LiOH

Uses.   Preparation of lithium salts.

Toxicity.   Highly toxic. Caustic. Causes burns. LC_{50} (4 h) inhalation rat 960 mg m^{-3}.

Lithium nitrate CAS No. 7790-69-4

$LiNO_3$

Uses.   Pyrotechnics. Rocket propellant. Ceramics.

Toxicity.   Irritant. Oxidizing. Harmful. Target organs blood and CNS.

Lithium peroxide CAS No. 12031-80-0

Li_2O_2

Uses.   Space technology; it absorbs carbon dioxide and liberates oxygen. Hardening agent for certain plastics.

Toxicity.   Oxidizing and corrosive. Harmful if injected. Irritant to skin and mucous membranes. Over-exposure can result in chemical pneumoconiosis or pulmonary oedema.

LUTETIUM CAS No. 7439-94-3

Lu

Lutetium, a member of the lanthanide series of elements, has an atomic number of 71, an atomic weight of 174.97, a specific gravity of 9.84 at 25°, a melting point of 1663° and a boiling point of 3395°. The stable oxidation state is +3. ^{175}Lu (97.4%) occurs naturally with ^{176}Lu (2.6% , $t_{1/2} = 3 \times 10^{10}$ years).

Lutetium is often present in yttrium minerals but *monazite* containing $\sim 0.003\%$ Lu is the commercial source. Relative abundance in the earth's crust is ~ 0.8 ppm. The pure metal is obtained by the reduction of anhydrous $LuCl_3$ or LuF_3 with sodium or calcium. The metal dust is a fire and explosion hazard.

Levels in human organs are very low. Lutetium and its compounds appear to be of low toxicity but since data are limited they should be handled with care.

Further reading
Gschneidner, K.A., Jr, and LeRoy, K.S. (eds), (1988) *Handbook on the Physics and Chemistry of the Rare Earths*, North-Holland, Amsterdam.
Kirk–Othmer ECT, 4th edn, Wiley, New York, 1995, Vol. 14, pp. 1091–1115.

Lutetium(III) hexachloride hexahydrate CAS No. 15230-79-2

$LuCl_3.6H_2O$

Uses.   Chemical agent.

Toxicity.   May be harmful by inhalation, ingestion or skin absorption. Causes skin and eye irritation. LD_{50} oral mouse 7100 mg kg^{-1}.

MAGNESIUM CAS No. 7439-95-4

Mg

Magnesium, a member of Group 2 of the Periodic Table, has an atomic number of 12, an atomic weight of 24.31, a specific gravity of 1.74 at 20°, a melting point of 649° and a boiling point of 1100°. The element consists of three natural isotopes: ^{24}Mg (78.99%), ^{25}Mg (10.00%) and ^{26}Mg (11.01%). The silvery white metal has a main oxidation state of +2.

Widely distributed in nature, the sixth most abundant element in the earth's crust, at 25 000 ppm. Most important minerals as a source of magnesium are *magnesite*, $MgCO_3$, *dolomite*, $MgCO_3.CaCO_3$, *brucite*, $Mg(OH)_2$, and *carnallite*, $KCl.MgCl_2.6H_2O$. Magnesium also occurs in many silicate minerals such as *olivine*, $(MgFe)_2SiO_4$ and *asbestos*, $(CaMg)_2SiO_4$. Magnesium chloride is present in sea water at ~ 1250 ppm and is a major source of the metal. The metal is extracted by electrolysis of molten salts. Typical environmental concentrations are marine algae 6–20 g kg^{-1}, plants 1–8 g kg^{-1}, fish and mammals 1 g kg^{-1} dry weight. Coal contains high magnesium concentrations.

Common foods have the following magnesium concentrations ($mg\,kg^{-1}$): nuts 2000, cereals 800, dairy products 180, fruits 70. An average adult ingests 240–480 $mg\,day^{-1}$ but intakes of 3.6–4.2 $mg\,kg^{-1}$ are considered adequate for healthy adults. The magnesium content of the human body ranges between 272 and 420 $mg\,kg^{-1}$ wet tissue. The total body content for a 70 kg human is about 35 g, with 60% residing in bone. Human blood contains 18–23 $mg\,l^{-1}$.

Uses.　Manufacture of lightweight alloys. Flares. Pyrotechnics. Incendiary bombs. Reducing agent in the production of other metals, e.g. titanium, uranium. Anti-corrosion coatings. Preparation of Grignard reagents.

Magnesium compounds are of variable toxicity, which is dependent on the anion. Magnesium is an essential element for plants (in the synthesis of chlorophyll) and animals. Many biological phosphates, including ATP, are associated with Mg^{2+} ions. Under normal circumstances magnesium shows low toxicity to all species but it is dependent on the anion.

Particles of metallic magnesium that perforate the skin or gain entry through cuts or scratches may produce a severe local lesion characterized by the evolution of gas and acute inflammation, frequently with necrosis, often referred to as 'chemical gas gangrene'. The lesion is slow to heal. The most serious industrial hazard of magnesium is fires with the danger of serious burns to workers.

In critical situations for humans, such as kidney failure or heart surgery, it is possible for blood magnesium levels to rise sharply through either ingested salts or intravenous overloading. Hypermagnesaemia may lead to prolonged drowsiness, decreased tone in muscles, bradycardia, decrease in blood pressure and finally cardiac arrest. Deficiency of magnesium in living organisms is a much more serious threat than toxicity. The characteristic symptoms for plants is chlorosis (insufficient chlorophyll production). Since magnesium is required by a wide range of metabolic enzymes involved in most fundamental body processes, lack of the element will produce a diverse list of clinical symptoms: gross muscular tremor, ataxia, tetany, hallucinations, confusion, delirium, low serum magnesium concentration, phlebothrombosis, etc. Treatment by intravenously administered magnesium sulphate (25–50 $mg\,kg^{-1}$ of body weight) is the usual therapy. Magnesium salts may be harmful by inhalation, ingestion or skin absorption. They may cause eye, skin and mucous membrane irritation.

Occupational exposure limits.　OSHA TLV (TWA) 15 $mg\,Mg\,m^{-3}$ total dust; respiratory fraction 5 $mg\,m^{-3}$.

Further reading
Kirk–Othmer ECT, 4th edn, Wiley, New York, 1995, Vol. 15, pp. 622–723.
King, R.B. (ed.) (1994) *Encyclopedia of Inorganic Chemistry*, Wiley, Chichester.
Birch, N.J. (1988) Magnesium, in: Seiler, H.G., Sigel H. and Sigel, A. (eds) (1988) *Handbook on the Toxicology of Inorganic Compounds*, Marcel Dekker, New York, pp. 397–403.

Magnesium(II) acetate tetrahydrate CAS No. 16674-78-5

$(CH_3CO_2)_2Mg.4H_2O$

Uses.　Dye fixative in textile printing. Deodorant. Disinfectant. Antiseptic.

Toxicity.　Poison by intravenous routes. LD_{50} intravenous mouse 111 $mg\,kg^{-1}$.

Magnesium(II) bromide CAS No. 7789-48-2; hexahydrate 13446-53-2

$MgBr_2$

Uses. Organic syntheses. Medicine (sedative).

Toxicity. N.a.

Magnesium(II) chloride CAS No. 7786-30-3; hexahydrate 7791-18-6

$MgCl_2$

Uses. Metallic magnesium production. Sorel cement.

Toxicity. LD_{50} oral rat $8100\,mg\,kg^{-1}$; oral mouse $7600\,mg\,kg^{-1}$.

Magnesium(II) fluoride CAS No. 7783-40-6

MgF_2

Uses. Ceramics. Glass. Single-crystal polarizing prisms. Lenses and windows.

Toxicity. Moderately toxic.

Magnesium(II) hydroxide CAS No. 1309-42-8

$Mg(OH)_2$

Uses. Medicinal, Milk of Magnesia.

Toxicity. LD_{50} oral rat $8500\,mg\,kg^{-1}$; intraperitoneal rat $2780\,mg\,kg^{-1}$.

Magnesium(II) nitrate hexahydrate CAS No. 13446-18-9

$Mg(NO_3)_2.6H_2O$

Uses. Pyrotechnics.

Toxicity. Severe irritant to eyes and skin. LD_{50} oral rat $5440\,mg\,kg^{-1}$.

Magnesium(II) oxide CAS No. 1309-48-4

MgO

Uses. Fire-resistant material. Additive to fertilizer and feed.

Toxicity. Inhalation may cause metal fume fever and leukocytosis in humans. Tumorigenic and possible carcinogen to laboratory animals. LD_{50} oral mouse $810\,mg\,kg^{-1}$.

Magnesium(II) sulphate CAS No. 7487-88-9; heptahydrate 10034-99-8

$MgSO_4$

Uses. Medicinal, Epsom Salts. In the anhydrous state as a drying agent.

Toxicity. LD_{Lo} intravenous woman $80\,mg\,kg^{-1}$.

Magnesium tungstate(VI) CAS No. 13573-11-0

$MgWO_4$

Uses. Fluorescent lighting. Solid-state masers.

Toxicity. May be harmful by all routes. Eye and skin irritant. Can cause dermatitis.

MALATHION CAS No. 121-75-5
(Karbofos, Mercaptothion).
Wide-spectrum insecticide, acaricide. Toxicity class III. LD_{50} rat $1200\,mg\,kg^{-1}$; LD_{50}
fish $0.1\text{-}0.3\,mg\,l^{-1}$; LD_{50} bee $0.71\,\mu g/bee$

MeO—P—S—CH—CO$_2$Et $C_{10}H_{19}O_6PS_2$ $M = 330.4$
       ‖
       S
   MeO   CH$_2$ CO$_2$Et

S-1,2-Bis(ethoxycarbonyl)ethyl-O,O-dimethylphosphorodithoate; m.p. 3°; b.p. 165°/
$0.7\,mmHg$; v.p. 4×10^{-5} Torr (30°); water solubility $145\,mg\,l^{-1}$ (25°).

The toxicity of impurities and transformation products may exceed that of
malathion itself, e.g.

MeS—P—S—CH—CO$_2$Et   MeO—P—S—CH—CO$_2$Et
       ‖   ‖
       O   O
   MeO   CH$_2$—CO$_2$Et   MeO   CH$_2$—CO$_2$Et

          Isomalathion   Malaoxon
      LD_{50} rat $113\,mg\,kg^{-1}$ LD_{50} rat $158\,mg\,kg^{-1}$

Malathion is largely detoxified in warm-blooded mammals through hydrolysis of
an ester group, but in pests the metabolism leads to the much more toxic malaoxon. It
has been found that malaoxon is formed to some extent during spraying and a level in
air of $62\,ng\,m^{-3}$ has been reported.

Malathion is used against pests on cotton, rice, potatoes, soft and stoned fruit,
vegetables and parasites of cattle, poultry, dogs, cats, human head and body lice and
stored grain. Its residues are widespread in water, soil, crops, animals and fish.

Maximum level in drinking water: UK $7\,\mu g\,l^{-1}$; EC $0.1\,\mu g\,l^{-1}$ (malathion). The
workplace limit in air is a TWA of $10\,mg\,m^{-3}$ (UK and USA).

Codex Alimentarius: acceptable daily intake $0.02\,mg\,kg^{-1}$ (1966). Maximum residue levels: wheat flour 2.0, root vegetables 0.5, citrus fruits $4.0\,mg\,kg^{-1}$. The levels in
the EC are similar but include malaoxon.

Indian government research showed that seven out of eight chilli samples had 100
times the permissible levels of malathion.

MANCOZEB CAS No. 8018-01-7
(Manzeb).
Fungicide. Toxicity class IV. LD_{50} rat $5\text{-}7\,g\,kg^{-1}$; LC_{50} fish $2\text{-}8\,mg\,kg^{-1}$ (48 h)

⁻S—C—NH.CH$_2$.CH$_2$.NH—C—S—Mn(Zn)⁺
    ‖   ‖
    S   S

Manganese ethylenebis(dithiocarbamate), a polymerized complex with the Zn salt;
m.p. 192–194° (dec.); water solubility $6\text{-}20\,mg\,l^{-1}$.

In addition to this mixed derivative of manganese and zinc, the dithiocarbamate
group includes the individual manganese derivative Maneb (CAS No. 12427-38-2)

and its zinc analogue Zineb (CAS No. 12122-67-7). Although this group is of low mammalian toxicity, they may give rise to ethylenethiourea, which does have adverse effects (see **Ethylenethiourea**).

Mancozeb is applied to field crops, fruit and vegetables. It may cause tumours and birth defects in animals at high dosage, but no standards have been set in the UK or USA.

Codex Alimentarius as dithiocarbamate. Maximum residue levels: potato 3.0 (EC 0.05) wheat 0.2, lettuce 5.0 (EC 5.0), tomato 3.0 mg kg^{-1}.

MANGANESE CAS No. 7439-96-5

Mn

Manganese, a member of Group 7 of the Periodic Table, has an atomic number of 25, an atomic weight of 54.94, a specific gravity of 7.20 at 20°, a melting point of 1246° and a boiling point of 2061°. Manganese is a reddish grey or silver, brittle, metallic element forming compounds in the oxidation states $+1, +2, +3, +4, +6$ and $+7$.

Natural manganese exists as only one isotope, ^{55}Mn. The isotope ^{54}Mn, whose sources are nuclear weapons testing and LW reactors, has a half-life of 312 days. Relatively abundant element with an abundance of ~ 1000 ppm of the earth's crust, the 12th most abundant element (only iron is a more abundant transition metal). Principal ores are *pyrolusite* (MnO_2) and *rhodochrosite* ($MnCO_3$). The metal is obtained by reduction of the oxide with sodium or aluminium. Anthropogenic sources of manganese particles are metal smelting and mining and sewage sludge. An estimated global amount of 10^6 t year^{-1} is released into the environment, so Mn is a common air contaminant. Typical concentrations in the environment are soil 500–1000 ppm, air 0.01–10 μg m^{-3}, drinking water 0.05 mg l^{-1}, sea water 0.03–0.8 μg l^{-1}, plants 1–700 mg kg^{-1} (ppm dry weight) (legumes, nuts, tea, cloves, etc. accumulate Mn). Daily adult human diet ~ 3 mg (RDA, 2.5–5.0 mg day^{-1}). Stored in the kidneys and liver, 1 mg kg^{-1}. Average total human body content (70 kg adult), 20 mg.

Uses. Alloys. Steel. Alkaline batteries. Ceramics. Glass. Dyes. Paints. Fertilizers. Antiknock agent for petrol engines.

Manganese is an essential trace element for all living creatures. At higher levels, Mn is toxic. Workers in manganese ore mills, smelting works, battery factories and manganese mines are vulnerable to Mn poisoning. Human systemic effects by inhalation of fumes and dust causing degenerative brain changes, changes in motor activity, muscle weakness and increase in incidence of upper respiratory infections and pneumonia. Exposure to high concentrations for as little as 3 months may produce the symptoms, but usually cases appear 1–3 years after exposure with the CNS most affected. Very few poisonings have occurred from ingestion. The neurotoxic effects of exposure to manganese can be acute and mainly psychiatric (known as manganese madness) or chronic with symptoms similar to Parkinson's disease, with muscular twitching varying from fine tremor in the hand to coarse rhythmic movements of the arms, legs and trunk, rigidity and bradykinesia (slow movements). Manganese increases dopamine autoxidation. This reaction, which seems to be irreversible, is probably one of the mechanisms behind the clinical picture of manganism with symptoms similar to Parkinson's disease. Apart from the effects on dopamine, manganese has also been shown to reduce levels of γ-aminobutyric acid and substance P in striatal cells, and also impairs energy metabolism by decreasing the adenosine-5-

triphosphate and increasing the lactate level. These findings indicate an impairment in oxidative metabolism. Manganese inhibits, in an irreversible way, glutathione-*S*-transferase, a finding which suggests that manganese can cause an interneural accumulation of cytotoxic compounds. The metal activates many enzyme reactions involved in the metabolism of organic acids, phosphorus and nitrogen. Manganese activates reduction of NO_2^- to ammonia. Also involved in photosynthesis. Constituent of some respiratory enzymes and other enzymes. It is a co-factor for the enzyme pyruvate carboxylase. Responsible for protein synthesis. In workers exposed to levels of less than $1\,mg\,m^{-3}$ ($18\,\mu mol\,m^{-3}$) an effect on performance was noted in several psychiatric tests. Some workers also showed an increase in neuropsychiatric disorders.

Deficiency of manganese in plants leads to diseases. In animals and birds, deficiency causes still births, low birth rates and growth impairment. Animals consuming a manganese-rich diet may show interference with haemoglobin formation. Excess of the metal in plants causes chlorosis.

Occupational exposure limit.   OSHA PEL (TWA) $1\,mg\,Mn\,m^{-3}$; STEL $3\,mg\,m^{-3}$. Manganese dust is flammable.

Toxicity.   A skin and eye irritant. Possible carcinogen. Experimental tumorigenic. Mutation data reported.

Further reading
ECETOX (1991) *Technical Report 30(4)*, European Chemical Industry Technology and Toxicology Centre, Brussels.
Environmental Health Criteria 17. Manganese, WHO/IPCS, Geneva, 1981.
Keen, C.L. and Leach, R.M. (1988) Manganese, in Seiler, H.G., Sigel, H. and Sigel, A. (eds), *Handbook on Toxicity of Inorganic Compounds*, Marcel Dekker, New York, pp. 405–415.
Kirk–Othmer ECT, 4th edn, Wiley, New York, 1995, Vol. 15, pp. 963–1055.
Savic, M. (1986) Manganese, in Friberg, L., Nordberg, G.F. and Voux, V.B. (eds), *Handbook on the Toxicology of Metals*, 2nd edn, 1986, Elsevier, Amsterdam, Vol. II, pp. 354–386.
SI 1991 No. 472 The Environmental Protection (Prescribed Processes and Substances) Regulations 1991, HMSO, London, 1991.

Manganese(II) acetate CAS No. 638-38-0; tetrahydrate 6156-78-1

$Mn(OOCH_3)_2$

Uses.   Oxidation catalyst. Textile dyeing. Paints and varnishes. Fertilizers. Animal food additive.

Toxicity.   Poison by ingestion. Mutation data. LD_{50} oral rat $3730\,mg\,kg^{-1}$.

Manganese(II) chloride tetrahydrate CAS No. 13446-34-9

$MnCl_2.4H_2O$

Uses.   Catalyst in chlorination of organic compounds. Paint drier. Dyeing industry. Fertilizers/feed additives.

Toxicity.   Poison. moderately toxic. Mutation, teratogenic and reproduction effects. LD_{50} oral rat $1484\,mg\,kg^{-1}$; intraperitoneal mouse $138\,mg\,kg^{-1}$.

Manganese(II) fluoride CAS No. 7782-64-1

MnF_2

Uses. Manufacture of special glass. Lasers.

Toxicity. Poison.

Manganese(IV) oxide CAS No. 1313-13-9

MnO_2

Uses. Dry cell batteries. Oxidizing agent for organic and inorganic compounds. Porcelain paint. Glass and enamels.

Toxicity. Poison. Neurological damage to central nervous system. LD_{50} subcutaneous mouse $422\,mg\,kg^{-1}$.

Manganese(II) nitrate CAS No. 10377-66-9; tetrahydrate 20694-39-7

$Mn(NO_3)_2$

Uses. Ceramics. Intermediate. Catalyst. Synthesis of MnO_2.

Occupational exposure limit. N.a. Fire and explosive risk with organic compounds.

Toxicity. N.a.

Manganese(II) sulphate monohydrate CAS No. 10034-96-5

$MnSO_4.H_2O$

Uses. Additive to fertilizers. Fertilizers. Paints and varnishes. Ceramics. Fungicide. Synthetic MnO_2. Catalyst.

Toxicity. Low toxicity.

(Methylcyclopentadienyl)manganesetricarbonyl CAS No. 12108-13-3

$MeCpMn(CO)_3$

Uses. Antiknock agent for petrol engines.

Toxicity. Poisonous by most routes.

MCPA CAS No. 94-74-6
(Metaxon, Agroxone).
Systemic herbicide. Toxicity class IV. LD_{50} rat $900\,mg\,kg^{-1}$; LC_{50} rainbow trout $232\,mg\,l^{-1}$

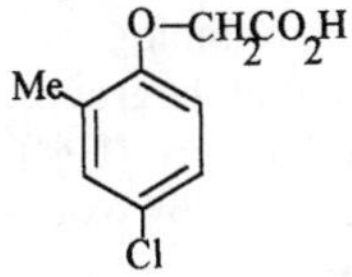

$C_9H_9ClO_3$ $M = 200.6$

4-Chloro-*o*-tolyloxyacetic acid; m.p. 119°; v.p. $2.3 \times 10^{-5}\,Pa$; water solubility $630\,mg\,l^{-1}$.

Its plant growth regulatory action was first reported in 1945 by Slade and it was introduced commercially in 1950 by Union Carbide. It was once the most commonly used phenoxyherbicide in the USA but has now been replaced by phosphorus compounds and carbamates; it was no longer detectable in leachate samples taken in 1992.

MCPA was used for the post-emergence control of broad-leaved weeds in cereals, rice and potatoes, under trees and on road verges. It is phytotoxic to vines, vegetables and cotton. It is degraded to 4-chloro-2-methylphenol, followed by ring hydroxylation and cleavage.

With 2,4-D and 2,4,5-T, it was responsible for widespread losses of broad-leaved crops and vegetables in the Tala valley of Natal. The damage arose from drifting spray and a ban on use was applied in 1989, however, legal action brought against 17 suppliers failed on the grounds that it should have been directed at the users.

The WHO recommends a maximum level in drinking water of $0.5 \, \mu g \, l^{-1}$.

MERCURY CAS No. 7439-97-6

Hg

Mercury, a member of Group 12 of the Periodic Table, has an atomic number of 80, an atomic weight of 200.59, a specific gravity of 13.55 at $20°$, a melting point of $-38.83°$, and a boiling point of $356.9°$. The element is a silver, heavy, mobile liquid and forms compounds in the $+1$ and $+2$ oxidation states. There are seven naturally occurring isotopes: ^{196}Hg (0.15%), ^{198}Hg (9.97%), ^{199}Hg (16.87%), ^{200}Hg (23.10%), ^{201}Hg (13.18%), ^{202}Hg (29.86%) and ^{204}Hg (6.87%).

The abundance of mercury in the earth's crust is about 0.08 ppm. It occurs mainly as native mercury and as HgS, *cinnabar*. Natural mercury arises from the earth's crust through volcanic eruptions, coal burning, smelting, chlor-alkali production plants and evaporation from oceans and amounts to more than $50\,000 \, t \, year^{-1}$. Typical concentrations in the environment are air $0-10 \, ng \, m^{-3}$, soil < 100 ppb and in UK ~ 40 ppb, fresh water 6 ppb $(ng \, l^{-1})$ (MAC drinking water 2 ppm), sea water < 1 ppm, plants and animals $1-100 \, \mu g \, kg^{-1}$. Some aquatic species can bioaccumulate mercury, as shown by the following bioconcentration factors: freshwater fish 63 000, marine fish 10 000 and marine invertebrates 100 000. Human daily intake (60 kg adult) $\sim 3 \, \mu g$; adult male (70 kg) 6 mg total body content.

Uses.   Wide applications in science, industry, agriculture and dentistry. Batteries. Electrical and control systems. Catalyst. $^{192}Hg(t_{1/2} = 5 \, h)$ is used in medical diagnostics.

A non-essential element, toxic to mammals. It is a cumulative poison because of its slow elimination from the body. Although it is not carcinogenic, mercury causes teratogenic and reproductive changes to laboratory animals. Mercury forms strong covalent bonds in biological systems, especially those containing sulphur, nitrogen and carbon. The toxicity of mercury derives from its ability to block biologically active molecules in proteins containing sulphydryl, amine, carboxy and other groups. Three major species of mercury occur in the environment: native mercury (Hg^0), Hg^{2+} and alkylmercury (RHg^+). The metabolism of mercury is achieved by oxidation/reduction between the three oxidation states.

The metal is fairly volatile even at ordinary temperatures; at $25°$ 1 m^3 of air holds 20 mg of mercury. Air saturated with mercury is very toxic. Acute human poisoning by inhalation of mercury vapour has usually occurred as a result of industrial accidents (fires in a mercury mine) or by gross violation of health and safety rules, e.g. electric welders with mercury-contaminated equipment. Acute mercury poisoning is characterized by a metallic taste, skin rashes, eye irritation, lung damage, nausea, diarrhoea and albuminuria. Chronic exposure to mercury at concentrations of $0.2–1.3 \, mg \, m^{-3}$ for 6 months showed the workforce to have deteriorated motor coordination, early onset of fatigue, impaired short-term memory and mercurial tremor. A blood mercury level of $0.2 \, \mu g \, g^{-1}$ produces neurological symptoms. About 90% of Hg in blood occurs in erythrocytes. Inorganic mercury salts are corrosive to the skin and mucous membranes, e.g chronic exposure by poisoning with mercury fulminate (an explosive) caused dermatitis.

Ingestion of salts causes pharyngitis, dysphagia, vomiting, abdominal pain, bloody diarrhoea, loosening of teeth, nephritis and hepatitis. Little absorption of liquid mercury in the stomach, but for inorganic salts ca 10% is absorbed. Mercury(II) is capable of being incorporated into the transfer DNA which plays a central role in protein biosynthesis. The most obvious genotoxic effect of inorganic mercury compounds is the induction of C-mitosis with inactivation of the mitotic spindle. Some evidence of carcinogenic activity of mercury(II) chloride, $HgCl_2$, in male rats. Transformation of mercury in soils and aquatic environments is initially to Hg^{2+}, followed by methylation to the highly toxic monomethyl- or dimethylmercury (usually anerobically).

Mercury and its compounds are absorbed by the roots and translocated. In excess, mercury becomes phytotoxic, causing stunted growth in roots and shoots (plant species dependent). Most of the mercury in fish occurs as $MeHg^+$, which is retained for years or until the fish is eaten. In humans, inhaled or ingested methylmercury readily penetrates biological membranes and concentrates in the liver, kidney and brain. In the foetal and young child brain it can cause irreversible damage to the central nervous system. Organomercurials, including methylmercury, have been shown to be about ten times more effective than $HgCl_2$ in inducing abnormal mitosis and single strand breaks in cellular DNA. No studies of the effects of high exposure levels of mercury or methylmercury on humans have been reported, but it is probably carcinogenic.

Mercury poisoning is a worldwide problem and some attempts have been made to reduce its use. For example, the amount of mercury in alkaline Mn batteries and zinc carbon batteries has been considerably reduced to a very small percentage. The dumping of mercury to landfill sites is banned in the USA. Alternative technology is available to replace the use of mercury in chlor-alkali plants.

Two of the most reported examples of mercury poisoning were due to the ingestion of $MeHg^+$. Between 1953 and 1960, a strange illness disabled 111 persons severely and killed 43 of them. In 1956, Minamata disease was discovered near Minamata Bay in southern Japan. It was shown that the disease was caused by the local population eating fish containing high levels of methylmercury chloride (CH_3HgCl). The source of the pollutant was identified as Chisso Corporation, a manufacturer of vinyl chloride who had discharged 200–600 t of Hg-polluted effluent into the Bay (Tsubaki and Irukayama, 1977). In 1964, a smaller but similar outbreak of Minimata disease was recognized in Niigata, Japan (26 poisoned, 5 died). In humans, the onset of the disease begins with numbness in the limbs and face, sensory disturbance and difficulties with hand movements. Progressively there is lack of coordination and speech, hearing and visual impairment followed by paralysis, finger deformity, convulsions and death. In Iraq, in the winter of 1971–72, ingestion of bread made from seed wheat treated with a $MeHg^+$ fungicide resulted in over 500 deaths, with over 60 000 exposed.

The use of organomercury compounds as insecticides and fungicides has led to several disastrous poisonings, some involving thousands of people. Other cases include: 1956, Iraq, over 100 poisoned from ethylmercury fungicide on seed, 14 died; 1960, Baghdad, 221 suffered fungicide poisoning; 1961, Pakistan, chronic Hg poisoning, 100 people, 4 died; 1969, New Mexico family, waste seed Hg poisoning; early 1960s, Sweden, drastic reduction in bird population due to methylmercury fungicide, with the use of this fungicide banned in 1966.

Mercury compounds may be fatal if inhaled, swallowed or absorbed through the skin. They may cause eye and skin irritations. Other symptoms include reproductive disorders and damage to the kidney and liver.

Mercury salts are used in skin-lightening creams and as antiseptics creams and ointments.

Occupational exposure limit. OSHA TLV (TWA) 0.05 mg Hg vapor m^{-3}, inorganic Hg salts 0.1 mg m^{-3}; MAC for drinking water 2 ppb.

Toxicity. LD_{Lo} oral man 1429 mg kg^{-1}. LD_{50} oral rat 166 mg kg^{-1}. LC_{50} (28 days) rainbow trout 0.005 mg l^{-1}.

Reference
Tsubaki, T. and Irukayama, K. (eds) (1977) *Minamata Disease, Methylmercury Poisoning in Minamata and Niigata, Japan*, Elsevier, Amsterdam, Tokyo.

Further reading
Berlin, M. (1986) Mercury, in Friberg, L., Nordberg, G.F. and Voux, V.B. (eds), *Handbook on the Toxicology of Metals*, 2nd edn, Elsevier, Amsterdam, Vol. II, pp. 387–435.
Boffetta, P., Merler, E. and Vainio, H. (1993) *Scand. J. Work. Environ. Health*, **19**, 1–7.
ECETOX (1991) *Technical Report 30(4)*, European Chemical Industry Technology and Toxicology Centre, Brussels.
Environmental Health Criteria 86. Mercury – Environmental Aspects, WHO/IPCS, Geneva, 1989.
Environmental Health Criteria 118. Inorganic Mercury, WHO/IPCS, Geneva, 1991.
IARC Monograph 58. *Mercury*, IARC, Geneva, 1991.
Kirk–Othmer ECT, 4th edn, Wiley, New York, 1995, Vol. 16, pp. 212–244.
SI 1991 No. 472 The Environmental Protection (Prescribed Processes and Substances) Regulations 1991, HMSO, London, 1991.
Wennberg, A. (1994) *Scand. J. Work Environ. Health*, **20**, 65–71.

Diethylmercury(II) CAS No. 593-74-8

$(C_2H_5)_2Hg$

Uses. Source of ethyl radicals. Ackylating agents.

Toxicity. LD_{50} oral rat 51 mg m^{-3}

Dimethylmercury(II) CAS No. 593-74-8

$(CH_3)_2Hg$

Uses. Source of methyl radicals. Ackylating agents.

Occupational exposure limit. USA TLV (TWA) 0.01 mg m^{-3}

Toxicity. Neurotoxic. LD_{50} oral rat 44 mg kg^{-1}.

Mercury(II) acetate CAS No. 1600-27-7

$Hg(OOCCH_3)_2$

Uses. Chemical intermediate and catalyst.

Occupational exposure limit. OSHA TLV (TWA) $0.025\,mg\,m^{-3}$.

Toxicity. Poison all routes. Mutation and reproductive data. LD_{50} oral rat $41\,mg\,kg^{-1}$.

Mercury(I) chloride CAS No. 10112-91-1

Hg_2Cl_2

Uses. Formerly used as a purgative and in teething powders for children. Repeated application has led to dementia, colitis, renal failure and death.

Toxicity. Poison. LD_{50} oral rat $210\,mg\,kg^{-1}$; LD_{50} oral human $5\,mg\,kg^{-1}$.

Mercury(II) chloride CAS No. 7487-94-7

$HgCl_2$

Uses. Catalyst in the manufacture of PVC. Wood preservation. Electroplating.

Toxicity. LD_{50} oral rat $1.0\,mg\,kg^{-1}$; LD_{50} oral human $29\,mg\,kg^{-1}$; LD_{Lo} oral human $86\,mg\,kg^{-1}$.

Mercury(II) nitrate monohydrate CAS No. 7783-34-8

$Hg(NO_3)_2.H_2O$

Uses. Analyical reagent. Nitration of aromatic organic compounds. Preparation of mercury fulminate, $Hg(CNO)_2$.

Toxicity. LD_{50} oral mouse $8\,mg\,kg^{-1}$.

Mercury(II) oxide CAS No. 21908-53-2

HgO

Uses. As cathode in dry cell battery.

Toxicity. Poison. LD_{50} oral rat $18\,mg\,kg^{-1}$; intraperitoneal mouse $4500\,\mu g\,kg^{-1}$.

Mercury(II) sulphate CAS No. 7783-35-9

$HgSO_4$

Uses. Catalyst in aldehyde synthesis. Catalyst in the dyeing industry. Battery electrolyte.

Toxicity. LD_{50} oral rat $57\,mg\,kg^{-1}$; oral mouse $25\,mg\,kg^{-1}$.

Methylmercury(II) choride CAS No. 22967-92-6

CH_3HgCl

Uses. RHgX-type compounds are used in the manufacture of urethane, vinyl acetate, fungicides and slimicides.

Toxicity. LD_{50} oral rat $10\,mg\,kg^{-1}$; LD_{50} oral human $5\,mg\,m^{-3}$. Symptoms of human toxicity are effects on the brain, sensory organs, vision and speech and involuntary movement.

METHANE CAS No.74-82-8

CH_4 $M = 16.0$

B.p. $-162°$; water solubility $24\,mg\,l^{-1}$; soluble in organic solvents.

The major natural source is natural gas which provides the methane for the chemical industry. This is largely converted into 'synthesis gas' by desulphurization, mixing with excess steam and passage over a nickel catalyst at $800°$ and 35 atm:

$$CH_4 + H_2O \longrightarrow CO + 3H_2$$
(excess)

This can be varied by the introduction of N_2 and removal of CO, leading to the synthesis of ammonia; this is the most important outlet for synthesis gas and consumes 5% of world natural gas supplies. Methanol, formaldehyde and acetic acid are also important products obtained from synthesis gas (Witcoff and Reuben, 1996).

Leakage of natural gas from industry is a major source of airborne methane, but significant quantities are released from rice paddy fields, landfill sites, anaerobic fermentation in swamps and rain forests and the voiding of cattle. Emissions from rice fields in the Szechzuan Province of China release methane at the rate of $60\,mg\,m^{-2}\,h^{-1}$. In the UK, methane emissions from sites adjacent to, or even below, building sites have caused serious explosion damage. In the village of Loscoe, Derbyshire, a bungalow built over a former tip was destroyed by a methane gas explosion. A housing estate of 15 000 units was built in Berkshire on the former Woodley Airfield in 1991, but in 1996 some of these properties bordering a dump site were threatened by methane release. These incidents are especially serious for householders because of the legal difficulties in apportioning blame between Local Authorities and others.

Diesel exhaust contains 17% of methane and petrol engines emit 24%, while municipal incinerator emissions contain 1.6–10 ppm of methane. It is important to control methane transfers to air as it is a significant greenhouse gas. Although carbon dioxide is more abundant (345 ppm) than the normal methane level (1.7 ppm), the latter has a greater potential for global warming. This is because its IR absorption maximum is near $3000\,cm^{-1}$, which coincides with a window where the effect of carbon dioxide and water vapour are minimal.

Another undesirable property of methane from an environmental point of view is its long lifetime, over 10 years. It reacts very slowly with ozone and the reaction with OH radicals has a half-life of 7000 days, this reaction is also important because it produces oxidized species which contribute to urban smog:

$$^{\bullet}OH + CH_4 \longrightarrow {}^{\bullet}CH_3 + H_2O$$

This sequence is typical of the alkanes as a group and homologues of methane give rise to higher aldehydes and thence to peroxynitrates (see **PAN**).

Reference

Witcoff, H.A. and Reuben, B.G. (1996) *Industrial Organic Chemicals*, Wiley, New York.

Further reading
Harrison, R.M. (1996) in *Pollution, Causes, Effects and Control*, 3rd edn, Royal Society of Chemistry, Cambridge, pp. 174–179.

METHANETHIOL CAS No. 74-93-1
LC_{50} inhalation mice $6.5 \, mg \, m^{-3}$

CH_3SH CH_4S $M = 48.1$

Methyl mercaptan; b.p. 60°; v.p. 1520 Torr (26°); dens. 0.866; v. dens. 1.66; pK_a 10.33. Solubility in water $23 \, g \, l^{-1}$ (20°) and forms a crystalline hydrate; soluble in diethyl ether, EtOH, MeOH. Highly flammable with an odour of rotting cabbage, threshold $3 \times 10^{-3} \, mg \, m^{-3}$ (1.5 ppb). The threshold value in water, commonly sewage effluent, is 0.02–2 ppb; the odour is most noticeable when sewage digestion is incomplete.

It is manufactured by the action of sodium hydrosulphide on dimethyl sulphate:

or by reaction between NaSH and methyl chloride.

Methanethiol occurs naturally in higher plants, e.g. radish and algae. It occurs in the sea along with other volatile sulphur compounds notably dimethyl sulphide, but also H_2S, COS, CS_2 and dimethyl disulphide.Wine contains methanethiol and its oxidation products dimethyl disulphide and dimethyl sulphide (up to $400 \, \mu g \, l^{-1}$, Rapp and Pretorius, 1990). The natural source of methanethiol in wine is methionine but it may taint the product by hydrolysis of the pesticide Orthene (otherwise known as Acephate), with associated loss of revenue.

orthene

In the reverse reaction methionine is obtained from MeSH for use as an animal feed. Other uses are as an odorant for gases and in the organic synthesis of methylthio compounds:

ametryn

a thiodicarbonate including the
methylthioethylideneamino group

The main odorants in sewage treatment (Bonnin *et al.*, 1990) are hydrogen sulphide, alkanethiols and dialkyl sulphides at the stages of thickening, thermal processing and de-watering, as indicated in the following table (all concentrators in $mg \, m^{-3}$)

	MeSH	EtSH	MeSMe	H$_2$S
At a covered thickener: mean	0.21	< 0.03	< 0.03	1.59
max.	1.17			6.68
Transported sludge: mean	0.05	< 0.03	< 0.03	0.013
max.	0.08			0.013

The toxic effects of methanethiol are similar to those of hydrogen sulphide but less severe. Low level exposure at $0.005\,\mathrm{mg\,m^{-3}}$ for 6 months led to changes in lung and heart function; higher doses cause narcosis and convulsions and human fatalities are on record.

The ACGIH has set the TLV (TWA) at 0.5 ppm ($0.98\,\mathrm{mg\,m^{-3}}$). The UK long-term limit is also 0.5 ppm.

Reference
Bonnin, C., Laborie, A. and Paillard, H. (1990) Odour nuisances created by sludge treatment: problems and solutions, *Water Sci. Technol.* **22**, 65–74.
Rapp, A. and Pretorius, P.J. (1990) *Foreign and Undesirable Flavours in Wine, Developments in Food Science*, Vol. **24**, Elsevier, Amsterdam.

Further reading
Whitham, G. (1995) *Organosulphur Chemistry*, Oxford University Press, Oxford.

METHANOL CAS No. 67-56-1
LD_{50} rat $5.6\,\mathrm{g\,kg^{-1}}$; LC_{50} (48 h) trout $8\,\mathrm{g\,l^{-1}}$

CH₃OH CH_4O $M - 32.0$

Methyl alcohol; b.p. 64.5°; v.p. 92 Torr (20°); dens. 0.79; v. dens 1.11; $\log P_{ow} - 0.77$; odour threshold $5\,\mathrm{mg\,m^{-3}}$.

First identified in 1661 by Boyle; a product from the distillation of wood. Methanol is flammable and miscible with water and most organic solvents; it forms many constant-boiling mixtures.

It can be obtained from lignitic coal, although the price is not at present competitive. The principal route is from synthesis gas, which is obtained by the steam reforming of hydrocarbons at about 800° and 35 atm. Synthesis gas is often obtained from methane:

$$CH_4 + H_2O \longrightarrow CO + 3H_2$$

but almost any hydrocarbon fraction can be used:

$$C_nH_m + nH2O \longrightarrow nCO + (n + m/2)H_2$$

then for methanol:

$$CO + 2H_2 \longrightarrow CH_3OH$$

When catalysed by Zn/Cr the reaction proceeds at 200–400° and 100 atm, but it is more economical to use a $Cu/Zn/Al_2O_3$ catalyst when at a temperature in the range 210–270°, a lower pressure of 50 atm is sufficient.

Methanol is used as a solvent for nitrocellulose, ethylcellulose, natural and synthetic resins and inorganic salts and as an anti-freeze. Major quantities are required for the manufacture of formaldehyde and as a methylating agent – here demand has escalated since 1990 owing to the increased use of methyl *tert*-butyl ether(MTBE) (CAS No. 1634-04-4) in motor fuels. Methanol was 21st in importance in the USA in 1994 when production was 10.8×10^9 lb; world production (1992) was 19.6×10^9 lb. The principal outlets were formaldehyde 38%, MTBE 19%, acetic acid 7%, solvents 4%, dimethyl terephthalate 3% and miscellaneous 29%.

The symptoms of methanol poisoning are similar to those of ethanol intoxification but milder; they include loss of judgement, slurred speech and visual impairment. Ingestion of methanol is life threatening and causes metabolic acidosis, rapid and shallow breathing and severe abdominal pain. As little as 8 g can cause blindness and the adult lethal dose is about 30 ml. Poisoning can follow intake through the skin. Methanol is metabolized to formaldehyde and formic acid in the liver and kidneys, but this is much slower than that of ethanol, some methanol being eliminated unchanged through the lungs and urine. Adult deaths have been recorded after exposure for 12 h at 4000–13 000 ppm. In the wood heel industry there was no evidence of injury to workers at levels of 160–780 ppm (213–1040 mg m^{-3}); vision is affected at levels in the range 1200–8300 ppm (1600–11400 mg m^{-3}). Methanol occurs in cigarette smoke at a level of 700 ppm and in the exhaust of a petrol engine at 100–600 ppb.

Methanol reacts with nitrogen peroxide in polluted air to form methyl nitrite-MeONO, which accelerates ozone formation in smog. This reaction can be put to good use since alkyl nitrites are more readily sensed by ECD. GC on 10% tricresyl phosphate will detect methanol as nitrite at 3 ppb and ethanol as EtONO at 0.6 ppb. GC analysis of methanol is important because of its use in gasoline; a level of 2 µg ml^{-1} in a 200 µl sample can be detected by FID using Carbowax supported on the bonded phase of a capillary column (Pollak and Kawagee, 1991). Methanol in wastewater may be determined by direct injection on to an Apiezon L column at 97° (Horna, 1988).

The ACGIH has set a TLV (TWA) skin at 200 ppm (262 mg m^{-3}).

References

Horna, A. (1988), Determination of methanol, *n*-butanol and toluene by direct aqueous injection chromatography, *J. Chromatogr.*, **457**, 372.

Pollak, G. and Kawagee, J.L. (1991) Determination of methanol in whole blood by capillary GC with direct on-column injection, *J. Chromatogr.*, **570**, 406–411.

METHOMYL CAS No. 16752-77-5

(Lannate, Methavin)

Systemic insecticide, acaricide. Toxicity class II. LD$_{50}$ rat 17 mg kg^{-1}; LC$_{50}$ rainbow trout 3.4 mg l^{-1}. Toxic to bees

CH$_3$.NH.C-O-N═C‹SMe ‹Me   C$_5$H$_{10}$N$_2$O$_2$S $M = 162.2$

S-Methyl-*N*-(methylcarbamoyloxy)thioacetimidate, a mixture of geometrical isomers including mainly the *Z*-form; m.p. 79°; v.p. 6.6 mPa (25°); water solubility 58 g l^{-1}.

Introduced by E.I. du Pont de Nemours, it is a cholinesterase inhibitor for the control of insects and spider mites on fruit, vines, olives, vegetables, cotton, tobacco, etc., and also for flies in animal houses. It is rapidly metabolized to acetonitrile and carbon dioxide.

Methomyl use requires Prior Informed Consent and it has been involved in incidents of misuse in Egypt where edible fruits were grown in the restricted space of plastic housing and tunnels. Aerial spraying at high doses has led to immunity in the pink bollworm, white fly and aphids. It is very toxic if swallowed or if the dust is inhaled. Workplace limits in air are set in UK and USA at TWA $2.5\,mg\,m^{-3}$

Codex Alimentarius acceptable daily intake $0.03\,mg\,kg^{-1}$. Maximum residue levels: wheat 0.5, potatoes 0.1, pineapple 0.2, sugar beet $0.1\,mg\,kg^{-1}$.

METHYL BROMIDE CAS No. 74-83-9

Soil sterilant, fumigant. Toxicity class II. LD_{50} rat $214\,mg\,kg^{-1}$; LC_{50} (2 h) inhalation, mouse $1.54\,g\,m^{-3}$; LC_{50} (96 h) fish $11-12\,mg\,l^{-1}$

CH_3Br $M = 94.9$

Bromomethane, Brom-o-gas, Desbrom, MBR-2, Terr-o-gas; b.p. 3.6°; v.p. (20°) $1893\,kPa$ (1420 Torr); dens. (20°) 3.97; v. dens. 3.27; log P_{ow} 1.19; Henry's law constant $0.533\,kPa\,m^3\,mol^{-1}$; UV maximum 202 nm. Water solubility $18\,g\,l^{-1}$ at 20°; below 4° it forms a crystalline hydrate, $CH_3Br.20H_2O$; freely soluble in most organic solvents.

Methyl bromide is produced by the reaction of methanol with hydrogen bromide, a reaction which is reversed in aqueous systems:

$$CH_3Br + H_2O \longrightarrow CH_3OH + H^+ + Br^-$$

In the pH range 3.0–8.0 this reaction is of S_N1 type; at pH 7.0 it has a half-life of 12 days.

World production for 1990 was 66 640 t but production losses are modest, 1000 kg in the USA annually, when compared with those from use as a fumigant. When applied 28 cm deep in soil at $50-125\,g\,m^{-2}$ almost all is lost to the air and it was estimated that in 1973 in the USA this amounted to $11 \times 10^6\,kg$. 2% of lachrymatory chloropicrin is added to enhance detection, otherwise the threshold is $65\,mg\,m^{-3}$.

The insecticidal properties were first reported in 1932 by Le Goupil. In addition to soil fumigation, it is used for books and buildings and for food crops, coffee, cocoa, timber, tobacco, etc., besides the protection of food in storage and during transportation in ships, trains and trucks.

It is formed naturally by ocean algal synthesis and is found in coastal waters with methyl chloride and methyl iodide, natural sources account for 60–80% of the total in the environment. Methyl bromide can diffuse through plastics and so enter a water body.

In humans acutely exposed at a level of $390-1950\,mg\,m^{-3}$ (100–550 ppm), non-fatal poisoning leads to adverse reactions of the nervous system, lungs, eyes, skin and kidneys. Human fatalities have occurred at levels above 8000 ppm; seven deaths were recorded in California during 1951–87. Usage is growing in Costa Rica where a worker was poisoned in a water melon plantation, but in the USA and the EC methyl bromide is gradually being withdrawn from use, although there is no single substitute – diethyl ether can be used as a fumigant, malathion for stored products and chloropicrin for soils.

Current concern about the use of methyl bromide arises because the fraction which reaches the stratosphere is photolysed to give bromine atoms, and owing to their longer lifetime these are 50 times more potent than chlorine atoms in destroying ozone.

The limit in workplace air is set at a TLV (TWA) of $20\,mg\,m^{-3}$ in the USA and the UK. The EPA limit for drinking water is $50\,\mu g\,l^{-1}$.

The *Codex Alimentarius* gives the following maximum residue levels of inorganic bromine in foods: cucumber 50, lettuce 100, tomato $75\,mg\,kg^{-1}$.

Further reading

Dinham, B. (1993) *The Pesticide Hazard*, The Pesticide Trust, ZED Books, London.

Environ. Health Criteria, World Health Organization, Geneva, 1995, Vol. 166.

METHYL ETHYL KETONE CAS No. 78-93-3
LD_{50} rat $800\,mg\,kg^{-1}$; LC_{50} (96 h) fish 2–$5.6\,g\,l^{-1}$

$$CH_3-C-CH_2CH_3 \qquad\qquad C_4H_8O \quad M = 72.1$$
$$\underset{O}{\overset{\|}{}}$$

2-Butanone, MEK, methylacetone; b.p. 80°; v.p. (20°) 77.5 Torr; v. dens 2.41; log P_{ow} 0.26. Water solubility $275\,g\,l^{-1}$; soluble in common organic solvents.

It has an odour like acetone and the threshold is 2 ppm ($5.9\,mg\,m^{-3}$). A natural flavour component in a range of foods, including cheese and fermented products, it has been identified in cigarette smoke and in volatile emissions from vetch and clover.

Methyl ethyl ketone is produced by the dehydrogenation of sec-butanol, typically at 400–550° with a ZnO catalyst, or by the oxidation of butane. In 1987 production in the USA was $305 \times 103\,t$ and in Western Europe $215 \times 10^3\,t$.

Use depends largely on its excellent solvent properties for protective coatings (65% in the USA), adhesives (15%) and magnetic tape (8%); also for de-waxing and as a chemical intermediate. The majority escapes to the atmosphere.

MEK is detectable in air by HPLC with UV detection at a level of $3\,\mu g\,m^{-3}$. It can be determined in water by direct-injection GC with FID to $40\,\mu g\,l^{-1}$.

The oral toxicity towards mammals is low. In humans severe toxic effects have followed 'glue sniffing' or 'huffing' the vapours of glues and lacquer thinners. Chronic huffers inhale fumes from as much as $500\,ml\,day^{-1}$ over sessions of several hours. The consequences include atrophy of muscles, weakness in the legs, damage to teeth and loss of weight. MEK potentiates the neurotoxic effects of *n*-hexane, when motor and sensory nervous response times are reduced, remission may take up to a year and be incomplete in serious cases.

Instances of occupational exposure have shown that MEK also potentiates the neurotoxic effects of methyl *n*-butyl ketone. In 1973, 1157 workers in an Ohio, USA, fabric factory were examined and 86 had symptoms of peripheral neuropathy, parathesia in the arms and legs and weakness in the hands and legs. MEK levels ranged between 85 and 760 ppm and MBK between 9 and 640 ppm.

The USA has set a TLV (TWA) at 200 ppm ($590\,mg\,m^{-3}$) and a STEL of 300 ppm.

METHYL ISOCYANATE CAS No. 624-83-9
LD_{50} rat $51\,mg\,kg^{-1}$

$$CH_3-N=C=O \qquad\qquad C_2H_3NO \quad M = 57.1$$

Isocyanatomethane, MIC, isocyanic acid methyl ester; b.p. 39°; v.p. (20°) 400 Torr; dens. 0.97. Very flammable, with a sharp unpleasant odour.

The most commonly used commercial production for isocyanates is the reaction of the corresponding amine with phosgene:

It is used principally in the production of insecticides and herbicides, e.g. aldicarb and carbaryl, and to some extent for flexible polyurethane foams (see **Toluene diisocyanates**). US output in 1984 was 35×10^6 lb.

Owing to its volatility and toxicity, the major risk from MIC is the inhalation of vapour, but skin penetration has caused death in animals. Mild exposure leads to irritation of the eyes, nose and throat. More severe exposure leads to bronchial irritation and breathing difficulties – these symptoms may only appear several hours after contact.

MIC was responsible for the worst recorded industrial accident at the Union Carbide plant at Bhopal on 3 December 1984, when a large storage tank discharged the vapour through the pressure relief valve. The excess pressure was caused by the entry of several tons of chloroform and water, which led to the release of gaseous products – methylamine and carbon dioxide:

A shanty town had been allowed to develop around the plant and its inhabitants were the worst affected – at least 2000 people died from lung failure and about 200 000 became ill. The incident was blamed on poor maintenance; the refrigeration system was not working and the scrubber was shut down, so permitting the escape of vapour. A related incident at the Union Carbide plant in the Kanawha Valley, West Virginia, was reported in 1985. This lead to the release of aldicarb oxime (2800 lb) and methylene chloride (700 lb) and 135 people were treated for irritation of the eyes, throat and lungs.

MIC is not carcinogenic but it can sensitize the bronchial tract in humans, when further exposure to low concentrations can cause asthmatic symptoms.

The ACGIH has set the TLV (TWA) at 0.02 ppm ($0.047 \, mg \, m^{-3}$).

Further reading
Kirk–Othmer ECT, 4th edn, Wiley, New York, 1996, Vol. 14, pp. 902–934.
Kletz, T. (1994) *What Went Wrong?*, 3rd edn, Gulf, Houston, TX, pp. 303–305.
Worthy, W. (1986) *Chem. Eng. News*, **64**, 9.

METOLACHLOR CAS No. 51218-45-2
(Dual, Pennant)
Herbicide. Toxicity class III. LD_{50} rat 2780 mg kg^{-1}; LC_{50} rainbow trout 2 mg l^{-1}

Me Me
CH-CH$_2$-OMe
N-C-CH$_2$-Cl $C_{15}H_{22}ClNO_2$ $M = 283.8$
Et O

2-Chloro-6'-ethyl-N-(2-methoxy-1-methylethyl)acet-o-toluidide;   b.p.100°/0.001 mmHg; v.p. 1.3×10^{-5} Torr (20°); water solubility $530 \, \text{mg} \, l^{-1}$ (20°).

Introduced by Ciba-Geigy and of major use in the USA on corn, soybeans, sorghum, sugar, potatoes with recent consumption in 14 States of the corn belt running at $17 \times 10^6 \, \text{kg year}^{-1}$.

The half-life in soil is about 20 days and following spring application the levels in storm run-off can reach $40 \, \mu\text{g} \, l^{-1}$ with a median of $1.3 \, \mu\text{g} \, l^{-1}$; in early summer local river levels can reach $6 \, \mu\text{g} \, l^{-1}$. A typical level is $160 \, \text{ng} \, l^{-1}$ in the main Mississippi, which drains some 40% of the whole USA and, flowing at a rate of $18\,499 \, \text{m}^{-3} \, \text{s}^{-1}$, carried about 200 t of metolachlor into the Gulf of Mexico (see **Cyanazine**).

In 1990 in Iowa 72% of all distributed herbicides consisted of alachlor, atrazine, cyanazine and metolachlor. The total usage of active compounds in that year was 22 000 t. Levels in May–June in surface water exceeded $1 \, \mu\text{g} \, l^{-1}$.

The WHO has set a maximum level for drinking water of $5 \, \mu\text{g} \, l^{-1}$; Canada specifies $50 \, \mu\text{g} \, l^{-1}$. No air standards have been set.

Further reading

Pereira, W.E. and Hostettler, F.D. (1993) Non-point source contamination of the Mississippi and its tributaries by pesticides, *Environ. Sci. Technol.*, **27**, 1542–1552.

METRIBUZIN CAS No. 21087-64-9
(Sencor, Lexone)
Selective systemic herbicide. Toxicity class III. LD_{50} rat $2000 \, \text{mg} \, \text{kg}^{-1}$; LC_{50} rainbow trout $64 \, \text{mg} \, l^{-1}$

O   NH$_2$
N
Me
Me-C-   -SMe $C_8H_{14}N_4OS$ $M = 214.3$
Me   N-N

4-Amino-6-*tert*-butyl-4,5-dihydro-3-methylthio-1,2,4-triazin-5-one; m.p. 126°; v.p. 0.06 mPa (20°); water solubility $1.2 \, \text{g} \, l^{-1}$.

Introduced in 1973 by Bayer and used for pre- and post-emergence control of grasses and broadleaved weeds in soyabeans, potatoes, sugar cane and cereals. It is cytotoxic to many other crops.

Total annual usage on soybeans in the USA (1990) was $5 \times 10^5 \, \text{kg}$. After spring application runoff levels in June in the Charles river, Iowa, reached $2 \, \mu\text{g} \, l^{-1}$. The maximum in storm water was $7.6 \, \mu\text{g} \, l^{-1}$, with median levels in the Mississippi at $13 \, \text{ng} \, l^{-1}$.

Metribuzin is rapidly photolysed in water with a half-life in ponds of about 7 days. Maximum levels in drinking water permitted in the USA and Australia are 175 and $5 \, \mu\text{g} \, l^{-1}$, respectively.

MEVINPHOS CAS No. 26718-65-0
(Phosdrin)
Broad-spectrum systemic insecticide, acaricide. Toxicity class I. LD_{50} rat $3\,mg\,kg^{-1}$; LC_{50} rainbow trout $17\,\mu g\,l^{-1}$

$C_7H_{13}O_6P$ $M = 224.1$

Methyl 3-(dimethoxyphosphinoyloxy)but-2-enoate; m.p. (E)-isomer 21°, (Z)-isomer 7°; v p 17 mPa (20°); miscible with water.

Introduced by the Shell Chemical, the technical product contains 60% (E)- and 20% (Z)-isomer. The minimum detectable level is 0.1 ppm.

Used for the control of chewing insects and spider mites on stone fruit, soft fruit, water melons, vegetables, potatoes, cotton, etc. It is rapidly hydrolysed to less toxic products, including dimethyl phosphate and phosphoric acid. No control standards have been set in water.

The limit in workplace air in the UK and USA (NIOSH) is a TWA of $0.1\,mg\,m^{-3}$.

Codex Alimentarius: acceptable daily intake $0.0015\,mg\,kg^{-1}$ (1972). Maximum residue levels (sum of both isomers) leaf vegetables 0.5-1.0, citrus fruit 0.2, other fruit $0.2-1.0\,mg\,kg^{-1}$.

MIREX CAS No. 2385-85-5
(Dechlorane)
Insecticide, fire retardant. LD_{50} rat $235\,mg\,kg^{-1}$; LC_{50} fish $0.2-30\,mg\,kg^{-1}$

$C_{10}Cl_{12}$ $M = 545.6$

Dodecachlorooctahydro-1,3,4-metheno-1H-cyclobuta(cd)pentalene; m.p. 485° (dec.); water solubility $70\,ng\,l^{-1}$ (24°).

Discovered in 1940 and first produced industrially at Niagara Falls (NY) in 1959.

It was introduced in south-eastern USA in 1958 to eradicate the fire ant, when 1.4×10^7 acres were treated up to 1965. It has also been used on leaf cutter ants (South America) and for mealybugs on pineapples (Hawaii).

The detection limit in water is $0.3\,pg\,l^{-1}$. The input into the Niagara river was $10\,kg\,year^{-1}$ during 1985-91, but levels declined from a peak of $200\,pg\,l^{-1}$ (1966) to

$25 \, \text{pg} \, \text{l}^{-1}$ (1990). Mirex has reached Grand Banks, Newfoundland via the St Lawrence river.

It is very persistent and the biomagnification is exceptionally high, being 12 000 in algae and over 30 000 in mussels. It has been detected in human milk in the range 0.2–$6 \, \text{mg} \, \text{kg}^{-1}$ and in the fat of residents in south eastern USA at an average level of $1.3 \, \text{mg} \, \text{kg}^{-1}$. Levels in human adipose tissue in south-eastern USA: 1971–72, 0.16–5.94, average $2.499 \, \text{mg} \, \text{kg}^{-1}$; 1975–76 average $0.29 \, \text{mg} \, \text{kg}^{-1}$.

No national standards have been set but owing to its persistence and bioconcentration it use was discontinued in 1976. However, residues were found in Venezuelan birds and their eggs in 1991 and more recently it was reported that levels in sediment in Casco Bay, Maine, were in the range of 0.04–0.66 ppb.

Further reading

Kutz, F.W., Wood, P.H. and Bottimore, D.P. (1991) Organochlorine pesticides and PCBs in human adipose tissue, *Rev. Environ. Contam. Toxicol.*, **120**, 55–61.

MOLYBDENUM CAS No. 7439-98-7

Mo

Molybdenum, a member of Group 6 of the Periodic Table, has an atomic number of 42, an atomic weight of 95.94, a specific gravity of 10.2 at 20°, a melting point of 2622° and a boiling point of 4639°. Molydenum is a cubic, silver-white metal forming compounds in the +2 to +6 oxidation states, but the +4 and +6 states predominate. The naturally occurring isotopes are ^{92}Mo (14.84%), ^{94}Mo (9.25%), ^{95}Mo (15.92%), ^{96}Mo (16.68%), ^{97}Mo (9.55%) and ^{98}Mo (24.13%) and ^{100}Mo (9.63%).

The crustal abundance of Mo is usually in the range 1–2.3 ppm, 53rd in relative order of abundance. It is widely distributed in nature. The most important ore is *molybdenite*, MoS_2, but it is also obtained as a by-product from copper and tungsten mining operations. The pure metal is obtained by hydrogen reduction of ammonium molybdate, NH_4MoO_4.

Coal burning in power stations is a source of molybdenum in air; coal contains 35–300 ppm of molybdenum. It is claimed that around $1000 \, \text{t} \, \text{year}^{-1}$ of Mo are released into the atmosphere through the burning of fossil fuels. Molybdenum is concentrated in sewage sludge ($\sim 15 \, \text{ppm}$). Molybdenum is found in soils 0.5–6 ppm; drinking water < 1 ppb; sea water 10 ppb ($10 \, \mu\text{g} \, \text{l}^{-1}$) as MoO_4^{2-}; milk $0.05 \, \text{mg} \, \text{kg}^{-1}$; ordinary plants 0.1–0.3 ppm dry weight, but some plants bioacccumulate Mo resulting in much higher values; fish $400–800 \, \text{ng} \, \text{g}^{-1}$ dry weight; adult human diet $\sim 100 \, \mu\text{g} \, \text{day}^{-1}$. Average total human body content (adult, 70 kg) < 5 mg.

Uses. Pigments/paints. Fuel additives/lubricants. Alloy with steel replacing toxic chromium. Fertilizers.

Essential to plants and animals, particularly for the growth of plants. Several important molybdenum-containing hydroxylases are known, e.g. the reduction of nitrate by the molybdenum-containing nitrate reductase enzyme according to the scheme

$NO_3^- \longrightarrow NO_2^- \longrightarrow$ amino acids $\longrightarrow NH_3$

nitrate   nitrite   ammonia

is a pathway for the formation of amino acids. Other examples include sulphite oxidase (sulphite/sulphate transformation in the liver, and arsenite oxidase which

oxidizes arsenite to arsenate). The biological fixing of nitrogen by legumes is catalysed by molybdenum-containing nitrogenases. Phosphorus and sulphur can exert a
major and synergistic influence on the availability and uptake of molybdenum by
plants. Copper is antagonistic to molybdenum. Molybdenum is a component of at
least six enzymes, nitrate reductase, nitrogenase, formate dehydrogenase, aldehyde
oxidase, xanthine oxidase and sulphite oxidase: the last three are found in animals.

Many pastures contain a low molybdenum content, which results in low growth in
ruminants. Improved rates of growth are achieved by molybdenum supplementation,
usually in the form of sodium/ammonium molybdate. Severe molybdenum deficiency
in adult patients gave symptoms of coma, tachycardia and night blindness. Low intake
of the metal over a long period may lead to oesophageal or stomach cancer. Higher than
normal levels can lead to molybdenum poisoning and to the condition molybdenosis.
Molybdenum compounds are highly toxic to laboratory animals. Subchronic exposure
to Mo may result in mild renal failure. Cattle are more affected than sheep, with horses
the least affected. Symptoms are severe diarrhoea and loss of weight. Prolonged over-
exposure results in anaemia, lameness and greying of dark hair. Increased levels of
molybdenum are detectable in the liver. This condition is associated with copper-
deficient soils, and made worse by increasing sulphur in the diet. Sufficient copper
supplementation can counteract almost all the disorders arising from Mo toxicity.

Fly ash from a Danish coal-fired power station used as road filler caused the fatal
poisoning of five horses. Autopsy showed high levels of molybdenum and low copper
levels. A farmer sued Rhone-Poulenc whose nearby factory produced high levels of
molybdenum, fluoride and sulphur, claiming that these elements were the cause of
eye defects in calves. Inhalation of molybdenum or molybdenum(VI) oxide, MoO_3,
dust by workers over a period of 3–5 years produced cases of pneumoconiosis.
Inhalation of dust from molybdenum-containing alloys or carbides can cause
'hard-metal lung disease.'

Occupational exposure limits. OSHA PEL (TWA) insoluble Mo compounds
$15\,mg\,Mo\,m^{-3}$; soluble Mo compounds $5\,mg\,m^{-3}$. Molybdenum dust is flammable.

Further reading

Abumrad, N.-N. (1984) Molybdenum – is it an essential trace element?, *Bull. N. Y.
Acad. Med.*, **60**, 163–171.

Braithwaite, E.P. and Haber, J. (1984) *Molybdenum – an Outline of its Chemistry and
Uses*, Elsevier, Amsterdam.

Couglin, M.P. (ed.) (1980) *Molybdenum and Molybdenum Enzymes*, Pergamon Press,
Oxford.

Friberg, L. and Lener, J. (1986) Molybdenum, in Friberg, L., Nordberg, G.F. and
Voux, V.B. (eds), *Handbook on the Toxicology of Metals*, 2nd edn, 1986, Elsevier,
Amsterdam, Vol. II, pp. 446–461.

King, R.B. (ed.) (1994) *Encyclopedia of Inorganic Chemistry*, Wiley, Chichester, Vol.
5, pp. 2304–2330.

Kirk–Othmer ECT, 4th edn, Wiley, New York, 1995, Vol. 16, pp. 940–963.

Rajagopalan, K.R. (1987) Molybdenum – an essential trace element, *Nutr. Rev.*, **45**,
321–328.

Wennig, R. and Kirsch, N. (1988) Molybdenum, in Seiler, H.G., Sigel, H. and Sigel,
A. (eds), *Handbook on Toxicity of Inorganic Compounds*, Marcel Dekker, New
York, pp. 437–447.

Molybdenum(V) chloride CAS No. 10241-05-1

$MoCl_5$

Uses. Chlorination catalyst. Intermediate for organomolybdenum compounds. Brazing and soldering flux.

Toxicity. Poison. Corrosive irritant to skin, eyes and mucous membranes.

Molybdenum(VI) oxide CAS No. 1313-27-5

MoO_3

Uses. As a reagent for chemical analysis. Flame-retardant additive.

Occupational exposure limit. N.a. MoO_3 dust can cause pneumoconiosis.

Toxicity. Possible carcinogen. LD_{50} oral rat 2689 mg kg^{-1}. LC_{50} (4 h) inhalation rat 5840 mg m^{-3}.

Molybdenum(IV) sulphide CAS No. 1317-33-5

MoS_2

Uses. Lubricant additive. Solid lubricant. Hydrogenation catalyst.

Occupational exposure limit. ACGIH TLV (TWA) 10 mg m^{-3}.

Toxicity. Irritant. Possible mutagen.

MONOCROTOPHOS CAS No. 6923-22-4
(Azodrin, Nuvacron)
Systemic broad-spectrum insecticide, acaricide. Toxicity class I. LD_{50} rat 15 mg kg^{-1}; LC_{50} rainbow trout 5 mg l^{-1}

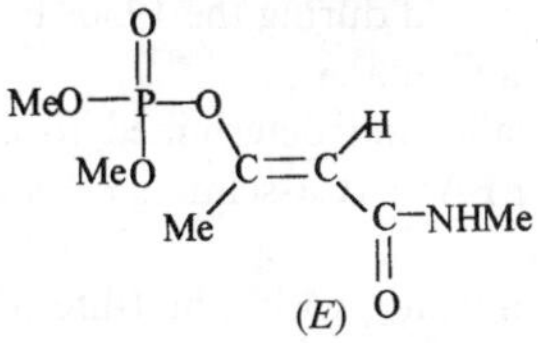

$C_7H_{14}NO_5P$ $M = 223.2$

Dimethyl(E)-1-methyl-2-(methylcarbamoylvinyl) phosphate; m.p. 25–30° (tech.), 55° (pure); v.p. 7×10^{-6} Torr (20°).

India uses large quantities for control of cotton pests; in 1988–89 the 655 t imported amounted to 45% of the total pesticides.

There have been numerous incidents of poisoning resulting from its misuse. Fifty migrants on a Transkei potato farm became ill after carrying water in a discarded container. In Paraguay (Conception) eight children suffered short term paralysis of the extremities of limbs; they lived close to a cotton crop treated with monocrotophos. Adults resident there suffered headaches, vomiting and profuse sweating. In the Parana state of Brazil in 1990 412 incidents of poisoning were recorded; 107(25%) were attributed to monocrotophos and 93(22%) to parathion.

Monocrotophos is very toxic if swallowed and it penetrates the skin.

The limit in workplace air in the USA (ACGIH) is a TWA of 0.25 mg m^{-3}; ADI 0.0005 mg kg^{-1} body weight (1991).

Monocrotophos is subject to Prior Informed Consent but to date no limits have been set in water or soil.

Codex Alimentarius: maximum residue levels include potato 0.05, milk $0.002\,mg\,kg^{-1}$.

MUSTARD GAS CAS No. 505-60-2

LD_{50} i.v. rat $0.7\,mg\,kg^{-1}$; inhalation rat TC_{Lo} $100\,\mu g\,m^{-3}$

$$S \big\langle \begin{array}{l} CH_2{-}CH_2Cl \\ CH_2{-}CH_2Cl \end{array}$$
$$C_4H_8SCl_2 \quad M = 159.1$$

Bis-2-(chloroethyl) sulphide, HD, 1,1'-thio-bis-(2- chloroethane); m.p. 14.5°; b.p. 98°/10 Torr; dens. 1.268 (25°); volatility $925\,mg\,m^{-3}$ (25°). Solubility in water $0.68\,g\,l^{-1}$ (25°); soluble in fat and most organic solvents.

It is possible that it was first synthesized by Despretz in 1822. It was manufactured during World War I and was used as a warfare agent. American production was then $18\,000\,kg\,day^{-1}$ by the reaction of ethylene with sulphur monochloride at 30°. In Germany it was produced by the sequence

$$CH_2{=}CH_2 \ + \ HOCl \longrightarrow \left[\begin{array}{cc} CH_2 & CH_2 \\ | & | \\ OH & Cl \end{array}\right] \xrightarrow{Na_2S} \ S\big\langle\begin{array}{l}CH_2{-}CH_2{-}OH\\CH_2{-}CH_2{-}OH\end{array} \xrightarrow{HCl} \text{Mustard Gas}$$

Mustard gas is primarily a vesicant, causing blistering of the eyes, skin and lungs; it is also a systemic poison which penetrates the skin. It is insidious in its action as symptoms only appear several hours after exposure. It was used during the 1930s by the Italians in Abyssinia (Ethiopia) and by the Japanese in China.

It may be collected from air by passage through decalin and determined to a level of $0.2\,ng\,ml^{-1}$ by GLC on Chromosorb W (2% FFAP) (Casselman *et al.*, 1973).

Mustard gas is rapidly decontaminated by hydrolysis in water, with a half-life of 5 min at 37°:

$$S\big\langle\begin{array}{l}CH_2{-}CH_2{-}Cl\\CH_2{-}CH_2{-}Cl\end{array} \xrightarrow{H_2O} \ 2\,HCl \ + \ HO{-}CH_2{-}CH_2{-}S{-}CH_2{-}CH_2{-}OH$$

This reaction is accelerated by alkali.

About 2700 American soldiers, who had been exposed to mustard gas during World War I, were found in the second following decade (1930–39) to have a higher mortality from pneumonia and tuberculosis than a control group with no exposure to mustard gas and which had not contracted pneumonia. A link with respiratory cancer was also established from a study of Japanese workers engaged in manufacture in an environment with mustard gas at 0.05–$0.07\,mg\,l^{-1}$. Of 172 deaths of former workers 48 were from cancer, including 28 from respiratory cancer.

Mustard gas reacts *in vivo* as an alkylating agent and attacks the proteins and nucleic acids of the lung, liver and kidneys. It also acts an antineoplastic agent through its effect on cancer cells and very many related compounds, but more important are the nitrogen mustards which, include the group

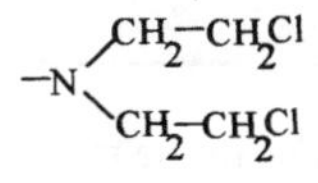

They have been investigated as anti-cancer agents; one useful example of this group is cyclophosphamide:

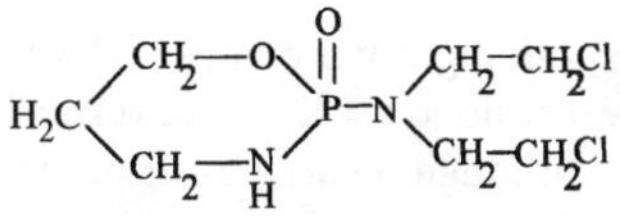

It is difficult to make the close distinction between the dose of an alkylating agent sufficient to give carcinostatis while avoiding damage to healthy host cells. The effectiveness of mustard gas as an alkylating agent is due to the water-soluble intermediate

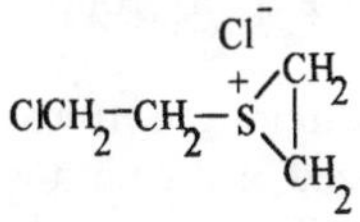

Other compounds in this group of warfare agents include 1,2-bis(2-chloro-ethylthio)ethane or Agent Q (CAS No. 3563-36-8) $ClCH_2CH_2-S-CH_2CH_2-S-CH_2CH_2-Cl$, prepared by the addition of two molecules of vinyl chloride to ethane-dithiol, and 1,2-bis(2-chloroethylthioethyl) ether or Agent T (CAS No. 63918-89-8), $[ClCH_2CH_2-S-CH_2CH_2]_2O$.

Reference
Casselman, A.A., Gibson, N.C.C. and Bannard, R.A. (1973), *J. Chromatogr.*, **78**, 317–327.

Further reading
IARC (1975) *Aziridines, N-, S- and O-Mustards and selenium*, IARC, Lyon, Vol. 9, pp. 181–216.
Kirk–Othmer ECT, 4th edn, Wiley, New York, 1993, Vol. 5, pp. 795–815.

NAPHTHYLAMINES

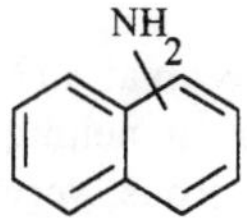

$C_{10}H_9N$ $M = 143.2$

	CAS No.	M.p. (°)	B.p (°)	V.p. (Torr) (106°)	Water solubility (gl⁻¹)
1-Napthylamine	134–32–7	50	301	1.0	1.7
2-Naphthylamine	91–59–8	113	306	1.0	10

1-Naphthylamine has an LD_{50} in the dog of $400\,\mathrm{mg\,kg^{-1}}$ and an LD_{50} in the rat of $779\,\mathrm{mg\,kg^{-1}}$. 2-Naphthylamine has an LD_{50} in the rat of $727\,\mathrm{mg\,kg^{-1}}$.

Both compounds are soluble in hot water and volatile in steam and also soluble in ethanol and diethyl ether. They have values of log P_{ow} of 2.4 and 2.5 and are therefore not expected to accumulate in fatty tissue.

They can be obtained by reduction of the corresponding nitro compound with iron–dilute HCl or by hydrogenation. Purification by distillation will reduce the level of 2- naphthylamine in a mixed product to 8–10 ppm, although on an industrial scale the level is about 0.5%. The best route to 2-naphthylamine is ammonolysis of 2-naphthol with $(NH_4)_2SO_3–NH_4OH$ at 150°.

1-Naphthylamine is used in the dyestuffs industry and for the synthesis of the rodenticide Antu (1-naphthyl-2-thiourea; CAS No. 86-88-4). 1-Naphthylphenyl-amine is used as an anti-oxidant in the rubber industry; it was originally held to be a human carcinogen, but earlier studies were invalidated by the presence of 4–10% of the 2-isomer. The IARC has classified 1-naphthylamine as a group 3 carcinogen.

2-Naphthylamine is a very dangerous substance and a group 1 carcinogen. It was formerly used as a dyestuffs intermediate but is no longer synthesized or used in the USA. It was banned in the UK in 1988 under the Control of Substances Hazardous to Health Regulations. It produced urinary bladder carcinomas in hamsters, dogs and non-human primates following oral administration.

Both of these amines have been administered orally to humans (MARC, 1986). Early studies on workers showed that 1-naphthylamine was absorbed at $10\,\mathrm{mg\,kg^{-1}}$ $\mathrm{day^{-1}}$ and only a small proportion was excreted in the urine. Both amines sensitize humans to dermatitis. Over the period 1900–52, 43% of those employed distilling 2-naphthylamine died of bladder cancer. Twelve coal tar dye workers from a group of 48 had bladder cancer diagnosed 8 years after exposure.

Reference
MARC Report (1986) Shuker, L.K., Batt, S., Rystedt, I. and Berlin, M., *The Health Effects of Aromatic Amines. A Review*, King's College, London.

Further reading
Kirk–Othmer ECT, 4th edn, Wiley, New York, 1966, Vol. 16, p. 987.

NEODYMIUM CAS No. 7440-00-8

Nd

Neodymium, a member of the lanthanide group of the Periodic Table, has an atomic weight of 60, an atomic weight of 144.24, a specific gravity of 7.007 at 20°, a melting point of 1016° and a boiling point of 3068°. It is a bright, silvery, lustrous, very reactive metal forming compounds in the +3 oxidation state. Natural neodymium consists of seven isotopes: ^{142}Nd (27.13%), ^{143}Nd (12.18%), ^{144}Nd (23.80%), ^{145}Nd (8.30%), ^{146}Nd (17.19%), ^{148}Nd (5.76%) and ^{150}Nd (5.64%).

The crustal abundance of neodymium is about 24 ppm. Found with other lanthanides in the minerals *monazite* and *bastnasite*. Neodymium levels in human organs are very low.

Uses. Up to 18% neodymium is present in misch–metal (cerium–lanthanum–neodymium) used as lighter flints. Neodymium compounds are used to colour glass and enamels purple. Laser component. Catalyst. Semiconductor dopant. Neodymium–iron–boron high-energy magnet.

No biological role. As with all rare earths, care is advised in handling neodymium and its compounds. It may be an anti-coagulant in blood. Neodymium compounds are of low to moderate toxicity for humans but are more toxic by injection routes. Neodymium compounds are eye and skin irritants.

Further reading
Gschneidner, K.A., Jr, and LeRoy, K.S. (eds), (1988) *Handbook on the Physics and Chemistry of the Rare Earths*, North Holland, Amsterdam.
Kirk–Othmer ECT, 4th edn, Wiley, New York, 1995, Vol. 14, pp. 1091–1115.

Neodymium(III) chloride hexahydrate CAS No. 13477-89-9

NdCl$_3$.6H$_2$O

Uses. Catalyst in the petroleum industry. Rubber manufacture.

Toxicity. LD$_{50}$ oral mouse 3962 mg kg^{-1}; intraperitoneal mouse 347 mg kg^{-1}.

NEON CAS No. 7440-01-9

Ne

Neon, a member of Group 18 (the noble gases) of the Periodic Table, has an atomic number of 10, an atomic weight of 20.18, a specific gravity (of neon gas) of 0.8999 at 0°/1 atm, a melting point of −248.67°, and a boiling point of −246.048° (1 atm). Only three isotopes occur naturally: ^{20}Ne (90.48%), ^{21}Ne (0.27%) and ^{22}Ne (9.25%). Its oxidation state is zero.

Neon is obtained by liquefaction and fractional distillation of air. In the atmosphere it is present to the extent of 1 part in 65 000 of air.

Uses. Neon advertising signs. High-voltage indicators. TV tubes. Gas lasers (with helium). Cryogenics. Electroluminescent plasma display screens.

Neon is a chemically inert element. It acts as an asphyxiant at high concentrations.

Further reading
King, R.B. (ed.) (1994) *Encyclopedia of Inorganic Chemistry*, Wiley, Chichester, 1994, Vol. 5, pp. 2660–2680.
Kirk–Othmer, ECT, 4th edn, Wiley, New York, 1995, Vol. 13, pp. 1–53.

NICKEL CAS No. 7440-02-0

Ni

Nickel, a member of Group 10 of the Periodic Table, has an atomic number of 28, an atomic weight of 58.71, a specific gravity of 8.90 at 20°, a melting point of 1455° and a boiling point of 2913°. It has five natural isotopes: ^{58}Ni (68.08%), ^{60}Ni (26.22%), ^{61}Ni (1.13%), ^{62}Ni (3.63%) and ^{64}Ni (0.93%). ^{61}Ni is used as a tracer in biological systems. A silvery-white, hard, malleable and ductile metal forming compounds mainly in the +2 oxidation state.

The abundance of nickel in the earth's crust is about 80 ppm, 23rd in order of relative abundance. Main ores are *pentlandite* $(Ni, Fe)_9S_8$ and *garnierite* $[(Ni, Mg)_6Si_4O_{10}(OH)_2]$. Nickel pollution occurs from metal ore smelters. World-wide production is around 8×10^5 t year^{-1}. Typical concentrations in the environment are soil 4.4–228 ppm; air 5–35 ng m^{-3}; sea water 0.1 ppb; drinking water < 10 µg l^{-1}; sewage sludge up to 0.5%; tobacco 1.3–4.0 mg Ni kg^{-1}; daily adult human intake 100–700 µg; bone ~ 0.5 ppm; blood 0.01–0.05 ppm; average total human body (70 kg, adult) content 10 mg.

Uses. Electroplating. Alloy production. Stainless steels. Nickel–chromium heating elements. Nickel–cadmium batteries. Electronic components. Raney nickel catalysts.

Poor uptake by plants, so little enters the human food chain. Generally, when the nickel concentration exceeds 50 ppm, plants may suffer from excess nickel and exhibit toxic symptoms of chlorosis and necrosis. Nickel-containing ureases and hydrogenases have important biological roles.

The evidence for the carcinogenicity of nickel compounds in humans arises from studies of workers employed in the high-temperature oxidation of nickel matte and nickel copper matte in electrolytic refining, in copper plants and in the extraction of nickel salts in the hydrometallurgical industry. The highest risk of lung and nasal cancer was found among nickel refinery workers exposed to nickel sulphate and to combinations of nickel sulphides and oxides. Metallic nickel, nickel monoxide, nickel hydroxide and crystalline nickel sulphides are carcinogenic in experimental animals. The evidence for the carcinogenicity of other nickel compounds in experimental animals is not conclusive.

Plant workers in nickel sinter factories showed an increased risk of lung and nasal cancer. Adverse effects of soluble nickel salts are found on kidney tubular function, at high exposure levels. As a possible mutagen, nickel can cause chromosome damage, both *in vivo* and *in vitro*. Nickel induces DNA damage by binding of the metal ion to DNA and nuclear proteins. Carbonyl nickel compounds give an increased risk of lung and nasal cancer (Sunderman, 1993). Cases of nickel poisoning have been reported in patients dialysed with nickel-containing dialysate; also in electroplaters who accidentally ingested water contaminated with nickel sulphate and nickel chloride.

Occupational exposure limits.   OSHA PEL (TWA) soluble Ni salts 0.05 mg m^{-3}.

Toxicity.   At least 29 million Europeans are allergic to nickel, which causes sensitization of the skin and contact dermatitis. For this reason, nickel is banned from use in articles which come into contact with skin (EC Directive 94/27/EC).

Reference
Sunderman, F.W. (1993) Biological monitoring of nickel in humans, *Scand. J. Work Environ. Health*, **19**, Suppl. 1, 34–38.

Further reading

ECETOX (1991) *Technical Report 33*, European Chemical Industry Ecology and Toxicology Centre, Brussels.

Environmental Health Criteria 118. Nickel. WHO/IPCS, Geneva, 1991.

Health and Safety Committee (1992) *Nickel and You*, HSC, London.

IARC (1987) *IARC Monograph*, Suppl. 7, p. 67.

King, R.B. (ed.) (1994) *Encyclopedia of Inorganic Chemistry*, Wiley, Chichester, Vol. 5, pp. 2384–2412.

Kirk–Othmer ECT, 4th edn, Wiley, New York, 1996, Vol. 17, pp. 1–42.

Nieboer E. (1992) *Nickel and Human Health*, Wiley, Chichester.

SI 1991 No. 472 The Environmental Protection (Prescribed Processes and Substances) Regulations 1991, HMSO, London, 1991.

Nickel(II) ammonium sulphate CAS No. 15699-18-0

$(NH_4)_2Ni(SO_4)_2$

Uses. Electroplating metals. Analytical reagent.

Toxicity. LD_{50} oral rat $400 \, mg \, kg^{-1}$

Nickel(II) chloride CAS No. 7718-54-9; hexahydrate 7791-20-0

$NiCl_2$

Uses. Nickel plating. Catalyst.

Toxicity. Toxic. Possible carcinogen. LD_{50} oral rat $157 \, mg \, kg^{-1}$; intraperitoneal mouse $48 \, mg \, kg^{-1}$; intravenous dog $40 \, mg \, kg^{-1}$.

Nickel(II) subsulphide CAS No. 11113-75-0

Ni_3S_2

Uses. Catalyst.

Toxicity. Tumorigenic. Possible carcinogen. LD_{50} oral rat $400 \, mg \, kg^{-1}$.

Nickel(II) sulphate CAS No. 7786-81-4; hexahydrate 10101-98-1

$NiSO_4$

Uses. Nickel plating. Mordant in dyeing and printing of fabrics. Jewellery manufacture.

Toxicity. LC_{50} rainbow trout $0.36 \, mg \, l^{-1}$; LD_{Lo} intravenous rabbit, dog, mouse $33–76.4 \, mg \, kg^{-1}$.

Nickel tetracarbonyl CAS No. 13463-39-3

$Ni(CO)_4$

Uses. Preparation of pure nickel.

Occupational exposure limit. US TLV (TWA) $0.05 \, mg \, m^{-3}$.

Toxicity. LC_{50} (20–30 min) inhalation cat, rat, mouse $0.067–0.24 \, mg \, l^{-1}$. LC_{50} (30 min) inhalation dog $2.5 \, mg \, l^{-1}$. Highly toxic. Symptoms are headache, vertigo, nausea, vomiting, insomnia and pulmonary symptoms. The liver, kidneys, spleen glands and brain are also affected.

NIOBIUM CAS No. 7440-03-1

Nb

Niobium, a member of Group 5 of the Periodic Table, has an atomic number of 41, an atomic weight of 92.90, a specific gravity of 8.57 at 20°, a melting point of 2477° and a boiling point of 4742°. Niobium is a steel-grey, lustrous, ductile and malleable metal and forms compounds in several oxidation states, +3, +4 and principally +5. One natural isotope of Nb exists, ^{93}Nb; other isotopes include $^{88-92}$Nb and $^{94-100}$Nb. Minute amounts (2×10^{-10} %) of ^{95}Nb ($t_{1/2} = 2.03 \times 10^{-4}$ years) are found in natural niobium.

The average crustal abundance of niobium is 24 ppm. It is found in the minerals *niobite, niobite-tantalite, pyrochlore, euxenite* and, its chief source, *columbite* [(Fe, Mn)Nb$_2$O$_6$]. Coal combustion leaves niobium in the coal slag residue at concentrations of 10–24 ppm. Some plants bioaccumulate niobium. Sea water contains ~ 0.005 ppb [as Nb(OH)$_6^-$]. Human adults have a body burden of ~ 1.5 mg Nb, distributed through all parts of the body. Bone < 0.05 ppm; blood 0.005 mg l^{-1}.

Uses   Iron–niobium alloy for stainless steels. Special alloys. As a getter to remove gases in electronic vacuum tubes. Superconducting niobium–titanium and niobium–tin alloys.

Not an essential nutrient for humans or animals. It is accepted that niobium and its compounds are probably of low toxicity to humans. The metal dust is an eye and skin irritant. Ingestion of niobium compounds and subsequent gastrointestinal absorption may cause kidney damage.

There are very few cases of niobium poisoning. There was increased risk of cancer in Norwegian niobium mine workers exposed to thoron–radon daughter, ^{95}Nb($t_{1/2} =$ 35 days). Dust exposure during manufacture of niobium carbide, NbC (component of hard metal), and niobium borides, NbB and NbB$_2$.

Occupational exposure limit.   Drinking water MAC 0.01 mg l^{-1}

Toxicity.   LD$_{50}$ oral rat 50 mg kg^{-1}.

Further reading
King, R.B. (ed.) (1994) *Encyclopedia of Inorganic Chemistry*, Vol. 5, pp. 2444–2462 Wiley, Chichester, 1994.
Kirk–Othmer ECT, 4th edn, Wiley, New York, 1996, Vol. 17, pp. 43–68.
SI 1991 No. 472 The Environmental Protection (Prescribed Processes and Substances) Regulations 1991, HMSO, London, 1991.
Wennig, R. and Kirsch, N. (1988) Niobium, in Seiler, H. G., Sigel, H. and Sigel, A. (eds), *Handbook on Toxicity of Inorganic Compounds*, Marcel Dekker, New York, pp. 469–473.

Niobium (IV) carbide CAS No. 12069-94-2

NbC

Uses.   Cemented carbide-tipped tools. Special steels. Preparation of Nb metal. Coating graphite used in nuclear reactors.

Toxicity.   N.a. Avoid dust.

Niobium(V) chloride CAS No. 10026-12-7

$NbCl_5$

Uses. Preparation of pure Nb. Intermediate for other niobium compounds.

Toxicity. LD_{50} intraperitoneal mice, $61\,mg\,kg^{-1}$ and $14\,mg\,kg^{-1}$. LD_{50} oral mouse $940\,mg\,kg^{-1}$. May evolve fumes of hydrogen chloride.

Niobium(V) oxide CAS No. 1313-96-8

Nb_2O_5

Uses. Intermediate. Electronics. Optical glass.

Toxicity. N.a. Avoid dust.

NITRATES CAS No. 14797-55-8

NO_3^-

Usually very soluble salts. Large amounts taken orally may have serious or fatal effects. The symptoms are dizziness, abdominal cramps, vomiting, bloody diarrhoea, weakness, convulsions and collapse. Small repeat doses may result in weakness, depression, headache and mental impairment.

NITRIC ACID CAS No. 7697-37-2

HNO_3 $\qquad\qquad\qquad\qquad M = 63.0$

Uses. Manufacture of nitrates and nitro compounds used as fertilizers, explosives, organic chemicals including dye intermediates.

Occupational exposure to nitric acid can lead to permanent eye and skin burns and lung damage.

Occupational exposure limits. US TLV (TWA) 2 ppm ($5.2\,mg\,m^{-3}$). UK short-term limit 4 ppm ($10\,mg\,m^{-3}$)

Toxicity. Oxidizing and corrosive substance. Teratogenic and reproductive effects in experimental animals. LD_{Lo} oral human $430\,mg\,kg^{-1}$. LC_{50} inhalation rat 244 ppm.

Further reading
Kirk–Othmer ECT, 4th edn, Wiley, New York, 1986, Vol.17, pp. 80–108.

NITRIDES CAS No. 18851-77-9

N_3^-

Nitrogen forms binary compounds with almost all elements of the Periodic Table. Nitrides may be classifed into four groups: 'salt-like', covalent, 'diamond-like' and metallic (or interstitial). Many nitrides react with water to form ammonia gas, which is a severe irritant to mucous membranes and can be a fire hazard. The metallic group of nitrides, of general formulae MN, M_2N and M_4N, are the most extensive and important commercially. These compounds are usually opaque, very hard (> 8 on the Moh scale), chemically inert, refractory materials, with a metallic lustre and conductivity. These properties permit nitrides to be used in crucibles, high-temperature reaction vessels, thermocouple sheaths and related applications. Several metal nitrides are used as catalysts. Occupational exposure to dusts takes place when these materials are manufactured. Similarities with borides and carbides are notable.

Further reading
King, R.B. (ed.) (1994) *Encyclopedia of Inorganic Chemistry*, Wiley, Chichester, Vol. 5, pp. 2498–2515.

NITRITES CAS No. 14797-65-0

NO_2^-

Uses. Curing meat. Sodium nitrite is used on a large scale for the synthesis of hydroxylamine. Manufacture of azo dyes and pharmaceuticals.

Further reading
Ullmann's Encyclopedia of Industrial Chemistry, 5th edn, VCH, Weinheim, 1991, Vol. A17, p. 331.

NITROGEN CAS No. 7727-37-9

N (stable as N_2)

Nitrogen, a member of Group 15 of the Periodic Table, has an atomic number of 7, an atomic weight of 14.0067, a specific gravity of $0.81 \, g \, l^{-1}$ at $-195.8°$, a melting point of $-209.86°$ and a boiling point of $-195.8°$. Nitrogen is a colourless, odourless gas and forms compounds in the $+3$ and $+5$ oxidation states. Natural nitrogen consists of only two stable isotopes, ^{14}N (99.63%) and ^{15}N (0.37%).

Nitrogen makes up 78% by volume of the air, and is obtained industrially from air by liquefaction and separation of air. Average total human body content (adult, 70 kg) 1.8 kg.

Uses. Manufacture of ammonia (Haber process) and nitric acid. Inert gas in the electronics industry. Liquid nitrogen is used as a refrigerant for food products and also the transport of food. In the space industry for the air purging of instruments and components. In the oil industry to force oil to the surface.

Nitrogen acts as an asphyxiant, displacing oxygen. Exposure can cause nausea, headache and vomiting. Skin contact with liquid nitrogen can result in severe frostbite or 'burns.' Nitrogen breathed at high presssure, as in deep sea diving, can lead to the syndrome of decompression sickness if decompression is too fast.

Toxicity. LD_{50} oral rat $740 \, mg \, kg^{-1}$. Selective carcinogen. IARC Group 2 (sufficient animal data for carcinogenicity).

Further reading
Kirk–Othmer ECT, 4th edn, Wiley, New York, 1996, Vol. 17, pp. 153–172.
King, B.R. (ed.) (1994) *Encyclopedia of Inorganic Chemistry*, Wiley, Chichester, Vol. 5, pp. 2516–2557.

NITROGEN HYDRIDES
Nitrogen forms several hydrides but only ammonia, NH_3 (and salts), hydrazine, N_2H_4 (and salts) and hydrogen azide, HN_3 (and salts) are of environmental import-

ance. Hydroxylamine NH_2OH, is closely related in structure and properties to both ammonia and hydrazine.

NITROGEN OXIDES

Nitrogen forms five oxides, alphabetically dinitrogen pentoxide, dinitrogen trioxide, nitrogen dioxide, nitric oxide and nitrous oxide.

Dinitrogen pentoxide CAS No. 10102-03-1

N_2O_5

Uses. Nitrating agent.

Toxicity. Chronic symptoms of respiratory disease are evident even at low concentrations ($< 5\,mg\,m^{-3}$).

Further reading
ECETOX (1991) *Technical Report 30(4)*, European Chemical Industry Technology and Toxicology Centre, Brussels.

Dinitrogen trioxide CAS No. 10544-73-7

N_2O_3

Uses. Oxidant in special fuel systems. Preparation of pure alkali metal nitrites.

Occupational exposure limits. N.a.

Toxicity. Similar to those of NO and NO_2.

Further reading
SI 1991 No. 472 The Environmental Protection (Prescribed Processes and Substances) Regulations 1991, HMSO, London, 1991

Nitrogen dioxide CAS No. 10102-44-0; **Dinitrogen tetroxide** 10544-72-6

$NO_2; N_2O_4$

Uses. Nitric acid production. Organic synthesis. The dimer N_2O_4 has been used extensively as the oxidant in rocket fuels for space missions.

Car exhausts are the principal source of the dioxide in air. In the stratosphere it forms acid rain:

$$\bullet OH + NO_2 + M(e.g.N_2) \longrightarrow HNO_3 + M$$

Occupational exposure limit. USA TLV (TWA) 3 ppm ($5.6\,mg\,m^{-3}$). UK short-term limit 5 ppm ($9.4\,mg\,m^{-3}$)

Toxicity. Toxic by inhalation (b.p. 21°). Experimental teratogen. Reproductive effects on experimental animals. Mutation data reported. An asphyxiant. Human systemic effects by inhalation: pulmonary changes, cough, dyspnea. LC_{50} (10 min) in halation mouse 1000 ppm. LC_{Lo} (1 min) inhalation human 200 ppm (see **Nitric oxide**).

Further reading
ECETOX (1991) *Technical Report 30(4)*, European Chemical Industry Technology and Toxicology Centre, Brussels.
Elsayed, N.M. (1994) Toxicology of nitrogen dioxide, *Toxicology*, **89**, 166–167.

SI 1991 No. 472 The Environmental Protection (Prescribed Processes and Substances) Regulations 1991, HMSO, London, 1991.

Norberg, S. *et al.* (1993) Health risk evaluation of nitrogen oxides (NO and NO_2), *Scand. J. Work Environ. Health*, **19**, 3–12.

Nitric oxide CAS No. 10102-43-9

NO $M = 30$

Uses.   Manufacture of nitric acid (Ostwald process). Bleaching rayon.

Occupational exposure limits.   US TLV (TWA) 25 ppm ($31\,mg\,m^{-3}$); UK short-term limit 35 ppm ($45\,mg\,m^{-3}$).

Toxicity.   Toxic gas. A severe eye, skin and mucous membrane irritant (60–150 ppm). Concentrations of 200–700 ppm may be fatal even after short exposures. Mutation data reported. LC_{50} (unknown duration) inhalation rat $1068\,mg\,m^{-3}$.

Car exhausts are the main source of NO as an atmospheric pollutant. In polluted environments the exact balance between nitric oxide and nitrogen dioxide (see above) is often not defined and the proportions change continuously in daylight through reactions such as

$$NO + O_3 \longrightarrow NO_2 + O_2$$
$$\bullet OH + NO_2 \longrightarrow HNO_3$$

Hence it is common practice to refer to the mixture of these gases as NO_x. Nitric acid formed as above contributes to the problem of acid rain (see **Sulphur dioxide**). Global emissions of NO_x in 1990 (see **Carbon monoxide**) were 32×10^6 t, divided between heavy trucks 50%, cars 42%, light trucks 6.5% and motor cycles 1.5%. In unpolluted air, oxygen and NO_x achieve a steady state including a little ozone, but the injection of CO and unburnt hydrocarbons into the atmosphere diverts the equilibrium:

$$NO_2 + h_v \rightarrow [O] + NO$$
$$[O] + O_2 \rightarrow O_3$$
$$O_3 + NO \rightarrow NO_2 + O_2$$

The key to the disturbance of this natural cycle is the increase in the level of NO_2 consequent upon its formation from radicals derived from unburnt hydrocarbon, without participation of ozone, e.g.

$$R + O_2 \rightarrow RO_2$$
$$RO_2 + NO \rightarrow RO + NO_2$$

when through the photolytic cleavage of the NO_2 further synthesis of ozone occurs, so distorting the natural system.

In the USA, the National Air Quality Standard (NAAQ) for oxides of nitrogen, expressed as NO_2, requires an annual arithmetic daily mean below 53 ppb ($100\,\mu g\,m^{-3}$). In developed countries the growth of traffic and unrestricted access to city centres have led to occasional episodes where limits have been exceeded.

In the UK at present the DOE levels (ppb) for NO_2 are described in general form as very good < 50, good 50–99, poor 100–299, very poor > 300. Winter conditions are the worst for air pollution; in 1993 NO_2 peaks above 200 ppb were recorded in Shefffield, Birmingham and Newcastle. In London in 1994 five sites recorded levels above 200 ppb, while a national survey gave a national kerbside average of 23 ppb with an urban background of 14 ppb (Bowen et al., 1995). The government response to a worsening situation is the proposal to introduce a new hourly average limit of 150 ppb by 2005. The present WHO guideline for a 1 h average limit is 210 ppb $(400 \, \mu g \, m^{-3})$.

Reference
Bowen, J.S., Broughton, G.F.S., Willis, P.G. and Clark, H. (1995) *Air Pollution in the United Kingdom 1993/4.* 7th Report, AEA Technology, Culham.

Further reading
Norberg, S. et al. (1993) Health risk evaluation of nitrogen oxides (NO and NO_2), *Scand. J. Work. Environ. Health*, **19**, 3–12.
King, R.B. (ed.) (1994) *Encyclopedia of Inorganic Chemistry*, Wiley, Chichester, Vol. 5, pp. 2482–2498.

Nitrous oxide CAS No. 10024-97-2

N_2O

Uses. General anaesthetics. Oxidant in chemical synthesis.

Stable gas formed naturally by blue–green algae and by *Rhizobium* bacteria active in the nodules of peas, beans and other legumes. In 1995 the N_2O level was ~ 0.3 ppm in air but it is expected to increase by 0.2% p.a. in future years. Nitrous oxide is slowly destroyed in the atmosphere with a residence time of 100 years:

$$N_2O + \text{hot}[O] \rightarrow 2NO$$

Occupational exposure limits. USA TLV (TWA) 5 ppm $(30 \, mg \, m^{-3})$. UK long-term limit MEL 100 ppm $(180 \, mg \, m^{-3})$.

Toxicity. An asphyxiant at high concentrations. LD_{50} oral rat 505 mg kg^{-1}. LD_{Lo} (2 h) inhalation humans 24 mg kg^{-1}. An experimental teratogen. Experimental reproductive effects. Mutation data reported.

Further reading
ECETOX (1991) *Technical Report 30(4)*, European Chemical Industry Technology and Toxicology Centre, Brussels.

NITROSAMINES

$$\begin{array}{c} R(H) \\ {\diagdown} \\ {}N-NO \\ {\diagup} \\ R_1 \end{array}$$

These are a formidable group of about 300 compounds of which over 70% induce cancer in experimental animals. Mice, rats and hamsters developed tumours after 1 week on a daily diet containing 0.4 mg kg^{-1}.

Two representative aliphatic members of the group are *N,N*-diethyl- and *N,N*-dimethylnitrosamine.

N, *N*-**Diethylnitrosamine** CAS No. 55-18-5

$(Et)_2N—NO$ $C_4H_{10}N_2O$ $M = 102.1$

b.p. 177°; dens. 0.94 (20°).

N, *N*-**Dimethylnitrosamine** CAS No. 62-5-9
LC_{50} (10 day) rainbow trout $1.77 \, g \, l^{-1}$; LC_{50} (4 h) rat 78 ppm

$(Me)_2N—NO$ $C_2H_6N_2O$ $M = 74.1$

B.p. 153°; dens. 1.0 (18°); v. dens. 2.56. Half-life in sea water and freshwater in daylight 2–3 days.

The acute toxicities in the rat (oral) are as follows:

Compound	$LD_{50} (mg \, kg^{-1})$
Dimethylnitrosamine	27–41
Diethylnitrosamine	216
n-Butylmethylnitrosamine	130
tert-Butylethylnitrosamine	1600

However, as is often the case these values are no guide to carcinogenicity – *N*-nitrosoazetidine:

$$CH_2—CH_2$$
$$| \quad \quad |$$
$$CH_2— N$$
$$\quad \quad \quad \diagdown NO$$

has a LD_{50} in the rat of over $1.6 \, g \, kg^{-1}$, yet multiple tumours were induced in rats fed for 46 days on a daily diet including only 5 mg of this nitroso compound.

Dimethylnitrosamine was once manufactured in the USA at the rate of $5 \times 10^5 \, kg \, year^{-1}$ as a precursor for the rocket fuel dimethylhydrazine, MeNHNHMe. Levels in air of up to $36 \, \mu g \, m^{-3}$ were recorded at a factory site in Boston, MA, and $1 \mu g \, m^{-3}$ was detected in the adjacent neighbourhood; production ceased in 1976.

Humans are also at risk from airborne nitrosamines from sulphur-rich chemicals used for the cross-linking of elastomers. For example, thiuram disulphides, $R_2N(C=S)S—S(C=O)NR_2$, insert sulphur cross-links and so harden the originally soft thermoplastic rubber. At the necessary temperatures these compounds react with NO_x and liberate the corresponding nitrosamine, thus tetraethylthiuram disulphide (R = Et) gives diethylnitrosamine; levels up to 270 ppb have been detected in tyre producing plants. Among British rubber workers in Birmingham during 1936–51 deaths from bladder cancer were almost double those in the general population.

Tobacco smoking is another source of airborne nitrosamines, formed by reaction between tobacco amines and the nitrite added as a preservative. Two of these products are *N'*-nitroso-*nor*-nicotine (**A**) and *N*-nitrosopyrrolidine (**B**):

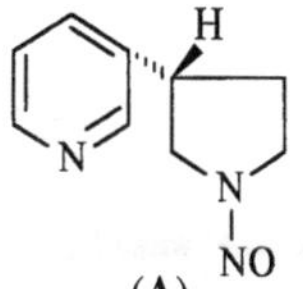

(A)    (B)

which are emitted at levels of 0.2–3.0 µg/cigarette and 110 ng/cigarette, respectively. Snuff (oral tobacco) contains dimethylnitrosamine ($37\,\mu g\,kg^{-1}$) and propylnitrosamine ($3.9\,mg\,kg^{-1}$). A disturbing feature is that cigarette production worldwide, with all the associated emissions, is on the increase. A United Nations survey (1994) showed UK production falling from 100×10^9 cigarettes (1982) to 12×10^9 (1991), but in China over the same period output rose from 942×10^9 to 1600×10^9 cigarettes.

Dimethylnitrosamine causes necrosis of the liver, internal bleeding, ascites and jaundice; it is carcinogenic in all animal species tested, causing liver, kidney and lung tumours. Mice, rats and hamsters developed tumours after 1 week on a daily diet containing $0.4\,mg\,kg^{-1}$. In the rat dimethylnitrosamine is metabolized to formaldehyde.

The formation of nitrosamines in acidic solution (pH 4.0) is well documented and raises concern about their production in the digestive system of humans. There is also a mechanism by which nitrosamines are formed via the action of nitrite ion on the glyceryl esters of fats:

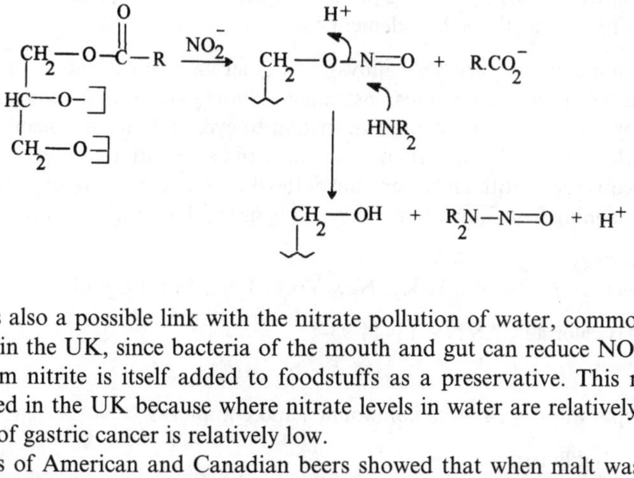

There is also a possible link with the nitrate pollution of water, commonly above EC limits in the UK, since bacteria of the mouth and gut can reduce NO_3^- to NO_2^- and sodium nitrite is itself added to foodstuffs as a preservative. This risk is not highly rated in the UK because where nitrate levels in water are relatively high, the incidence of gastric cancer is relatively low.

Analysis of American and Canadian beers showed that when malt was dried by direct heating, dimethylnitrosamine was formed and was detected in the product in the range $1.5–5.9\,\mu g\,l^{-1}$, however, with indirect drying, levels in the beer fell to about one tenth of this, $0.07–0.16\,\mu g\,l^{-1}$ (Scanlan et al., 1990).

A Swedish review showed that dimethylnitrosamine was present at levels up to $1.2\,mg\,kg^{-1}$ in bacon, smoked fish and whisky. In Chinese food diethylnitrosamine was found at $4.0\,mg\,kg^{-1}$ in dried seafood and at $2.6\,mg\,kg^{-1}$ in meat.

The ACGIH has rated dimethylnitrosamine as a group A2 carcinogen – a suspected human carcinogen at levels relevant to human exposure.

Reference
Scanlan, R.A., Barbour, J.F. and Chappel, C.I. (1990) *J. Agric. Food Chem.*, **38**, 442–443.

Further reading
IARC (1991) *Symposium 105. Relevance to Human Cancer of N- Nitroso compounds, Tobacco Smoke and Mycotoxins*, O'Neill, I.K. and Bartsch, H. (eds), IARC, Lyon.

NITROUS ACID CAS No. 7782-77-6

HNO_2 $M = 47.0$

Uses. Catalyst. Reducing agent.

Toxicity. Weak acid. Rapidly decomposed in water to nitric oxide and nitric acid.

OSMIUM CAS No. 7440-04-2

Os

Osmium, a member of Group 8 of the Periodic Table, has an atomic number of 76, an atomic weight of 190.2, a specific gravity of 22.57 at 20°, a melting point of 3033°, and a boiling point of $\sim 5000°$. The metal is a lustrous, bluish white, extremely hard inert element and forms compounds in the $+3, +4, +6$ and $+8$ oxidation states. Natural osmium consists of seven isotopes: ^{184}Os (0.02%), ^{186}Os (1.58%), ^{187}Os (1.6%), ^{188}Os (13.3%), ^{189}Os (16.1%), ^{190}Os (26.4%) and ^{192}Os (41.00%).

A rare element occurring at 0.005 ppm of the earth's crust. Found in platinum- and nickel-bearing ores as the native element.

Uses. Manufacture of very hard alloys, with other members of the platinum group. Manufacture of fountain pen nibs, instrument pivots and electrical contacts.

Poison by the intravenous route. An irritant to eyes and mucous membranes. The principal effects are ocular disturbances and an asthma condition caused by inhalation. It can also cause dermatitis and ulceration of the skin. There is only one reported case of osmium poisoning from inhalation, which gave rise to dermatitis and bronchitis.

Further reading
Kirk–Othmer ECT, 4th edn, Wiley, New York, 1996, Vol. 19, pp. 347–406.

Osmium(III) chloride CAS No. 13444-93-4

$OsCl_3$

Uses. Preparation of osmium compounds and complexes.

Toxicity. Highly toxic.

Osmium(VIII) oxide (Osmium tetroxide) CAS No. 20816-12-0

OsO_4

Uses. Contrast agent in electron microscopy.

Occupational exposure limits. OSHA TLV (TWA) 0.0002 mg Os m^{-3}.

Toxicity. Highly toxic. Poison by ingestion, inhalation and intraperitoneal routes. Humans system effects by inhalation. Lacrimation and other eye effects, including dimming of the cornea. Damage to the respiratory tract. LD_{50} intraperitoneal rat $14\,100\,\mu g\,g^{-1}$

OXYGEN CAS No. 7782-44-7

O (Stable as O_2)

Oxygen, a member of Group 16 of the Periodic Table, has an atomic number of 8 an atomic weight of 15.999, a density of $1.429\,g\,l^{-1}$ at $0°$, a melting point of $-218.8°$, and a boiling point of $-182.96°$. Its usual oxidation state is -2. Natural oxygen consists of three isotopes: ^{16}O (99.762%), ^{17}O (0.038%) and ^{18}O (0.200%).

Oxygen forms 21% of the atmosphere by volume, it is obtained by liquefaction and fractional distillation of air. The element and its compounds make up 49.2% by weight (492 000 ppm) of the earth's crust. Average total human body content (adult, 70 kg) 45 500 g.

Uses. Oxygen is used on a massive scale to prepare thousands of organic and inorganic compounds. Steel making. Metal cutting.

Liquid oxygen is a burns and fire hazard. Finely divided elements react explosively with liquid oxygen.

Further reading
Kirk–Othmer ECT, 4th edn, Wiley, New York, 1996, Vol. 17, pp. 919–940.

OZONE CAS No. 10028-15-6

O_3

Found in the atmosphere at ~ 0.05 ppm at sea level. Air pollutant. Obtained when an electric discharge passes through the oxygen of the air. Ozone is a gas at room temperature; m.p. $-193°$; b.p. $-111.9°$; density $2.144\,g\,l^{-1}$ at $0°$.

Uses. Water disinfectant. Bleaching textiles, waxes and oils. In organic synthesis.

Ozone is produced in the upper atmosphere (stratosphere) in two steps. The double bond in dioxygen ($500\,kJ\,mol^{-1}$) is dissociated by UV radiation having a wavelength of less than 240 nm. The oxygen atoms combine with O_2 and via a third-party catalyst, e.g. N_2, forms ozone:

$$O_2 + h\nu \rightarrow 2O$$
$$O + O_2 + M \rightarrow O_3$$

Ozone in the lower atmosphere (troposphere) is a source of the hydroxyl radical (HO•) through the reaction of oxygen atoms derived from the photolysis of ozone, and water:

activated $O + H_2O \rightarrow 2HO•$

The hydroxyl species is responsible for the oxidation of many compounds, especially those of carbon and sulphur. Some of the products are toxic and may be carcinogenic.

High ozone concentrations come from nitrogen oxides produced in car exhausts and are apparent in cases of photochemical smog as found in cities such as Los Angeles and Tokyo. High concentrations are toxic and harmful to animal and plant life. Ozone is a probable carcinogen.

Occupational exposure limits.   USA TLV (TWA) 0.05 ppm (0.1 mg m^{-3}). UK long-term limit 0.3 ppm (0.6 mg m^{-3})

Further reading
Alloway, B.J. and Ayres, D.C. (1997) *Chemical Principles of Environmental Pollution,*
    2nd edn, Blackie, London.
Kirk–Othmer ECT, 4th edn, Wiley, New York, 1996, Vol. 17, pp. 953–994.

PALLADIUM CAS No. 7440-05-3

Pd

Palladium, a member of Group 10 of the Periodic Table, has an atomic number of 46, an atomic weight of 106.42, a specific gravity of 12.02 at 20°, a melting point of 1555° and a boiling point of $\sim$ 3000°. It is a steel-white metal and forms compounds in the +2, +3 and +4 oxidation states. Only six stable isotopes occur naturally: ^{102}Pd (1.02%), ^{104}Pd (11.14%), ^{105}Pd (22.33%), ^{106}Pd (27.33%), ^{108}Pd (26.47%) and ^{110}Pd (11.73%).

Palladium is found along with platinum group metals and is also associated with nickel–copper ores. Rare element, 0.5 ppb in the earth's crust.

Uses.   Finely divided palladium is a catalyst for hydrogenation/dehydrogenation reactions. Jewellery. Electrical contacts. Surgical instruments.
    Little is known about the toxicity of palladium apart from the highly toxic metal fumes but it is generally thought to be moderate. Finely divided palladium is a fire hazard.
    Palladium salts are considered harmful by inhalalation, ingestion and absorption through the skin. Cause eye and skin irritation. May affect the lungs, blood, reproduction. Tumorigenic to experimental animals.

Further reading
Kirk–Othmer ECT, 4th edn, Wiley, New York, 1996, Vol. 19, pp. 347–406.
ECETOX (1991) *Technical Report No 30(4)*, European Chemical Industry Technology and Toxicology Centre, Brussels.
SI 1991 No. 472 The Environmental Protection (Prescribed Processes and Substances)
    Regulations 1991, HMSO, London, 1991.

Palladium(II) chloride CAS No. 7647-10-1

PdCl$_2$

Uses.   Photography. Indelible ink.

Toxicity.   LD$_{50}$ oral rat 2700 mg kg^{-1}; LD$_{Lo}$ intravenous rabbit 18.6 mg kg^{-1}.

PAN CAS No. 2278-22-0
LC$_{50}$ inhalation rat 95 ppm (4 h)

$$CH_3—C—O—O—NO_2 \qquad\qquad C_2H_3NO_5 \quad M = 121.1$$
$$\underset{O}{\overset{\|}{}}$$

Peroxyacetyl nitrate, acetyl nitroperoxide. Highly explosive phytotoxic and lachrymatory component of smog whose effect is felt within 5 min by humans on exposure to it at 5 ppm.
    PAN was first detected in smog in 1956 by Haagen-Smit and Fox. It is generated in air by the influence of sunlight on volatile organic compounds and nitrogen peroxide.

With the formation of oxygenated species, including aldehydes (see **Methane**), through the action of the •OH radical the essential acetyl radical is produced:

$$CH_3-C\overset{O}{\underset{H}{\diagup}} \quad \xrightarrow[-H_2O]{•OH} \quad CH_3-\overset{•}{C}=O \xrightarrow{O_2} CH_3-C\overset{O}{\underset{O-O^•}{\diagdown}} \xrightarrow{NO_2} CH_3-C\overset{O}{\underset{O-O-NO_2}{\diagdown}}$$

This reaction is induced by natural hydrocarbons such as plant terpenes but the natural PAN levels fall below 1 ppb. In European cities the highest levels are usually reached in summer in the afternoon, following oxidation of hydrocarbons released earlier in the day, when NO_x production is also significant. Under extreme conditions the concentration of PAN may reach 20–30 ppbv, although its effect is less serious than that of ozone. In the Paris basin the highest levels were found in summer at 33.6 ppb, but in Tokyo observations from 1971 to 1985 show peaks at 10–22 ppb in winter. French forests have been damaged by PAN imported from Strasbourg and Germany, which raised the background level of 0.2 ppb to 1–2 ppb.

PAN is scavenged by the NO released by vehicles and so maximum levels are reached some 30–100 km away from the immediate source. Its concentration may be monitored by Fourier transform IR spectrometry, and also by GC on a polar phase below its decomposition temperature at 20–30° when EC detection is sensitive to 1 pptv.

Further reading
Ciccioli, P. (1993) in Bloemen, H.J. Th. and Burn, J. (eds), *Chemistry and Analysis of VOCs in the Environment*, Blackie, Glasgow.
Haagen-Smit, A.J. and Fox, M.M. (1956) *Ind. Eng. Chem.*, **48**, 1484–1487.
Ridley, B.A. (1991) Recent measurements of oxidised nitrogen compounds in the troposphere, *Atmos. Environ. A*, **25**, 1905–1926.

PARAQUAT CAS No. 4685-14-7
(Gramoxone, Cekuquat) 1910-42-5.
Herbicide Toxicity class II. LD_{50} rat, cat, guinea pig 30–112 mg kg^{-1}; LC_{50} rainbow trout 32 mg l^{-1}

$C_{12}H_{14}Cl_2N_2$ $M = 257.2$

1,1'-Dimethyl-4,4'-bipyridinium dichloride; decomposes at 300°; water solubility 700 g l^{-1}.

First described in 1958 by Brian and introduced by ICI Plant Protection for the control of broad-leaved weeds in fruit orchards and on plantation crops, vines, olives, tea, sugar beet, etc., and also on aquatic weeds. It is widely used in developing countries as it is cheap. ICI still produce 90% of the product (1993) but factories are also active in Texas, Japan, Brazil, India and Malaysia. It is very strongly absorbed by clays and is rapidly deactivated on contact with soil.

Paraquat is an eye irritant. Accidental ingestion by humans has led to fatal poisoning with severe injury to the lungs (fibrosis). It is a contaminant in marihuana at levels averaging 450 ppm and inhalation presents a risk to regular users.

Dermal exposure via soaked clothing can be fatal and 40 such deaths occurred in Negri Sembilan during 1989. Lesser consequences are inflamed eyelids, dermatitis and falling out of nails. In 1989 there were 171 pesticide-related suicides in Negri Sembilan. Paraquat is a common means of suicide elsewhere in Malaysia and in Sri Lanka – 1 teaspoonful of the concentrate will kill, there is no antidote and death can take up to 3 weeks

The limit in air in the UK and USA is TWA $0.1 \, \text{mg m}^{-3}$. Various standards apply to drinking water: in the UK(DOE) $0.3 \, \mu\text{g l}^{-1}$, USEPA $30 \, \mu\text{g l}^{-1}$, the WHO gives $0.3 \, \mu\text{g l}^{-1}$.

Codex Alimentarius: acceptable daily intake $0.004 \, \text{mg kg}^{-1}$ (1986). Maximum residue levels: polished rice, 0.5, milk 0.01, potatoes 0.2 (EC 0.05), vegetables $0.05 \, \text{mg kg}^{-1}$.

Further reading
Pestic. News (1992) (16) 4–6.

PARATHION CAS No. 56-38-2 (Et ester) CAS No. 298-00-0 (Me ester) **(Thiophos)**
Non-systemic, wide-range insecticide, acaricide. Toxicity class I

	Et ester	Me ester
LD$_{50}$ rat	$2 \, \text{mg kg}^{-1}$	$6 \, \text{mg kg}^{-1}$
LC$_{50}$ rainbow trout	$1.5 \, \text{mg kg}^{-1}$	$2.7 \, \text{mg kg}^{-1}$

$C_{10}H_{14}NO_5PS \quad M = 291.3$

O,O-Diethyl-*O-p*-nitrophenyl thiophosphate; m.p. 6°; b.p. $162°/0.6 \, \text{mmHg}$; v.p. $0.89 \, \text{mPa}$ (20°); water solubility $2.4 \, \mu\text{g ml}^{-1}$ (25°).

The ethyl ester was discovered by Schrader in 1946 and was introduced as a pesticide by ICI Plant Protection in 1952.

Parathion is effective against aphids and red spider mite and is used for their control on cereals, fruit, vines, hops and vegetables. The methyl ester ($C_8H_{10}NO_5PS$, $M = 263.3$) came to be preferred because it retained the activity but was less dangerous to mammals. The effectiveness depends on the conversion by the mixed function oxidases of insects into the highly toxic oxon form (see **Malathion**).

The half-life of parathion in water is 108 days and that of paraoxon is 144 days. During spraying in California under foggy conditions in 1986, levels up to $39 \, \mu\text{g l}^{-1}$ were detected.

The largest consumption of pesticides consumed per acre in all of Asia is in the Indian State of Uttar Pradesh where cotton is the main crop. Here parathion and endrin were sprayed every 4 days at levels of $0.25–1.0 \, \text{litre acre}^{-1}$. In 1987 this overdosing led to pest resistance and subsequent bankruptcy of plantation owners.

The ethyl ester is Prior Informed Consent listed with a USEPA limit in drinking water of $9 \, \mu\text{g l}^{-1}$ and a workplace air limit of $0.2 \, \mu\text{g l}^{-1}$ in the UK and USA.

Codex Alimentarius: acceptable daily intake $0.005\,mg\,kg^{-1}$ (1967). Maximum residue levels: citrus fruits 1.0, vegetables $0.7\,mg\,kg^{-1}$.

PARTITION COEFFICIENT

The *n*-octanol–water system mimics the membrane lipid–water barrier and the partition coefficient (K_{ow} = concentration of substance in the octanol phase / concentration in the water phase) between these two phases gives an indication of the tendency for toxins to cross cell membranes and so produce adverse effects. The *n*-octanol is a surrogate for fish lipid. The measured K_{ow} values for organic compounds range widely from 10^{-3} to 10^7 and hence it is usual to quote log K_{ow} (log P_{ow}):

for $K_{ow} = 0.01$ log $K_{ow} = -2.00$, showing little affinity for a lipid membrane;

for $K_{ow} = 100$ log $K_{ow} = 2.00$, showing considerable affinity for a lipid membrane.

For di-*n*-octyl phthalate $K_{ow} = 1.6 \times 10^4$ when measured at low solute concentration at 20°; log $K_{ow} = 4.204$ and this compound will readily cross a cell membrane. Note that at equilibrium the octanol phase contains $2.3\,mol\,l^{-1}$ of water and the aqueous phase contains $4.5 \times 10^{-3}\,mol\,l^{-1}$ of octanol, therefore K_{ow} can be reliably obtained by comparing solubilities in the pure solvents.

Log K_{ow} values for same substrates over a range of polarities are glucose -3.24, piperazine -1.17, ethanol -0.31, CCl_4 2.83, styrene 2.95, cyclohexane 3.44 and pyrene 4.88. The first three compounds are unlikely to cross the cell wall barrier in fish.

Further reading

Samiullah, Y. (1990) *Prediction of the Environmental Fate of Chemicals*, Elsevier/British Petroleum, Amsterdam.

PENDIMETHALIN CAS No. 40487-42-1
(Herbadox, Prowl)
Herbicide. Toxicity class III. LD_{50} rat $1250\,mg\,kg^{-1}$; LC_{50} rainbow trout $0.14\,mg\,l^{-1}$; LD_{50} bee $50\,\mu g/bee$

$C_{13}H_{19}N_3O_4 \quad M = 281.3$

N-(1-Ethylpropyl)-2, 6-dinitro-3, 4-xylidine; m.p.58°; v.p. 4.0 mPa (25°); water solubility $0.3\,mg\,l^{-1}$.

First reported by Sprankle in 1974 and introduced by American Cyanamide for the control of annual grasses and broad- leaved weeds in cereals, cotton, potatoes, fruit, brassicas, etc. The C-4 methyl substituent is metabolized by oxidation to a hydroxymethyl group and then to a carboxylic acid.

It is imported into Brazil. No international standards have been set.

PENTACHLOROPHENOL CAS No. 87-86-5

Non-selective insecticide, fungicide herbicide. LD_{50} rat $50\,mg\,kg^{-1}$; LD_{Lo} oral human $29\,mg\,kg^{-1}$; LC_{50} (48 h) rainbow trout $93\,\mu g\,l^{-1}$

$C_6HCl_5O \quad M = 266.3$

PCP, Penchorol; m.p. 191°; b.p. 310° (dec.); v.p. 1.2×10^{-4} Torr (100°) v.p. 40 Torr (211°); dens. 1.99; $\log P_{ow} 3.32$ at pH 7.2. Weak acid, $pK_a 4.71$, solubility in water $80\,mg\,l^{-1}$. Soluble in acetone ($215\,g\,l^{-1}$), diethyl ether, benzene. The sodium salt crystallizes from water as the monohydrate, solubility in water $330\,g\,l^{-1}$ (25°). This is used as a general disinfectant and as a slimicide in the manufacture of pulp and paper.

Pentachlorophenol is produced by the chlorination of phenol catalysed by $AlCl_3$ or $FeCl_3$ at 100–180°. It was introduced as a timber preservative in 1936 and is now applied to wood as an ester (e.g. the laurate); also used for the control of termites and as a fungicide and molluscicide. Peak production in the USA was 22×10^6 kg, when it was estimated that 3–6 million workers were exposed to it.

The airborne mean concentration in wood treatment plants ranged up to $300\,\mu g\,m^{-3}$. Incineration of 2,3,4,6-tetrachlorophenol and PCP produced polychloro-dibenzdioxins (PCDD) at $10\,\mu g\,g^{-1}$ of phenate. These compounds and polychloro-dibenzfurans (PCDF) are formed as by-products in the processing of PCP. Levels of Cl_7CDF in the range 130–$197\,pg\,g^{-1}$ were found in the blood plasma of workers who had been employed over 10 years in the industry, their average blood level was $4.7\,\mu g\,ml^{-1}$ (Teshka, et al. 1992).

Phenolic pollutants may be determined by HPLC using UV or electrochemical detection. The UV detection limits for 2, 4, 6- trichlorophenol and PCP are 2.0 and 0.5 ppb, respectively (Baldwin and Debowski, 1988). PCP may also be determined directly, without derivatization, by GLC on a capillary column coated with trifluoro-propylsilicone and methylsilicone (Reigner et al., 1990).

PCP dust at levels over $1\,mg\,m^{-3}$ irritates the eyes and may cause permanent damage; it is also absorbed through the skin. The dermal LD_{50} rat is $96\,mg\,kg^{-1}$. Pentachlorophenol affects mitochondrial electron transport so as to increase meta-bolic rate and raise the body temperature; in severe cases involving sawmill workers coma and death have resulted. Additional risks arise from the PCDD/PCDF impu-rities; in unregulated PCP (1977) levels could be significantly toxic: octa-CDD < 3600, hepta-CDD < 180, hexa-CDD < 23, octa-CDF < 210, hepta-CDF < 320, hexa-CDF < 36 ppm. The British Wood Preserving and Dampproofing Association subsequently set the following limits: octa-CDD < 800, hepta-CDD < 200, hexa-CDD < 8, octa-CDF < 200, hepta-CDF < 100, hexa-CDF < 20 ppm.

The sodium salt of PCP was used in China to control snail-borne schistosomiasis. It was subsequently found to have a total TCDD/TCDF content of 1000 ppb with a TEQ of 30 ppb. A control group had a lipid OCDD level of 117 ppb, whilst adults exposed to the spray had lipid OCDD in the range 40–748 ppb. In 1980 Brazil imported 427 t of pentachlorophenol.

The ACGIH has set a TLV (TWA) for vapour contact with the skin at $0.5\,mg\,m^{-3}$ and the WHO guideline for drinking water is $10\,\mu g\,l^{-1}$

References

Baldwin, D.A. and Debowski, J.K. (1988) Determination of phenols by HPLC down to ppt levels, *Chromatographia*, **26**, 186–190.

Reigner, B.G., Rigod, J.F. and Tozer, T.N. (1990) Simultaneous assay of pentachlorophenol and its metabolite tetrachlorohydroquinone by GC without derivatisation, *J. Chromatogr.*, **533**, 111–114.

Teshka, K., Kelly, S.J., Wiens, N., Hertzman, C., Dimich-Ward, H., Ward, J.E.H. and van Oostdam, J.C. (1992) Dioxins and furans in residents of a forest industry region of Canada, *Chemosphere*, **25**, 1741–1751.

Further reading

Chem. Safety Data Sheets (1991) **4b**, *Toxic Chemicals*, Royal Society of Chemistry, Cambridge.

Chemosphere (1988) **17**, 627–631.

PERCHLORIC ACID AND PERCHLORATES CAS No. 7601-90-3; 14797-73-0

$HClO_4$, ClO_4^-

Perchloric acid is a poison by ingestion. A serious irritant to the eyes, skin and mucous membranes. Powerful oxidizing agent. Inorganic perchlorates are irritant to the human body, avoid skin contact. Organic perchlorates are shock sensitive and explosive compounds. Perchlorates are powerful oxidizing agents.

Toxicity. LD$_{50}$ oral rat $1100\,mg\,kg^{-1}$.

PERMETHRIN CAS No. 52645-53-1
(Ambush, Kafil).
Insecticide. Toxicity class II (Ambush). LD$_{50}$ rat 430–$4000\,mg\,kg^{-1}$; LC$_{50}$ rainbow trout $2.5\,\mu g\,l^{-1}$; LD$_{50}$ bees > $50\,\mu g/bee$

$C_{21}H_{20}Cl_2O_3$ $M = 391.3$

(3-Phenoxyphenyl)methyl-3-(2,2-dichloroethenyl)-2,2-dimethylcyclopropane carboxylate; technical mixture m.p. 34–35°, pure *cis*-isomer 65°, *trans*-isomer 47°; v.p. $0.045\,mPa$ (25°); water solubility $0.2\,mg\,l^{-1}$

First reported by Elliott *et al.* in 1973 and introduced by ICI Agrochemicals. It has non-systemic action against pests in cotton, fruit, vines and vegetables at rates in the range 25–$200\,g\,ha^{-1}$. It is also used for control of flies and as a wool preservative. Permethrin is rapidly degraded through hydrolysis of the ester linkage.

Toxicity values vary depending on the *cis/trans* ratio. No standards have been set.

Codex Alimentarius acceptable daily intake $0.05\,mg\,kg^{-1}$. Maximum residue levels: wheat flour 0.2, milk 0.05, potatoes 0.05, citrus fruit 0.5, other fruit 0.05–$1\,mg\,kg^{-1}$.

PESTICIDES

Pesticides were first used to kill and therefore control organisms which reduce crop yields or which are hazardous to health, e.g. the malarial mosquito and the housefly. In use, the toxic risk they present must be taken into account, but these are outweighed by the social and financial benefits to farmers and consumers worldwide.

The earliest applications go far back into history – sulphur was first used in the years B.C. to control insects while compounds of arsenic and mercury were in early use by the Chinese to control body lice. Natural pesticides such as pyrethrum did not come into use in Europe until the late seventeenth century. Bordeaux Mixture for the control of diseases on fruit trees was discovered in the late nineteenth century.

Other substances in use early in this century include Paris Green, ferrous and lead sulphate, nicotine and various coal tar derivatives. However, the problems of pollution from pesticides and the need for their control only arose with the introduction and greatly increased scale of use of the synthetic organic pesticides (Conway and Pretty, 1991). These included the following classes: herbicides, insecticides, fungicides and acaricides for the control of mites and ticks.

Immediately after the Second World War, the most common compounds were organochlorines. The dramatic success of DDT in controlling lice which carry malaria, yellow fever and typhus enouraged an escalation in the use of this group including lindane, endrin, dieldrin and the chlorophenoxy herbicides 2, 4- D, 2, 4, 5-T and MCPA.

Unfortunately, owing partly to instances of gross over- application and to their resistance to natural degradation, unacceptably high levels of DDT, lindane and endrin accumulated and entered food chains. As a result, they were implicated in the decline in the populations of predatory birds. Examples of such incidents are the death of grebes at Clear Lake, California, in 1954 and the poisoning of trout and salmon at Prince Edward Island in the 1960s.

High levels of organochlorine pesticides in air and groundwater and their detection in human milk contributed to public pressure for limitation on their use. Aldrin and dieldrin are now banned in developed countries, DDT was banned by the USEPA as early as 1972, followed by 2, 4, 5-T in 1983. DDT was banned in the UK in 1984 but organochlorines continue to be manufactured and used in Third World countries.

The dilemma for regulators is illustrated by DDT, which is not toxic to humans and while evidently undesirable environmentally is effective in control of diseases endemic in the Third World.

With the decline in the use of organochlorines their place has increasingly been taken by the less persistent but more expensive synthetic pyrethroids, carbamates and organophosphorus compounds (*Pesticide Manual*, 1994). The latter group poses a further problem for regulators since many are highly toxic to mammals including humans, and proper precautions must be taken to avoid poisoning of workers in the factory and the field (Dinham, 1993).

In the UK there is current concern about the exposure of workers to sheep dips where the persistent organochlorines lindane and dieldrin have been supplanted by non-persistent but toxic organophosphates. There is also an ongoing investigation into 'Gulf War Syndrome', where it is alleged that in 1990–91 the health of service personnel was affected by the unmonitored distribution of organophosphates, including diazinon, fenitrothion and malathion. The use of an unapproved form of malathion for delousing captured Iraqi troops has been criticized (*Daily Telegraph*, 1996).

Details of the toxicology and control levels of individual pesticides may be found in a dictionary (Richardson and Gangolli, 1992–95) and levels set for residues in foodstuffs are given by the World Health Organization (*Codex Alimentarius*, 1994).

References
Codex Alimentarius (1994) FAO/WHO, Rome.
Conway, G.R. and Pretty, J.N. (1991) *Unwelcome Harvest*, Earthscan, London.
Daily Telegraph, 12 and 16 December 1996.
Dinham, B. (1993) *The Pesticide Hazard*, The Pesticide Trust, London.
Richardson, M.L. and Gangolli, S. (eds) (1992–95) *Dictionary of Substances and Their Effects*, Royal Society of Chemistry, Cambridge, Vols 1–7.
The Pesticide Manual (1994) 10th edn, British Crop Protection Council, Royal Society of Chemistry, Cambridge.

Further reading
International Code of Conduct in the Distribution and Use of Pesticides, FAO, Rome, 1990.

PHENOL CAS No. 108-95-2
LD_{50} rat 400 mg kg^{-1}; LC_{50} (96 h) sole 14 mg l^{-1}; LC_{50} (48 h) rainbow trout 7.7 mg l^{-1}

C_6H_6O $M = 94.1$

Hydroxybenzene, carbolic acid; b.p. 181°; m.p. 41°; v.p. (20°) 0.36 Torr. Water solubility 67 g l^{-1}, entirely soluble above 68°; dissociation constant in water $(20°, K_a) = 1.28 \times 10^{-10}$; soluble in most organic solvents.

Sampled by absorption in NaOH solution or from air on Tenax or carbon. Detection limit 2 µg m^{-3} by GC-FID and 0.3 µg m^{-3} by GC-MS. With HPLC and UV detection at 275 nm the limit is 50 µg m^{-3}, this may be improved by diazo coupling:

Phenol is a constituent of coal tar and of manure where it occurs at up to 55 µg kg^{-1} dry weight in air (The Netherlands).

The major industrial route is from cumene (isopropylbenzene; see **Acetone**) and production losses are estimated at 0.16 g kg^{-1} produced. World production in 1990 was 4.8×10^6 t year^{-1}, including USA 1.5, EC 1.2, Japan 0.4 and Eastern Europe 0.9×10^6 t.

The principal use (ca. 40%) is the synthesis of phenolic resins as binding materials for chipboard, insulation board, paints and moulds. The phenol content in these materials ranges between 2 and 50%. It is also used (ca 20%) for the synthesis of caprolactam, an intermediate for nylon 6, for 2,2-bis-1-hydroxyphenylpropane (bisphenol A) another resin component, for chlorophenols including pentachloro-phenol, and for aspirin.

Phenol is formed by the action of OH radicals on benzene. It has been detected in incinerator emissions, especially from wood burning; in the USA 109×10^6 t of fuel wood gave rise to 7200 t of phenols and in car exhaust phenol itself is found up to a level of 2.0 ppm. It occurs in sidestream smoke at ca. 0.4 mg/cigarette. In 1992 the UK produced 6.4×10^6 t of coke by carbonization of 9.0×10^6 t of coal at 1300° and emissions of phenols were estimated at 5000 9000 t year^{-1}.

Phenol occurs in ferrous foundry leachate. Levels of 460 µg l^{-1} have been detected in leachate near landfill and 40% of active US sites showed positive. Petroleum refinery effluent typically contains 200 µg l^{-1}. In the UK, surface water levels were 0.60 µg l^{-1} (river Dee) and in the range 0.1–16 µg l^{-1} (Mersey estuary); the Hayashida river in Japan had a phenol level of 306 µg l^{-1} arising from discharge by the leather industry. Phenol occurs in rainwater as the major phenolic component at levels in the range 2–8 µg l^{-1} (USA).

Phenol is degraded in sewage within 8 h, its lifetime in freshwater is < 1 day and in soil it survives for 2–5 days. The quarterly average level in air in Santa Clara (USA) was 0.13 µg m^{-3} and in urban Osaka it was 2.1–3.9 µg m^{-3}; near a US resin factory the level varied between means of 66–306 µg m^{-3}.

Phenol is incorporated into antiseptics, disinfectants, lotions and ointments, although dressings containing 5–10% of phenol are no longer in use as they can cause necrosis of the skin and tissue. Its solutions are corrosive to the eyes and skin and the vapour irritates the respiratory tract; inhalation may lead to anorexia, weight loss, vertigo and the passing of dark urine.

The odour threshold in some individuals may be as high as 5 ppm, whilst the taste threshold in water is 0.3 mg l^{-1}. Pollution of a river in North Wales by accidental release led to levels in the range 4.7–10.3 µg l^{-1} for several days with a significant rise in gastrointestinal illness; chlorophenols appeared when this water was chlorinated. Ingestion of 4.8 g caused death of the patient but others have survived larger doses.

Occupational exposure ocurred in physicians in the last century when phenol was widely used as an antiseptic; other cases were found in Bakelite production workers. A major mortality study, involving 360 000 man years of workers in five companies producing phenol – formaldehyde resins, did not reveal any significant variation from that of the general population. There was no statistically meaningful increase in cancer of the kidney, or oesophagus or in Hodgkin's disease. The USEPA has classified phenol as a group D substance – data inadequate for assessing carcinogenic potential.

The occupational limit (TWA) set by the ACGIH is 19 mg m^3 and the EC limit for drinking water is 0.5 mg l^{-1}.

Further reading
Environ. Health Criteria, World Health Organization, Geneva, 1994, Vol. 161.

PHOSGENE CAS No. 75-44-5

$Cl_2C{=}O$ CCl_2O $M = 98.9$

Carbonyl chloride; m.p. $-118°$; b.p. 8°; v.p. 1180 Torr (20°); v. dens. 3.4. Phosgene has a characteristic hay-like odour, threshold 0.6 mg m^{-3} (0.14 ppm). It is slightly soluble in water, which hydrolyses it slowly at 20° to CO_2 and HCl, it is soluble in benzene, acetic acid, etc.

First prepared in 1812 by Davy by the reaction of CO and chlorine in the light. Manufactured by passing these reagents over activated carbon; emissions are controlled on-site by scrubbing with sodium hydroxide solution or water; alternatively, it may be burnt to CO_2 and water. Phosgene is usually produced and used on the same site as its high toxicity is a potential risk during transport.

It is widely used as an intermediate for the preparation of isocyanates for the manufacture of polyurethanes and polycarbonate, also for the synthesis of pharmaceuticals and pesticides. The demand in the USA in 1994 was 1.18×10^6 t.

Phosgene may be formed when chlorinated solvents are heated in air. At levels above the odour threshold $(0.6 \, \mathrm{mg \, m^{-3}})$ it becomes choking and unpleasant. There is little warning of risk and the symptoms of poisoning may be delayed by 2–24 h, these include burning in the throat and chest with breathing difficulty in severe cases, followed by death from pulmonary oedema. Experience in World War I led to an estimated human LC_{50} value of 400 ppm for 2 min.

A visual indication of phosgene vapour is the yellow–orange coloration of paper impregnated with diphenylamine and p-dimethylaminobenzaldehyde; this test is positive at the threshold limit level. The ACGIH has set a TLV (TWA) of 0.1 ppm $(0.40 \, \mathrm{mg \, m^{-3}})$.

Further reading
Kirk–Othmer ECT, 4th edn, Wiley, New York, Vol. 18.
Reichert, D., Sphengler, U. and Henschler, D.J. (1979) *Chromatographia*, **179**, 181.

PHOSPHORUS CAS No. 7723-14-0

P

Phosphorus, a member of Group 15 of the Periodic Table, has an atomic number of 15, an atomic weight of 30.9738, a specific gravity (white P) of 1.82 at 20°, a melting point (white P) of 44.1° and a boiling point (white P) of 280°. Phosphorus exists in four allotropic forms: white (or yellow), red and black (or violet). The two stable oxidation states are +3 and +5. Naturally occurring phosphorus consists of only one stable isotope, ^{31}P. The radioactive isotope ^{32}P is used as a tracer.

Phosphorus is widely distributed in combination with many minerals. Average concentration in the earth's crust is 1000 ppm. Numerous phosphate minerals are known, including the various *apatites*, $Ca_5(PO_4)_3X$, where X can be OH^- (*hydroxyapatite*), F^- (*fluorapatite*) and the less common Cl^- (*chlorapatite*). Phosphate rock, which contains tricalcium phosphate, is an important source of the element (worldwide production $159 \, \mathrm{t \, year^{-1}}$).

The phosphorus cycle is important for all living matter. Constituent of RNA and DNA. Typical concentrations in humans are bone $\sim 70\,000$ ppm; blood $350 \, \mathrm{mg \, l^{-1}}$; daily intake $1–2 \, \mathrm{g \, day^{-1}}$; total body mass (70 kg adult) 780 g.

Uses. Manufacture of phosphoric acid. Steels. Phosphor bronze. Rat poison. Tracer bullets. Incendiaries. Fertilizers.

Ordinary phosphorus is a waxy white solid. It takes fire spontaneously in air, burning to phosphorus(V) oxide. In a confined space it will remove oxygen and cause asphyxiation. Dangerous fire hazard when exposed to heat, flame or chemical oxidants. Six workers died as a result of a fire in a phosphorus flame-retardant plant in Charleston, South Carolina (17 June 1991).

Human poison by ingestion. Symptoms include cardiomyopathy, cyanosis, vomiting and sweating. Toxic amounts affect the liver and can cause severe eye damage. The lethal dose for humans when taken internally (as P_4) is $\sim 100\,mg$. Chronic exposure by inhalation can cause anaemia, gastrointestinal effects and brittleness of the long bones. The most common symptom is necrosis of the jaw (phossy-jaw). Red phosphorus is also a human poison

Occupational exposure level. USA PEL (TWA) $0.1\,mg\,m^{-3}$

Further reading
Emsley, J. and Hall, D. (1986) *The Chemistry of Phosphorus*, Harper and Row, New York.
Kirk–Othmer ECT, 4th edn, Wiley, New York, 1996, Vol. 18, pp. 719–796.

Phosphates CAS No. 14265-44-2

PO_4^{3-}

Alkali metal phosphates are important commercially. They are powerful irritants. Sodium triphosphate is used in detergents but its use is declining owing to the need to control eutrophication. Calcium phosphates are used as fertilizers. Phosphates are used in the production of special glasses, such as those used for sodium lamps. Bone ash (calcium phosphate) is used to produce fine chinaware and to produce monocalcium phosphate used in baking powder. Organophosphates are often highly toxic pesticides.

Further reading
Kirk–Othmer ECT, 4th edn, Wiley, New York, 1996, Vol. 18, pp. 669–718.

Phosphides CAS No. 14901-63-4

P^{3-}

Prepared from a metal and phosphorus. They should be kept away from water or acids since they react to form the very toxic phosphine. Dangerous fire hazard and moderate explosion hazard by chemical reaction.

Further reading
Environmental Health Criteria 73. Phosphine and Selected Metal Phosphines, WHO/IPCS, Geneva, 1988.

Phosphites CAS No. 14901-63-4

PO_3^{3-}

Uses. Powerful reducing agents.

Phosphoric acid CAS No. 7664-38-2

H_3PO_4

Uses. Pickling agent for steel. Thin film of iron(III) phosphate prevents corrosion. Catalyst. Flame proofing. Produced in multi-million tonne quantities.

Toxicity. Irritant to eyes, respiratory tract and mucous membranes.

Phosphorous acid CAS No. 10294-56-1

$P(OH)_3$

Uses. Preparation of phosphites.

Toxicity. Corrosive.

Phosphorus (III) chloride CAS No. 7719-12-2

PCl_3

Uses. Manufacture of phosphoric acid, H_3PO_3. Starting material for the herbicide glyphosate.

Occupational exposure limit. OSHA TLV (TWA) 0.5 ppm ($3 \, mg \, m^{-3}$)

Toxicity. Severe burns to skin, eyes and mucous membrane. LD_{50} oral rat $550 \, mg \, kg^{-1}$. LC_{50} (4h) inhalation rat $660 \, mg \, kg^{-1}$.

Phosphorus (V) chloride CAS No. 10026-13-8

PCl_5

Uses. Manufacture of chlorophosphazenes. Catalyst.

Occupational exposure limit. OSHA TLV(TWA) $1 \, mg \, m^{-3}$

Toxicity. Corrosive and toxic compound. Skin burns heal slowly and painfully. LD_{50} oral rat $660 \, mg \, kg^{-1}$. LC_{50} (10 min) inhalation mouse 120 ppm.

Phosphorus hydride: CAS No. 7803-51-2
Phosphine

PH_3

Uses. Chemical synthesis. Doping agent in semiconductors. Fumigant.

Occupational exposure limits. US TLV (TWA) 0.3 ppm ($0.42 \, mg \, m^{-3}$). UK short-term limit 0.3 ppm ($0.4 \, mg \, m^{-3}$).

Toxicity. Toxic and flammable gas. LC_{50} (4 h) inhalation rat 11 ppm. LC_{Lo} inhalation human 1000 ppm. Death is usually due to respiratory paralysis. Fumigant applicators exposed to phosphine above recommended standards showed chromosome changes. Twelve people developed nausea and one died when phosphine was emitted from a warehouse near to their apartment block. Metal workers at a Norwegian shipyard exposed to $\sim$ 1ppm phosphine reported symptoms of nausea, dizziness, dyspepsia and chest tightness.

Further reading
Environmental Health Criteria 73. Phosphine and Selected Metal Phosphides. WHO, Geneva, 1988.

Phosphorus(V) oxide CAS No. 1314-56-3

P_4O_{10}

Uses. Dehydrating agent. Manufacture of $POCl_3$. Intermediate in preparation of phosphate esters.

Toxicity. Irritating. LC_{50} (1 h) inhalation mouse $271 \, mg \, m^{-3}$.

Phosphorus (V) oxychloride CAS No. 10025-87-3

$POCl_3$

Uses. Manufacture of alkyl and aryl orthophosphate triesters.

Toxicity. Severe burns to skin, eyes and mucous membranes. Inhalation can cause pulmonary oedema. LD_{50} oral rat $380\,mg\,kg^{-1}$.

Phosphorus (III) sesquisulphide CAS No. 1314-85-8

P_4S_3

Uses. 'Strike anywhere' matches.

Toxicity. LD_{50} oral rabbit $100\,mg\,kg^{-1}$. Flammable.

Phosphorus(V) sulphide CAS No. 1314-80-3

P_4S_{10}

Uses. Lubricating oil additives. Insecticide. Ore flotation agents.

Toxicity. LD_{50} oral rat $390\,mg\,kg^{-1}$.

Phosphorus(V) sulphochloride CAS No. 3982-91-0

$PSCl_3$

Uses. Manufacture of insecticides such as parathion.

Toxicity. Poison.

PIRIMIPHOS-METHYL CAS No. 29232-93-7
(Actellic)
Insecticide, acaricide. Toxicity class II (diethyl).

	O,O-diethyl	*O,O*-dimethyl
LD_{50} rat	$200\,mg\,kg^{-1}$	$2050\,mg\,kg^{-1}$
LC_{50} rainbow trout	$0.02\,mg\,l^{-1}$	$0.64\,mg\,l^{-1}$

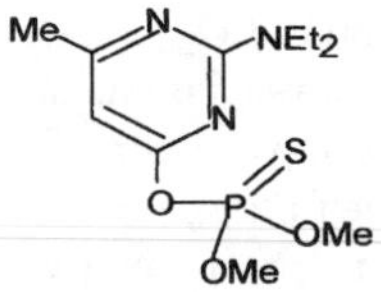

$C_{11}H_{20}N_3O_3PS$ $M = 305.3$

O-2-Diethylamino-6-methylpyrimidin-4-yl-*O,O*-dimethyl   phosphorothioate;   m.p. 18° (tech.) v.p. $2\,mPa$ (20°); water solubility $9\,mg\,l^{-1}$.

Introduced by ICI Plant Protection Division and used to control pests in warehouses, animal houses, grain stores, domestic and commercial premises and also on vegetables, cereals, fruit, vines, etc. The half-life in plants is < 2 days and it is *N*-de-ethylated to the mono-*N*-ethyl compound. The *O,O*-dimethyl compound is preferred to the *O,O*-diethyl analogue since it is less toxic and is placed in the USEPA class III.

Pirimiphos residues have been found amongst other pesticides in 25–90% of samples from 11 Egyptian Governates.

Codex Alimentarius acceptable daily intake $0.03\,mg\,kg^{-1}$. Maximum residue levels (for the Me ester): white bread 0.5, milk 0.05, potatoes 0.05, citrus fruits $2.0\,mg\,kg^{-1}$.

PLASTICIZERS
These compounds are the most widespread of all chemical pollutants – this results from their inclusion in a wide range of plastics in concentrations of up to 50% by weight of the finished product.

The major pollutant is di-(2-ethylhexyl)phthalate with US production (1984) 136×10^3 t and estimated global release 402 t year^{-1}. Other significant compounds are:

Compound	World production (1973) (t year^{-1})	Est. global release (t year^{-1})
Di-*n*-butyl phthalate	230×10^3	230
Diethyl phthalate	7.7×10^3	7.7
Dimethyl phthalate	2.2×10^3	6.5

The general structure of these phthalate esters is

Diethylhexyl phthalate (DEHP) accounts for 50% of all production but only a small proportion of this is added to food-contact plastics. The major plasticizer in contact with food is di-(2-ethylhexyl) adipate (DEHA), which is added to PVC to make flexible clingfilm, although phthalates and phosphates are also used to a lesser extent.

The thermoplastics account for 80% of the total market and of these the most important are polyethylene (PE), polypropylene (PP), polystyrene (PS) and PVC. This last is most receptive to the addition of plasticizers; the physical properties of the others are usually varied during the polymerization process by changing the make-up of the components, although phthalates and adipates are used in acrylic polymers.

In the UK in 1987 13 500 t of plasticized PVC was in contact with food, leading to the following estimates of dietary intake (mg/person/day): DEHA 16, dibutyl phthalate (DBP) 1.9 and dicyclohexyl phthalate (DCHP) 1.4, with others < 0.5. Some details of the properties of DEHA and DCHP follow; for more about DEHP, see **Diethylhexyl phthalate**.

Di-(2-ethylhexyl) adipate (DEHA) CAS No. 103-23-1

$$C_{22}H_{42}O_4 \quad M = 370.6$$

b.p. 417°, v.p. 26 Torr (200°); water solubility < 200 mg l^{-1}.

About 2.5% of adipic acid production goes into the productionof DEHA. It occurs in rivers in the USA in the concentration range 0.02–30 μg l^{-1}. It is not used in food packaging in the USA but is used otherwise as a solvent and plasticizer in bath oils, eye shadow, foundation creams and moisturizers.

The LD$_{50}$ for the rat is 9–15 g kg^{-1}. It has been classed as a group 3 carcinogen by the IARC.

Dicyclohexyl phthalate (DCHP) CAS No. 84-61-11

$$C_{20}H_{26}O_4 \quad M = 330.6$$

m.p. 58–65°; b.p. 218°/5 Torr.

These phthalates are produced by the reaction of the appropriate alcohol with phthalic anhydride and in the USA over 50% of this product is used for the manufacture of plasticizers.

Others

Other plasticizers include esters of benzene hexacarboxylic acid, e.g. tris(2-ethylhexyl) trimellitate:

This has a low vapour pressure and is added to electric cable insulation.

Phosphates are also added to PVC to build up its fire resistance. Two of these are tri-n-butyl phosphate, $(BuO)_3P{=}O$ (CAS No. 126-73-8), 1 kt of which was produced in the USA in 1983 with an annual global release estimated at 3 t, and triphenyl phosphate, $(PhO)_3P{=}O$ (CAS No. 115-86-6), with US production (1983) 14 kt and an estimated global release of 42 t.

Further reading

Verrier, P. (1993) in Wickson, E.J. (ed.), *Handbook of PVC Formulations*, Wiley, New York, p. 784.

PLATINUM CAS No. 7440-06-4

Pt

Platinum, a member of Group 9 of the Periodic Table, has an atomic number of 78, an atomic weight of 195.09, a specific gravity of 21.45 at 20°, a melting point of 1768°, and a boiling point of 3825°. It is a silvery white, malleable, ductile noble metal and forms compounds in the +2, +3 and +4 oxidation states. Natural platinum consists of six isotopes: ^{190}Pt (0.010%, $t_{1/2} = 6 \times 10^{11}$ years), ^{192}Pt (0.79%, $t_{1/2} \approx 10^{15}$ years), ^{194}Pt (32.9%), ^{195}Pt (33.8%), ^{196}Pt (25.3%) and ^{198}Pt (7.21%).

The element occurs native with other metals such as iridium, osmium, palladium, ruthenium and rhodium. *Sperrylite* $(PtAs_2)$, occurring with nickel ores, is an important commercial source. Concentration in the earth's crust is 0.001–$0.005\,mg\,kg^{-1}$.

Uses. Jewellery. Wire and vessels for laboratory use. Electrical contacts. Dentistry. Platinum–cobalt alloys. Contact process in the manufacture of sulphuric acid. Catalytic converters for car exhausts. Corrosion-resistant metal.

Exposure to platinum salts is mainly confined to occupational environments, particularly platinum metal refineries and catalyst manufacturing plants. Symptoms of over-exposure include corticaria, skin dermatitis and breathing disorders varying from sneezing to severe asthma. From modern day catalytic converters, Pt emission is ~ 2–$39\,ng$ per kilometre travelled with a particle size of 4–9 µm. Specific complexes of Pt such as *cis*-diamminedichloroplatinum(II) (cisplatin) are used therapeutically as anti-cancer drugs.

A reported case involved self-poisoning by ingestion of 10 ml of photographic toner containing 600 mg of potassium hexafluoroplatinate(IV). Symptoms included acute oliguria, kidney failure, acidosis, fever, cramps and gastroenteritis. The patient recovered in 6 days.

Occupational exposure level. OSHA PEL (TWA) $1\,mg\,m^{-3}$ (metal); $0.002\,mg\,Pt\,m^{-3}$ (soluble salts). Finely divided platinum is flammable.

Further reading

ECETOX (1991) *Technical Report 30(4)*, European Chemical Industry Technology and Toxicology Centre, Brussels.

Environmental Health Criteria 125. Platinum, WHO, Geneva, 1991.

Kirk–Othmer ECT, 4th edn, Wiley, New York, 1996, Vol. 19, pp. 347–406.

SI 1991 No. 472 The Environmental Protection (Prescribed Processes and Substances) Regulations 1991, HMSO, London, 1991.

Wiltshaw, E. (1979) Cisplatin in the treatment of cancer. *Platinum Met. Rev.*, **23**, 90–98.

Platinum(II) chloride CAS No. 10025-65-7

$PtCl_2$

Uses. Catalyst.

Toxicity. LD_{50} oral rat $18\,mg\,kg^{-1}$.

POLONIUM CAS No. 7440-08-6

Po

Polonium, a member of Group 16 of the Periodic Table, has an atomic number of 84, an atomic weight of ~ 210, a specific gravity of 9.32 at $20°$, a melting point of $254°$ and a boiling point of $962°$. Its principal oxidation states are $-2, +2, +4$ and $+6$. Polonium has a larger number of isotopes than any other element.

Polonium is a very rare natural element. Uranium ores contain $\sim 100\,\mu g\,t^{-1}$ of the element. It is obtained artificially by bombarding natural bismuth (^{209}Bi) with neutrons to give ^{210}Bi, the parent of ^{210}Po (half-life $= 138.39$ days). Polonium-210 is very dangerous to handle. A 1 ng amount emits as many α-particles as 5 g of radium. Damage to human tissue arises from the absorption of the energy of α-particles into tissue. Maximum permitted concentration for the human body is only $0.03\,\mu Ci$, which is equivalent to $6.8 \times 10^{-12}\,g$ of the element. MAC for polonium in air is $\sim 2 \times 10^{-11}\,\mu Ci\,cm^{-3}$. Weight for weight, polonium is about 2.5×10^{11} times as toxic as hydrogen cyanide.

Further reading

Eisenbud, M. (1987) *Environmental Radioactivity. From Natural, Industrial and Military Sources*, 3rd edn, Academic Press, New York.

King, R.B. (ed.) (1994) *Encyclopedia of Inorganic Chemistry*, Wiley, Chichester.

Kirk–Othmer ECT, 4th edn, Wiley, New York, 1996, Vol. 20, pp. 871–906.

Thompson, R. (ed). (1995) *Industrial Inorganic Chemicals. Production and Uses*, Royal Society of Chemistry, Cambridge.

Ullmann's Encyclopedia of Industrial Chemistry, 5th edn, VCH, Weinheim, 1993, Vol. A22, pp. 499–591.

POLYBROMOBIPHENYLS (PBBs)

This group should be compared with the PCBs (see **Polychlorinated biphenyls**) as they are also a complex mixture of isomers. There are a possible 209 structural isomers but the hexabromobiphenyls are the major components of the commercial product, which was used as a flame retardant for synthetic fibres and thermoplastics, including polyurethane foam.

3,3',4,4',5,5'-Hexabromobiphenyl CAS No. 60044-26-0

$$C_{12}H_4Br_6 \quad M = 627.5$$

M.p. 249° (benzene).

2,2',4,4',5,5'-Hexabromobiphenyl CAS No. 59080-40-9
TD_{Lo} mouse $360\,mg\,kg^{-1}\,day^{-1}$ for 6–15 days

$$C_{12}H_4Br_6 \quad M = 627.5$$

M.p. 72°; solubility in water $11\,\mu g\,l^{-1}$, soluble in acetone, benzene. IARC group 2B.

Octabromobiphenyl CAS No. 27858-07-7
LD_{Lo} rat $2\,g\,kg^{-1}$.

$$C_{12}H_2Br_8 \quad M = 785.4$$

In Michigan in May 1973, serious pollution resulted from the accidental mixing of the PBB-based fire retardant Firemaster FF-1 (295 kg) with animal feed, killing thousands of cattle, hogs and sheep and also millions of chickens. Over 10 000 residents were exposed to PBBs which they had ingested from meat, milk and eggs. Levels in human serum ranged between 0 and $1900\,\mu g\,l^{-1}$. US production ceased after the Firemaster incident.

Further reading
Landrigan, P.J., Wilcox, K.R., Silva, J., Humphrey, H.E.B., Kauffman, C. and
   Heath, C.R. (1979) Cohort study of Michigan residents exposed to PBBs, *Ann.*
   N.Y. Acad. Sci., **320**, 284–294.

POLYCHLOROBIPHENYLS (PCBS)
(Aroclor Kanechlor, Clophen)

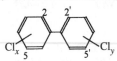

A group of 209 compounds prepared by the direct chlorination of biphenyl catalysed by $FeCl_3$. The process has the character of a radical substitution and a complex mixture of compounds with different chlorine numbers is produced after a short reaction time.

Because the possible isomers are permuted between 10 available positions, the number of monochloro compounds (3) = the number of nonachloro compounds (3, the possible places for the unsubstituted H atoms). The number of dichlorocompounds (15) equates with that of the octachlorocompounds (15), and so on. The total isomer numbers are:

monochloro = 3
dichloro   = 15
trichloro   = 25
tetrachloro = 42

```
pentachloro =  38
hexachloro  =  42
heptachloro =  25
octachloro  =  15
nonachloro  =   3
decachloro  =   1
Total          209
```

A direct consequence is that the industrial process produces complex, inseparable mixtures of isomers. This did not limit the applications of the PCBs, since these depend on properties common to the whole group. Commercial mixtures separated by distillation were made available by Monsanto and other companies and the fractions were coded for reference purposes; thus Aroclor 1242 implies (12) that the starting material is biphenyl and (42) has a chlorine content of 42%. Those of widest use were Aroclor 1242 and 1254; note that the highest possible chlorine content is found in $C_{12}Cl_{10} = 71\%$.

Aroclor 1242 CAS No. 53469-21-9
B.p. 325–366°; dens. 1.38–1.39.

The principal component PCBs are dichloro (16%), trichloro (49%) tetrachloro (25%) and pentachloro (8%).

Aroclor 1254 CAS No. 11097-69-1
B.p. 365–390°; dens. 1.49–1.51.

The principal component PCBs are tetrachloro (21%), pentachloro (48%), hexachloro (23%) and heptachloro (6%).

Modern GC techniques coupled with MS detection now make it possible to resolve these complex mixtures (Erickson, 1992).

A range of physical properties is shown by the individual isomers:

2-Chlorobiphenyl CAS No. 2051-60-7

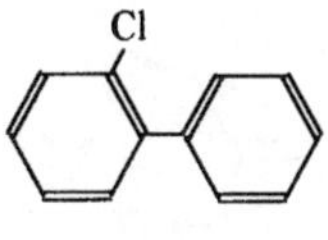

$C_{12}H_9Cl$ $M = 188.6$

B.p. 274°; m.p. 33°.

2,4,6-Trichlorobiphenyl CAS No. 35693 -92-6

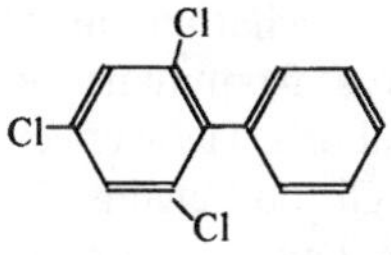

$C_{12}H_7Cl_3$ $M = 257.5$

B.p. 172°/15 Torr; m.p. 62°.

2,2′,4,4′5,5′-Hexachlorobiphenyl CAS No. 35065-27-1
TD_{Lo} Oral rat $50\,mg\,kg^{-1}$

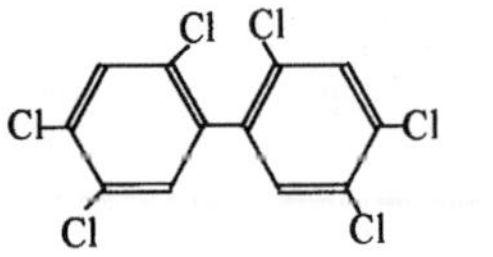

$C_{12}H_4Cl_6$ $M = 360.8$

An experimental teratogen. It is the major hexachlorobiphenyl in a group forming 22% of all PCBs in human tissue. The 3,3′,4,4′,5,5′-hexachloro isomer has teratogenic and reproductive effects – oral mouse TD_{Lo} $40\,mg\,kg^{-1}$.

A mixture of pentachlorobiphenyls had an LD_{50} in the rat of $3.58\,g\,kg^{-1}$.

In early use the PCBs had the advantages of non- flammability and stability. They were used as insulating liquids (askarels), in synthetic rubber, plasticizers, flame retardants, printing inks, paints, wallpapers and carbonless copy paper. These diverse applications led to their widespread distribution.

PCBs were first identified as environmental pollutants in 1966 by Soren Jensen as co-occurring with DDT isomers in a GLC analysis of human blood. Subsequently they were found at 90 ppm, a toxic level, in the body fat of polar bears, and they were held responsible for the decline of seal populations through 1950–70 in the Baltic and Dutch Wadden Sea. PCBs are associated with the decline in the populations of the European otter in the years following 1950.

The metabolism of PCBs is illustrated for 2,2′,5,5′-tetrachlorobiphenyl:

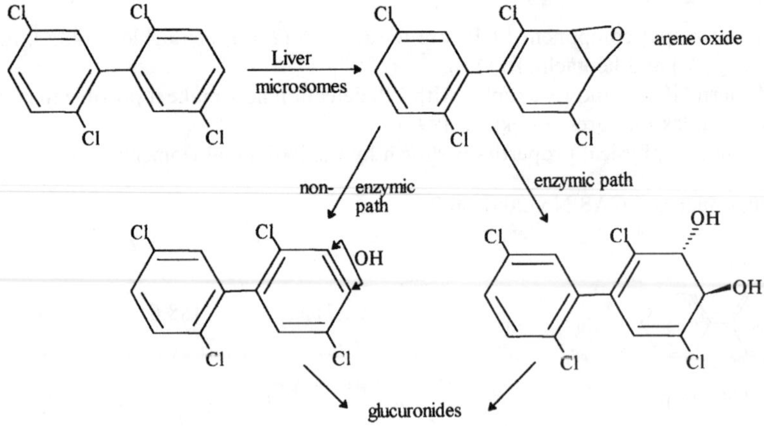

Those with higher chlorine numbers are particularly persistent and so are still detected despite the ending of production in the USA in 1976 and what amounted to a worldwide ban in the years after 1977. The need for a ban arose from numerous examples of adverse effects on animals such as those mentioned above. These effects arose because of the very high biomagnification of PCBs through food chains, for example the decline of the herring gull in the Great Lakes. In Lake Ontario the biomagnification followed the sequence-phytoplankton 250, mysid 45 000; smelt 835 000, herring gull 25×10^6 (Colborn et al., 1996).

Other recent examples come from samples taken in 1991 of the rural air at Ulm, Germany, when PCB levels of 170 pg m^{-3} were found, equivalent to those elsewhere in Europe and over the Great Lakes. Further evidence of the worldwide spread and persistence of these compounds comes from samples taken in 1992 of the milk from five seal species, namely Australian (44 ppm), Arctic (232 ppm), Californian (370 ppm) and Antarctic (10 ppm).

PCBs occur in everyone's body fat along with some 200 other contaminants. More seriously in the USA and Europe a breast-fed child receives five times the allowable daily allowance for a 150 lb adult. Dioxins also enter humans in this way.

A serious episode of human poisoning occurred in 1968 in southwest Japan, the Yusho incident, and a similar event took place in Taiwan towards the end of 1978. The toxic effects arose because of the contamination of rice oil with PCBs and PCDFs. In Taiwan some 2000 people were affected with the illness. It has since been concluded that the dermatological outbreak, which included chloracne, was largely caused by the PCDFs.

References
Colborn, T., Dumanowski, D. and Myers, J.D. (1996) *Our Stolen Future*, Little, Brown, Boston.
Erickson, M.D. (1992) *Analytical Chemistry of PCBs*, Lewis, Boca Raton, FL.

Other reading
Jensen, S. (1966) *New Sci.*, **32**, 612.
Kirk–Othmer ECT, 4th edn, Wiley, New York, 1993, Vol. 6, p. 127.

POLYCHLORODIBENZODIOXINS (PCDD) AND POLYCHLORODIBENZOFURANS (PCDF)

PCDD   PCDF

These dioxins and furans were produced inadvertently during chemical manufacture and processing and by fires and incineration of residues. They are associated with other 'dioxin-like' substances, which include PCBs, polychloroterphenyls, polyhalo-diphenyl ethers and polyhalonaphthalenes. The toxicity of these compounds is due to their ability to bind to a cellular receptor, the Ah receptor, when they are translocated to the nucleus where they bind to DNA to induce abnormality.

Like the PCBs, these substances are persistent in the environment and significant residues remain, despite increased control and regulation of potential sources. Their analytical chemistry and toxicology are also complicated by the wide range of isomers:

Numbers of isomers

	PCDD	PCDF		PCDD	PCDF
Monochloro	2	4	Pentachloro	14	28
Dichloro	10	16	Hexachloro	10	16
Trichloro	14	28	Heptachloro	2	4
Tetrachloro	22	38	Octachloro	1	1

Total isomers: PCDD 75; PCDF 135

PCDDs have no useful function and they are formed by the condensation of chlorophenols, especially when temperatures rise above the preparative optimum. This was the cause of the notorious Sevéso incident on 10 July 1976 at the ICMESA plant near Milan, when an overheated kettle discharged vapour which contaminated an area of 1430 ha including an intensely polluted zone A of 110 ha. The discharge included solvent glycols, alkali and 2,4,5-trichlorophenol, the major product, formed by the hydrolysis of 1,2,4,5-tetrachlorobenzene. Once the reaction temperature exceeds 220°, the undesirable condensation becomes significant:

a pre-dioxin

2,3,7,8-tetrachlorodibenzo-p-dioxin

This product is the most toxic of the whole group; the guinea pig is most susceptible with an LD_{50} of only $0.6\,\mu g\,kg^{-1}$. This compound is teratogenic in rats and mice at doses as low as 0.5–$3.0\,\mu g\,kg^{-1}$, but there is a wide variation amongst mammals for dogs are much less at risk and humans fall in the middle range of animal sensitivity.

Extensive studies have failed to link TCDD directly with human mortality, the principal effect is to induce the unsightly, painful and persistent form of acne known as chloracne. A significant moderating feature is that PCDD and PCDF are strongly adsorbed on sediments and in this form they are less toxic than when applied directly or in solutions.

The toxicity of individual compounds is related to that of 2,3,7,8-TCDD as unity by their toxic equivalent factor (TEF). 1,2,3,7,8-Penta-CDD (PCDD) has a TEF of 0.5, as has the most dangerous of the PCDFs. This latter group is formed from PCBs following their hydroxylation either by cytochrome-induced metabolism or by the •OH radical:

2,3,4,7,8-pentachloro-dibenzofuran

Humans are most at risk from long-term intake of low levels in food. The principal original sources of PCDD and PCDF arise from: incineration of waste, especially from municipal incinerators; exhaust emissions from leaded petrol engines which contain dichloroethane as a scavenger; the bleaching of wood pulp with chlorine; and other industrial activities such as wire stripping and metal refining.

The largest source of dioxins in the UK is the emission by British Steel of gases formed from residual cutting oils during the sintering of iron and steel: this is estimated to be 590 g toxic equivalents per year (Environmental Data Services, 1995). The toxic equivalent value (TEQ) derives from the TEF and is the sum of the effects of the components of a mixture scaled in proportion to 2,3,7,8-TCDD. For example, a sample of cow's milk included OCDD at 0.215 ng kg^{-1}, its TEF is 0.001 and therefore it contributed 0.0002(15) to the TEQ of 0.034 ng kg^{-1}. The major contribution came from 1,2,3,7,8-PCDD at a level of 0.012 ng kg^{-1}, a TEF of 0.5 and therefore a TEQ of 0.006 ng kg^{-1}.

There have been many studies of TCDD/TCDF in foodstuffs. Levels (ppt = ng kg^{-1}) in Russian samples collected (Schecter *et al.*, 1992) during 1988–90 were:

Compound	Lamb lipid	Swiss cheese	Fish
2,3,7,8-TCDD	0.27	ND	0.3
2,3,7,8-TCDF	0.41	0.7	9.4
Hexa-CDD	1.43	2.27	0.44
Hexa-CDF	1.36	0.6	1.78
Octa-CDD	6.5	22.0	17
Octa-CDF	7.2	ND	24
Total PCDD	10	34	19
Total PCDF	13	5	51

Notice the relatively high levels in fish; this may account for high adipose tissue levels in Japan, where a fish diet is favoured. Here, despite drinking water levels in fg ml^{-1}, 2,3,7,8-TCDD adipose levels reach 1486 ppt.

The Canadian forest industry includes 18 pulp mills using chlorine for bleaching and 95 sawmills using chlorophenolic fungicides (see **Pentachlorophenol**) and their release to Canadian rivers is a significant source which was monitored by analysis of PCDD/PCDF in the adipose tissue of residents in British Columbia (Teshka *et al.*, 1992):

Levels of PCDD/PCDF in adipose tissue (pg g^{-1})

2,3,7,8-TCDD	1.8–9.2	2,3,7,8-TCDF	0.46–7.5
1,2,3,4,7,8-Hexa-CDD	2.8–33	1,2,3,4,7,8-Hexa-CDD	1.2–27
1,2,3,6,7,8-Hexa-CDD	33–313	1,2,3,6,7,8-Hexa-CDF	2.9–26
Octa-CDD	67–1333	Octa-CDF	1.4–7.0

The mean values are significantly higher than those for Canadians resident in areas remote from the forestry industry (pg g^{-1}):

	BC residents	Other Canadians
2,3,7,8-TCDD	9.5	5.5
All hexa-CDD	145	59
Octa-CDD	1097	631

The use of capillary GC–MS allows the highly sensitive and selective analysis of incinerator emissions and almost all the PCDD/PCDFs with chlorine numbers 4–6 have been detected from this source. It is significant that samples of cow's milk from European farms contain a mean TEQ of 0.045 ppb with OCDD the major component where in New Zealand, where there is no municipal incineration, the background TEQ is lower at 0.006 ppb. Levels in human milk from dispersed locations in Western Europe were in the narrow range 30–40 ppb.

A summary of major accidental releases of PCDD during 1949–94 is available (Alloway and Ayres, 1997).

Media comment frequently implies that dioxins are carcinogenic in humans. There was a report in 1979 by Hardell and Sandstrom that raised the possibility of induction of a rare soft tissue cancer. This was later discounted by the Evatt Commission adjudicating cases in Australia and also by Sir Richard Doll.

Studies conducted by the USEPA, the US Veterans Association and NIOSH showed that 2,3,7,8-TCDD causes cancer in animals, but in 1985 no firm conclusion could be drawn about human risk, although the compound was declared a 'probable source of human cancer.' Ten years later, despite considerable further work, the position was unchanged (USEPA, 1994).

In assessing cancer risk, one model assumes that there is no safe threshold level, because induction occurs through a series of irreversible stages. Based on this model, the EPA arrived at an acceptable daily human intake of $0.006\,\mathrm{pg\,kg^{-1}\,day^{-1}}$; on the same basis, the US Food and Drug Administration has proposed a level an order of magnitude higher. Owing to the wide range of dioxin sources, the average citizen of an industrialized country is inevitably exposed to the higher level of $1–3\,\mathrm{pg\,day^{-1}}$.

Other authorities have evaluated a level based on the theory that a threshold dose does exist, below which there is no risk of cancer. Using this model, Canada and the WHO propose an acceptable daily intake of $10\,\mathrm{pg\,kg^{-1}\,day^{-1}}$, more than three orders of magnitude higher than that of the USEPA.

In setting a control level, regulators commit industry to the cost of control. This cost increases with the level of control in an exponential manner, so that if the cost for 60% control is $\$75\,\mathrm{t^{-1}}$ the cost for 90% control is $\$350\,\mathrm{t^{-1}}$. To enforce the most stringent of the above levels would bear very heavily on the community.

References

Alloway, B.J. and Ayres, D.C. (1997) *Chemical Principles of Environmental Pollution*, 2nd edn, Chapman & Hall, London, p. 299.

Teshka, K., Kelly, S.J., Wiens, N., Hertzman, C., Dimich-Ward, H., Ward, J.E.H. and Oostdam, J.C. (1992) Dioxins and furans in residents of a forest industry region of Canada, *Chemosphere*, **25**, 1741–1751.

USEPA (1994) *Public Review Draft of the Health Assessment Document for 2,3,7,8-tetrachloro-p-dioxin (TCDD) and related Compounds*, EPA/600/ BP-92/001a,b,c, USEPA, Cincinatti, OH.

Further reading

Environmental Data Service (1995) *ENDS Rep.*, **240**, 3–5.

Schechter, A., Furst, C., Grachev, M., Beim, A. and Koptug, V. (1992) Levels of dioxins, dibenzofurans and selected other compounds in food from Russia, *Chemosphere*, **25**, 2009–2015.

Startin, J.R. (1991) in Creaser, C.S. and Purchase, R. (eds), *Food Contaminants*, Royal Society of Chemistry, Cambridge, pp. 21–40.

POLYCYCLIC AROMATIC HYDROCARBONS (PAH)

This is a group of lipophilic substances that are ubiqitous in the environment. They are almost insoluble in water and are commonly sorbed on to airborne particles. They enter the environment from the following sources: tobacco smoke, incomplete combustion of coal, especially lignite, gas and oil-fired heating systems, wood fires, exhaust from petrol and diesel engines and the run-off from bitumen road surfaces. Industrial workers most exposed to risk are those engaged in coke production and iron and steel founding and those handling bitumen.

Some 45 individual PAH have been evaluated, all with four or more fused rings. Only a small selection can be shown here and these were chosen from those which are significantly carcinogenic and of most concern to the regulatory agencies:

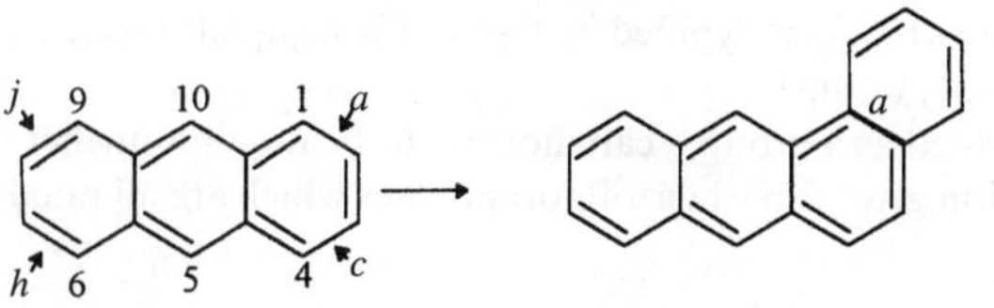

anthracene inactive

The point of ring fusion is indicated alphabetically

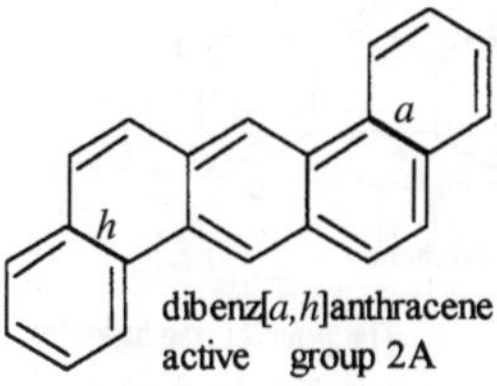

benz[*a*]anthracene active group 2A

dibenz[*a*,*h*]anthracene
active   group 2A

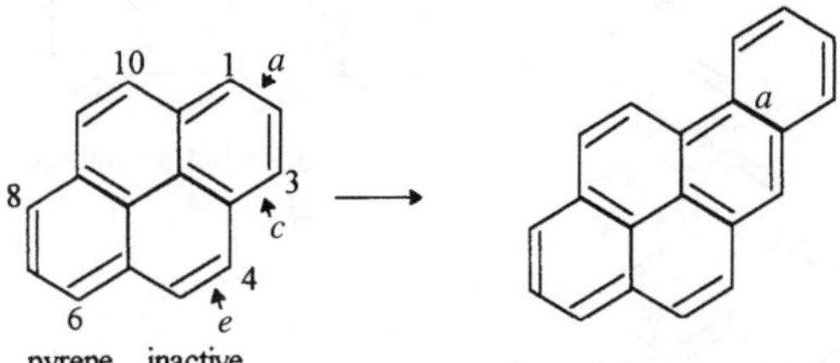

pyrene   inactive

benzo[*a*]pyrene   strongly active
group 2A

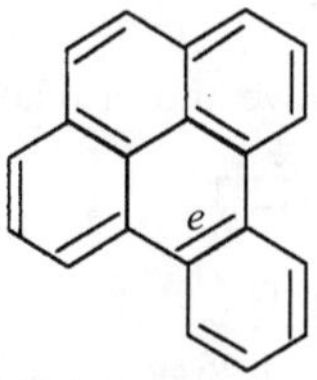

benzo[*e*]pyrene   weakly active
group 3

Benzo[a]pyrene is regarded as the most dangerous of this group as it is strongly carcinogenic in animals and is widely distributed. It becomes active after epoxidation *in vivo* to give the ultimate carcinogenic metabolite, the 7,8-diol-9,10-epoxide, which associates with cellular macromolecules – DNA, RNA or protein. The EC directive on drinking water limits total PAH to $0.2\,\mu g\,l^{-1}$. Details of those which present the greatest risk (IARC, 1983) are summarized:

Compound	Formula (M)	CAS No.	M.p. (°)	B.p. (°)	Log P_{ow}
Benz[a]anthracene	$C_{18}H_{12}$(228.3)	56–55–3	159	435	–
Dibenz[a,h]anthracene	$C_{22}H_{14}$(278.3)	53–70–3	267	524	6.75
Benzo[a]pyrene	$C_{20}H_{12}$(252.3)	50–32–8	177	312	6.35
Benzo[e]pyrene	$C_{20}H_{12}$(252.3)	192–97–2	179	250	–

Their water solubilities (25°) are typified by those of benz[a]anthracene, $14\,\mu g\,l^{-1}$, and dibenz[a,h]-anthracene, $0.5\,\mu g\,l^{-1}$.

Fluoranthene is rated as a group 3 carcinogen but fusion of a further benzene ring at the b, j or k position gives three benzofluoranthenes which are all rated as group 2B carcinogens:

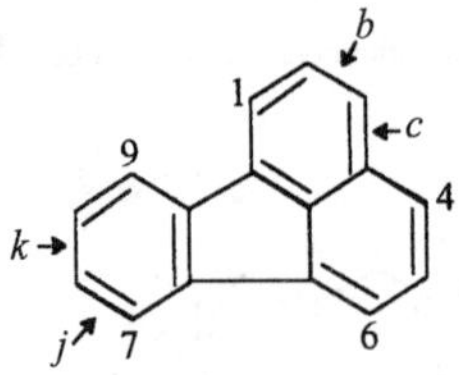

fluoranthene

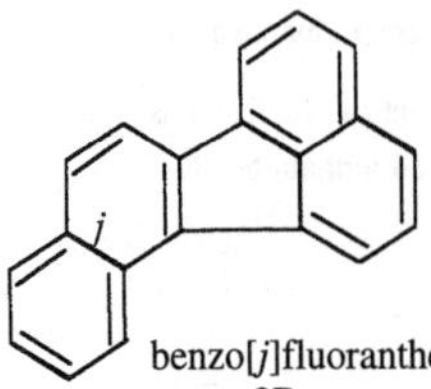

The point of ring fusion is indicated alphabetically

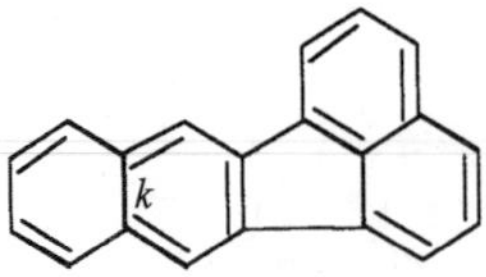

benzo[j]fluoranthene
group 2B

benzo[k]fluoranthene
group 2B

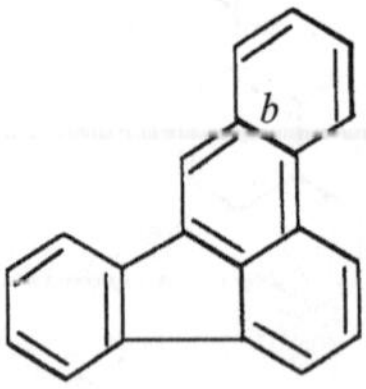

benzo[b]fluoranthene
group 2B

Further data on the above and of related compounds follow.

Anthracene CAS No. 120-12-7

$$C_{14}H_{10} \quad M = 178.2$$

m.p. 218°; b.p. 342°; dens. 1.25, log P_{ow} 4.45. Solubility in water $1.24\,mg\,l^{-1}$, soluble in EtOH, benzene, CCl_4.

There is no TLV; IARC group 3; emitted in vehicle exhaust at 534–642 $\mu g\,l^{-1}$; found in drinking water in the range 1–60 $ng\,l^{-1}$.

Benz[a]anthracene CAS No. 56-55-3
LD_{Lo} mouse 10 mg kg^{-1}
There is no TLV. IARC group 2A. Has a half-life of 2.9 h in air in summer and 7.8 h in winter. Strongly adsorbed on sediments and occurs in gasoline, bitumen and crude oil; found in domestic effluent at 0.19–0.32 ppb.

Dibenz[a,h]anthracene CAS No. 53-70-3
LD_{Lo} mouse 10 mg kg^{-1}
IARC group 2A. Present in the sludge of wood preservatives, coal tar and vehicle exhausts; found in meat, vegetables and cereals.

Pyrene CAS No. 129-00-0
LD_{50} rat 800 mg kg^{-1}

$$C_{16}H_{10} \quad M = 202.3$$

IARC class 3. M.p. 151°, b.p. 404°, dens. 1.27; $\log P_{ow}$ 5.00. Solubility in water 150 $\mu g\,l^{-1}$; soluble in acetone, benzene, diethyl ether. Emitted from vehicle exhaust. Occurs in surface water at 2–3.7 $ng\,l^{-1}$, in rainwater at 6–28 $ng\,l^{-1}$ and in char-broiled steak at 18 $\mu g\,kg^{-1}$.

Benzo[a]pyrene CAS No. 50-32-8
LD_{Lo} mouse 500 mg kg^{-1}; LD_{50} i.p. mouse 250 mg kg^{-1}.
IARC group 2A. In a Montreal study of 3726 cancer patients during 1979–85, 75% had been occupationally exposed to PAH, but it was concluded that there was no clear evidence of enhanced risk from 19 different cancer types (Krewski *et al.* 1990).

The groups exposed in the Quebec-based study included: from coal – miners, railway workers, roofers, ferrous foundry workers; from petroleum – truck drivers, commercial travellers, mechanics and repairmen; from wood – farmers, cooks, firefighters. Found in surface water up to 13 000 $ng\,l^{-1}$, in edible fats up to 62 $\mu g\,kg^{-1}$ and in char-broiled steak at 8 $\mu g\,kg^{-1}$.

Benzo[e]pyrene CAS No. 192-97-2
IARC group 3. Slightly soluble in water; soluble in acetone.
Found in surface water at 3–30 $ng\,l^{-1}$, in city air at 1–37 $ng\,m^{-3}$ and in char-broiled steak at 6 $\mu g\,kg^{-1}$.

Fluoranthene CAS No. 206-44-0
LD_{50} rat 2 g kg^{-1}

$$C_{16}H_{10} \quad M = 202$$

M.p. 111°; b.p. 375°; v.p. 10×10^{-6} Torr (20°); $\log P_{ow}$ 5.20. Solubility in water 0.2 mg l^{-1}, soluble in acetic acid, benzene, EtOH.

Found in the tars and exhaust from fossil fuels. Detected in rain at 5.6–1460 $ng\,l^{-1}$, in surface water at 4.7–6.5 $ng\,l^{-1}$, in drinking water at 2.6–132 $ng\,l^{-1}$ and in char-broiled steak at 20 $\mu g\,kg^{-1}$.

Cigarette smokers are exposed to risk from PAH. The following table gives the levels in μg per 100 cigarettes from mainstream (M) and sidestream smoke (S):

Compound	Level in smoke	Level in a smoky room
Anthracene	2–23 (M)	–
Benz[a]anthracene	0.4 (M)	–
Dibenz[a,h]-anthracene	0.4 (M)	–
Dibenz[a,j]-anthracene	1.1 (S)	6
Fluoranthene	1–27 (M)	99
Pyrene	5–27 (S)	2–66
Benzo[a]pyrene	0.5–8 (S)	3–760
Benz[e]pyrene	0.2–2.5(S)	3–18

Other dangerous products are methyl and heterocyclic nitrogen analogues of PAH and also their nitro derivatives.

Owing to their strong UV absorption and fluoresence the PAH are readily analysed by HPLC, although the greatest selectivity is obtained by capillary GC (Bartle, 1991)

References

Bartle, K.D. (1991) in Creaser, C.S. and Purchase, R. (eds), *Food Contaminants*, Royal Society of Chemistry, Cambridge, pp. 41–60.

IARC (1983) *Polynuclear Aromatic Compounds, Part 1: Chemical, Environmental and Experimental Data*, IARC, Lyon, Vol. 32.

Krewski, D., Siemiatycki, J., Naton, L., Dewar, R. and Gerin, M. (1990) Cancer risks due to occupational exposures to PAH: a preliminary report, *Environ. Sci. Res.*, **39**, 343–352.

POTASSIUM CAS No. 7440-09-7

K

Potassium, a member of Group 1 (alkali metals) of the Periodic Table, has an atomic number of 19, an atomic weight of 39.10, a specific gravity of 0.862 at 20°, a melting point of 63.3° and a boiling point of 759°. It is a ductile, silvery-white, very reactive metal and forms compounds in the +1 oxidation state. Natural potassium comprises of three isotopes: ^{39}K (93.258%); ^{41}K (6.73%) and the radioactive ^{40}K (0.012%, $t_{1/2} = 1.28 \times 10^9$ years). Natural radiation from this isotope is not usually harmful at this low concentration.

Potassium is the seventh most abundant element in the earth's crust and makes up $\sim 2.5\%$ (25 000 ppm) by weight. It is found in silicate minerals such as the feldspars and micas. Commercial sources of potassium are the more soluble minerals *sylvite*, KCl, *carnallite* $KCl.MgCl_2.6H_2O$, and *langbeinite*, $K_2Mg_2(SO_4)_2$. Soluble mineral deposits of potassium salts are referred to as potash deposits. The metal is obtained by electrolysis of the hydroxide or by reduction of KCl with sodium metal. Like Na^+, the potassium ion K^+ is a major ion in natural waters, ~ 380 ppm in sea water and ~ 1.5 ppm in river water. Potassium salts enter the body normally through ingestion. Typical human concentrations are; bone 2050 ppm, blood 1600 mg l^{-1}, daily intake 1.4–7 g. The amount of potassium stored in the human body (adult, 70 kg) varies from 140 to 250 g.

Uses. Sodium/potassium alloy as a heat transfer agent in nuclear reactors. Reducing agent. $^{43}K(t_{1/2} = 22\,h)$ is used in heart diagnosis.

Potassium is one among ca. 25 elements thought to be essential for healthy animal and plant life. As an essential element for plant growth, the greatest demand for potassium is in the manufacture of fertilizers. Many salts of K are of industrial importance and some are produced in greater than million tonne quantities (total production worldwide $32.1 \times 10^6\,t\,year^{-1}$). Anthropogenic sources of potassium include solid emissions from metallurgical works, industrial effluent discharge to rivers, runoff of fertilizer containing water from land to rivers and lakes, etc. The toxicity of K compounds is often that of the anion, not of potassium. Toxic dose for humans is $\sim 6\,g$; lethal dose $\sim 15\,g(K_2CO_3)$.

Occupational hazards. Reacts violently with water, liberating flammable hydrogen. Solid metal causes burns. Metal fumes are destructive to mucous membranes, skin and eyes. MAC (drinking water) $12\,mg\,l^{-1}$.

Further reading
ECETOX (1991) *Technical Report 30(4)*, European Chemical Industry Technology and Toxicology Centre, Brussels.
Kirk–Othmer ECT, 4th edn, Wiley, New York, 1996, Vol. 18, pp. 1047–1092.
SI 1991 No. 472 The Environmental Protection (Prescribed Processes and Substances) Regulations 1991, HMSO, London, 1991.
Ullmann's Encyclopedia of Industrial Chemistry, 5th edn, VCH, Weinheim, 1993, Vol. A22, pp. 31–103.

Potassium antimony tartrate trihydrate CAS No. 28300-74-5

$K_2[Sb_2(C_2O_3H)_4].3H_2O$

Uses. Antischistomal drug (infection caused by *Schistosoma japonica*). Mordant in leather and tanning industry. Once used as an emetic (tartar emetic).

Toxicity. Chronic over-exposure may damage kidneys, liver, lungs, blood and central nervous system. LD_{50} oral rat, rabbit $115\,mg\,kg^{-1}$; LD_{50} subcutaneous mouse $55\,mg\,kg^{-1}$.

Potassium arsenite CAS No. 10124-50-2

$K_3AsH_3O_3$

Uses. Manufacture of mirrors.

Toxicity. Poison. Irritant. LD_{50} oral rat $14\,mg\,kg^{-1}$.

Potassium bromide CAS No. 7758-02-3

KBr

Uses. Photography. As a source of bromine in organic syntheses.

Toxicity. N.a. Large doses can cause depression of the central nervous system.

Potassium carbonate CAS No. 584-08-7

K_2CO_3

Uses. Manufacture of glass, optical lenses, colour TV tubes and fluorescent lamps. It is also used in chinaware, textile dyes and pigments.

Toxicity.   Severe irritant. LD_{50} oral mouse $2570\,mg\,kg^{-1}$.

Potassium chloride CAS No. 7447-40-7

KCl

Uses.   Preparation of other potassium salts.

Toxicity. LD_{Lo} (2 d) oral infant $938\,mg\,kg^{-1}$; oral adult human $20\,mg\,kg^{-1}$; LD_{50} oral rat $2600\,mg\,kg^{-1}$.

Potassium chlorate CAS No. 3811-04-9

$KClO_3$

Uses.   Oxidizing agent. Explosives. Dyestuffs.

Toxicity.   Irritant to mucous membranes. Absorption in body can lead to cyanosis. Affects blood, liver and kidneys.

Potassium chromate CAS No. 7789-00-6

K_2CrO_4

Uses.   Oxidizing agent. Analytical reagent. Corrosion inhibitor. Pigments. Dyestuffs.

Occupational exposure limit.   OSHA PEL-CL $0.1\,mg\,CrO_3\,m^{-3}$.

Toxicity.   Highly toxic carcinogen, mutagen and teratogen. Affects lungs, kidneys, liver and blood. LD_{50} oral mouse $180\,mg\,kg^{-1}$; intramuscular rat $11\,mg\,kg^{-1}$.

Potassium cyanide CAS No. 151-50-8

KCN

Uses.   Catalyst. Organic synthesis. Electroplating.

Occupational exposure limit.   ACGIH TLV (TWA) $5\,mg\,CN\,m^{-3}$ (skin).

Toxicity. LD_{50} oral human $2857\mu g\,kg^{-1}$; LD_{50} oral rat $5\,mg\,kg^{-1}$. Affects blood and heart.

Potassium dichromate CAS No. 7778-50-9

$K_2Cr_2O_7$

Uses.   Oxidizing agent. Catalyst. Pigment in dyestuffs. Corrosion inhibitor.

Toxicity.   Carcinogen. Mutation data. Lungs and kidneys affected. LD_{50} intraperitoneal mouse $37\,mg\,kg^{-1}$.

Potassium fluoride CAS No. 7789-23-3

KF

Uses.   Manufacture of aluminium. Flux welding. Fluorinating agent in organic chemistry.

Occupational exposure limit.   ACGIH TLV (TWA) $2.5\,mg\,F\,m^{-3}$.

Toxicity. LD_{50} oral rat $245\,mg\,kg^{-1}$.

Potassium hexachloroplatinate(IV) CAS No. 16921-3-5

K_2PtCl_6

Uses. Electroplating.

Toxicity. Harmful by inhalation. Prolonged exposure may cause asthma.

Potassium hexafluorozirconate(IV) CAS No. 16923-95-8

K_2ZrF_6

Uses. Welding flux.

Toxicity. Harmful by all routes. Irritant to mucous membranes and upper respiratory tract. LD_{50} oral mouse $98\,mg\,kg^{-1}$.

Potassium hydroxide CAS No. 1310-58-3

KOH

Uses. Liquid soaps. Preparation of K compounds.

Occupational exposure limit. ACGIH TLV-CL $2\,mg\,m^{-3}$.

Toxicity. Extremely destructive to soft tissue. LD_{50} oral rat $273\,mg\,kg^{-1}$.

Potassium iodide CAS No. 7681-11-0

KI

Uses. Photography.

Occupational exposure limit. ACGIH TLV-CL 0.1 ppm I ($1\,mg\,I\,m^{-3}$).

Toxicity. Irritant.

Potassium nitrate CAS No. 7757-79-1

KNO_3

Uses. Pyrotechnics. Curing meats. Manufacture of glass. Fertilizers.

Toxicity. Absorption into the body leads to the formation of methaemoglobin, which in sufficient concentration causes cyanosis. Affects blood and the central nervous system. LD_{50} oral rat $3750\,mg\,kg^{-1}$; LD_{50} oral rabbit $1901\,mg\,kg^{-1}$.

Potassium nitrite CAS No. 7758-09-0

KNO_2

Uses. See **Nitrites**.

Toxicity. LC_{50} (2 h) inhalation mouse $85\,g\,m^{-3}$; LD_{50} oral rabbit $200\,mg\,kg^{-1}$

Potassium permanganate CAS No. 7722-64-7

$KMnO_4$

Uses. Oxidizing agent. Disinfectant. Analytical reagent.

Toxicity. LD_{Lo} oral human $143\,mg\,kg^{-1}$; LD_{50} oral rat $1090\,mg\,kg^{-1}$. Affects central nervous system, blood, kidneys and lungs.

Potassium sulphate CAS No. 7778-80-5

K_2SO_4

Uses. Analytical reagent. Fertilizers. Catalyst.

Toxicity. LD_{50} oral rat $6.6\,g\,kg^{-1}$ body weight. LD_{Lo} oral woman $750\,mg\,kg^{-1}$. Health examinations of workers employed in the manufacture of $K_2SO_4.KMg(SO_4)_3$ fertilizer showed substantial overall morbidity rate in terms of both sick individuals and number of days absent due to respiratory diseases caused by dust.

PRASEODYMIUM CAS No. 7440-10-0

Pr

Praseodymium, a member of the rare earth series of metals, has an atomic number of 59, an atomic weight of 140.91, a specific gravity of 6.773 (α-form) at $20°$, a melting point of $931°$, and a boiling point of $3512°$. The element is a soft, silvery, malleable and ductile metal forming compounds in the $+3$ and $+4$ oxidation states. Natural praseodymium exists as one stable isotope, ^{141}Pr (100%).

Monazite and *bastnasite* are the two main sources of praseodymium. Average concentration in the earth's crust is 9.5 ppm. The pure metal can be obtained by the calcium reduction of the anhydrous chloride or fluoride. Very low concentrations in human organs.

Uses. Misch metal ($\sim 5\%\,Pr$). Carbon-arc lights. Salts are used to colour glass (in certain mixtures to give a clean yellow colour) and enamels. Together with neodymium ('didymium') to colour welder's goggles. 'Didymium' used as a catalyst in the oxidation of hydrogen chloride to chlorine.

No biological role. Little is known about the toxicity of praseodymium and its salts so care should taken in handling . The metal as fibrous dust is carcinogenic. Symptoms of exposure may include burning sensation, coughing, wheezing, laryngitis, shortness of breath, headache, nausea and vomiting.

Further reading
Gschneidner, K.A. Jr, and LeRoy K.S. (eds.) (1988) *Handbook on the Physics and Chemistry of the Rare Earths*, North-Holland, Amsterdam.
Kirk–Othmer ECT, 4th edn, Wiley, New York, 1995, Vol. 14, pp. 1091–1115.

Praseodymium(III) chloride hexahydrate CAS No. 17272-46-7

$PrCl_3.6H_2O$

Uses. Chemical agent.

Toxicity. LD_{50} oral mouse $2987\,mg\,kg^{-1}$; subcutanous rabbit $351\,mg\,kg^{-1}$. Ingestion may cause sleepiness, anorexia and muscle weakness. Affects lungs and thorax.

Praseodymium(III) nitrate pentahydrate CAS No. 15878-77-0

$Pr(NO_3)_3.5H_2O$

Uses. Chemical agent.

Toxicity. LD_{50} oral rat $1859\,mg\,kg^{-1}$.

PROMETHIUM CAS No. 7440-12-2

Pm

Promethium, a radioactive member of the rare earth series of metals, has an atomic number of 61, an atomic weight of 145, a specific gravity of 7.26 at 20°, a melting point of 1042° and a boiling point of $\sim$ 3000°. It exhibits an oxidation state of +3.

Obtained by neutron bombardment of neodymium. Promethium is completely missing from the earth's crust but has been identified in the spectrum of star HR[465] in Andromeda.

Promethium salts luminesce in the dark owing to their radioactivity. Promethium-147, with a half-life of 2.5 years, is a commonly observed isotope since traces occur in uranium ores. Promethium-145 is the longest lived isotope, with a half-life of 17.7 years and a specific activity of 940 Ci g^{-1}.

Further reading
Gschneidner, K.A., Jr, and LeRoy, K.S. (eds.) (1988) *Handbook on the Physics and Chemistry of the Rare Earths*, North-Holland, Amsterdam.
Kirk–Othmer ECT, 4th edn, Wiley, New York, 1995, Vol. 14, pp. 1091–1115.

PROPACHLOR CAS No. 1918-16-7
LD$_{50}$ rat 1050 mg kg^{-1}; LC$_{50}$ (48 h) rainbow trout 0.28 mg l^{-1}; LC$_{50}$ (96 h) catfish 0.23 mg l^{-1}

$C_{11}H_{14}ClNO \quad M = 211.5$

Ramrod, acylide, 2-chloro-N-isopropylacetanilide; m.p. 77°; water solubility 580 mg l^{-1} (20°); soluble in most organic solvents except for aliphatic hydrocarbons.

Introduced in 1965 by Monsanto and formed by the reaction of chloroacetyl chloride with N-isopropylaniline. Applied at 4–6 kg ha^{-1} in water (150–300 l) as a pre- or early post-emergence herbicide, it is effective against annual grasses and some broad-leaved weeds. It undergoes microbial degradation and the half- life in soil is 2–3 weeks.

Propachlor may be detectd by TLC to a limit of 0.02 mg kg^{-1} and by GC-ECD to 4 ng kg^{-1}.

In the USA the acceptable residue limit in foodstuffs is 0.02–0.3 mg kg^{-1}. The German guideline for water is 0.01 mg l^{-1}.

Further reading
Environ. Health Criteria, **147**, World Health Organization, Geneva, 1993, Vol. 147.

PROPANIL CAS No. 709-98-8
(STAM, Surcopur)
Selective contact herbicide. Toxicity class III. LD$_{50}$ rat > 2.5 g kg^{-1}; LC$_{50}$ carp 8 mg l^{-1} (48 h)

$$C_9H_9Cl_2NO \quad M = 218.1$$

3,4-Dichloropropionanilide; m.p. $93°$; v.p. $0.026\,mPa$ $(20°)$; water solubility $130\,mg\,l^{-1}$ $(25°)$.

Introduced by Rohm & Haas. It inhibits photosynthesis and is used against broad-leaved weeds and grasses in rice and wheat. It is phytotoxic and therefore incompatible with carbamates and organophosphates. Propanil is rapidly degraded in soil to 3,4-dichloroaniline and CO_2. Costa Rica imported 537 t in 1989 and 391 t in 1990.

The No Effect Limit in rats is high at $400\,mg\,kg^{-1}$ of diet over a 2 year period and no standards have been set.

PROPYLENE CAS No. 115-07-1

$$CH_3CH = CH_2 \qquad\qquad C_3H_6 \quad M = 42.1$$

1-Propene, m.p. $-185°$; b.p. $-48°$; flash point $108°$. Extremely flammable; an asphyxiant which forms an explosive ozonide. Solubility in water 44 ml of gas per 100 ml; in ethanol 1250 ml of gas per 100 ml.

Produced as a by-product of ethylene or by the steam cracking of natural gas liquids. US production in 1990 was 11×10^6 t year^{-1} and in 1994 28.8×10^9 lb; seventh in the list of bulk chemicals. World consumption in 1992 was North America 5.6×10^6, Western Europe 5.9×10^6, Japan 2.6×10^6 and others 6.0×10^6, total 20.1×10^6 t. Steam cracking reduces the partial pressure of the hydrocarbon fraction by a free radical mechanism at temperatures in the range 870–1170 K. The steam prevents further reaction between the products, which include ethylene, propylene, 1,3-butadiene and benzene.

The principal uses for propylene are to make polypropylene, acrylonitrile, isopropyl alcohol and propylene oxide. The greatest environmental risk arises from evaporation from the manufacture of polypropylene by the slurry process – here propylene monomer and hydrogen are fed into a series of stirred pressurized vessels containing a hydrocarbon diluent and Ziegler–Natta catalyst, based on $TiCl_3$ with AlR_3 as cocatalyst. Air pollution arises at several points, such as the polypropylene drying and solvent recovery areas. Modern reactors have reduced, but not eliminated, these losses by internal recycling of propylene and with fugitive testing of valves and unions.

Since its discovery by Ziegler and Natta in 1954, polypropylene, a stereoregular polymer, has found widespread applications. These include carpet yarn, slit tape, non-woven fabric, fencing and moulded products including domestic appliances, furniture, packaging and the interiors of cars and trucks.

After exposure to propylene at 6.4 % by volume mild intoxication of humans was observed after 2–25 min. At 35% vomiting and vertigo ensued and after 2 min in 50% propylene subjects were anaesthetized.

Propylene itself presents no carcinogenic risk but the action of cytochrome P450 mixed function oxidase may produce propylene oxide (see **Propylene oxide** for structure). This product is a CNS depressant and has been classed as a group 2 carcinogen by the IARC. The ACGIH has set a TLV (TWA) of 20 ppm (48 mg m^{-3}) for the oxide, while regarding propylene as a simple asphyxiant with no other physiological effect.

See also **Liquified petroleum gas**.

Further reading
Chemical Safety Data Sheets, 5, Flammable Chemicals, Royal Society of Chemistry, Cambridge, 1992.
Kirk–Othmer ECT, 4th edn, Wiley, New York, 1996, Vol. 20, pp. 249–270.
Moore, E.P. (1996) *Polypropylene Handbook*, Hanser, Munich.

PROPYLENE OXIDE CAS No. 75-56-9
LD$_{50}$ rat 520 mg kg^{-1}

C_3H_6O $M = 58.1$

1,2-Epoxypropane, propene oxide; b.p. 34°; v.p. 59 kPa, 445 Torr (20°); water solubility 405 g l^{-1}

Determined to a limit of 60 ng m^{-3} by GC–MS or to 2 μg m^{-3} by absorption on Tenax followed by GC–FID.

The properties are similar to those of ethylene oxide and the principal use is in the synthesis of polyether polyols, e.g. $CH_3CH(OH)CH_2OCH_2CH_2OCH_2CH(OH)CH_3$ and propylene glycol for polyester fibres. It has minor uses as an antimicrobial and fumigant for foodstuffs and as a stabilizer for methylene chloride.

Propylene oxide enters the environment by evaporation and loss during production and it was estimated that in the USA 600 t were lost in this way in 1981. It is converted into glycols by reaction with OH radicals in air and in sea water it reacts with a half-life of 4 days to form chloropropanol and 1,2-propanediol.

Humans are exposed largely at the workplace. Propylene oxide alkylates haemoglobin, as does ethylene oxide, and has been found to be carcinogenic in rats and mice; it also causes chromosomal aberrations in mammals. It is rated as a group IIB carcinogen and a possible carcinogen in man,

See also **Ethylene oxide**.

Further reading
Environ. Health Criteria, World Health Organization, Geneva, 1985, Vol. 56.

PROTACTINIUM CAS No. 7440-13-3

Pa

Protactinium, a member of the actinide series of elements, has an atomic number of 91, an atomic weight of 231.036, a specific gravity of 15.37 at 20°, a melting point of 1572°, and a boiling point of 4227°. It forms compounds with oxidation states of +4 and +5. Protactinium has over 20 radioactive isotopes, the most common of which is ^{231}Pa with a half-life of 32 500 years. Pa is capable of being produced in gram quantities from uranium fuel elements.

The element occurs in *pitchblende* to the extent of about 1 part of ^{231}Pa to 10^7 parts of ore. This isotope is part of the ^{235}U radioactive decay series. Protactinium-234 is part of the ^{238}U radioactive decay series with a half-life of 6.75 h.

Unlikely to be a hazard from normal concentrations but this element is a high-energy (5.0 MeV) α-emitter and is a radiological hazard similar to polonium.

See also **Actinides**.

Further reading

Eisenbud, M. (1987) *Environmental Radioactivity. From Natural, Industrial, and Military Sources*, 3rd edn, Academic Press, New York.

PYRAZOPHOS CAS No. 13457-18-6
(Afugan)
Systemic fungicide with protective and curative action. Toxicity class II. LD_{50} rat 150-770 mg kg^{-1}; LC_{50} rainbow trout 0.5 mg kg^{-1}. Not toxic to bees

$$C_{14}H_{20}N_3O_5PS \quad M = 373.4$$

O-6-Ethoxycarbonyl-5-methylpyrazolo[1,5-*a*]pyrimidin-2-yl-diethylphosphorothioate; m.p. 52°; v.p. 0.22 mPa (50°); water solubility 4.2 mg l^{-1} (25°).

First reported by Smit in 1969 and introduced by Hoechst. Pyrazophos has a systemic action against fungi on apples, stone and soft fruit, strawberries, cereals, hops and vines.

The No Effect Limit is 5 mg kg^{-1} of diet over a 2 year period. The Codex MRL proposals are pending (1995).

PYRETHRINS
These naturally occurring insecticides are obtained largely from the plant *Chrysanthemum cinerariaefolium*. Their activity was known to the Chinese in the first century AD, but commercial production did not begin until 1850 and the first evidence of their structure was obtained in 1920 by Ruzicka and Staudinger.

The four principal components are esters formed by combination of either of the acids chrysanthemic or pyrethric with one or the other of the alcohols pyrethrolone and cinerolone:

About half of the current world production comes from Kenya; other major sources are Russia and Japan. Each flower head contains 2–4 mg (ca 2%) of the active compounds and these are extracted with light petroleum.

Key requirements for an active structure are a bulky group like *gem*-dimethyl two places removed from the ester function and unsaturated centres at the two extremities of the molecule. The activity is also related to the absolute confuguration. The *cis* and *trans* esters of the dextrorotatory (*R*) acids shown above are 20–50 times more active than the enantiomeric (*S*) compounds.

The highly active pyrethrolyl ester of 1(*R*)- chrysanthemic acid meets the structural criteria for activity and has an LD_{50} of 0.3 μg/insect, with an LD_{50} of 5 mg kg^{-1} in rats by the intravenous route. It is toxic to bees and the LC_{50} for the catfish is 114 mg kg^{-1}.

There have been numerous synthetic analogues of pyrethrin, which were extensively researched by Elliot and Janes at the Rothamstead Research Station. For further details of these insecticides, see **Allethrin, Deltamethrin, Permethrin**.

RADIUM CAS No. 7440-14-4

Ra

Radium, a member of Group 2 (alkaline earths) of the Periodic Table, has an atomic number of 88, an atomic weight of 226.025, a specific gravity of 5.5 at 20°, a melting point of 700°, and a boiling point of 1140°. The pure metal is brilliant white and forms compounds with an oxidation state of +2. Radium exists as at least 30 isotopes. Radium-228 with a half-life of 5.8 years occurs naturally as a decay product of thorium-223. Radium-226 is the longest lived isotope with a half- life of 1600 years and is naturally occurring from the ^{238}U decay chain. One gram of ^{226}Ra undergoes 3.7×10^{10} disintegrations per second and (= 1 Ci) via α-decay to yield ^{222}Rn ($t_{1/2} = 3.8$ days).

Found in the minerals *pitchblende, carnotite* and *uraninite*. There is about 1 g of radium in 7 t of pitchblende. Radium can be obtained from uranium processing. Concentration in the earth's crust is < 0.1 ppm. One gram of radium produces about 0.0001 ml (STP) of radon gas per day. Radium-226 and its daughter products are responsible for a major fraction of the dose received by humans from the naturally occurring internal emitters, are a radiological hazard and should be ventilated, when stored, to prevent the build-up of radon. The recommended MAC for total body content is 0.1 µg and exposure to 6 mrem year^{-1}. Radium is chemically similar to calcium and is absorbed from the soil by plants and passed up the food chain to humans. Bones contain about 89% (ppt) of the total body radium. In an adult skeleton containing 1000 g of Ca this corresponds to a median body burden of 23 pCi. The radium content of surface waters is low (0.1–0.5 pCi l^{-1} $\approx$ 1 µg of radium).

Uses. Self-luminous paints. Neutron sources. Cancer treatment.

A human carcinogen. Approximately 60 known deaths resulted from the use of radium in luminizing compounds (e.g. luminous dial watches) in the USA between 1912 and 1961.

Further Reading

ToxFAQS (1990) *Radium*, ATSDR, Atlanta, GA.

Eisenbud, M. (1987) *Environmental Radioactivity. From Natural, Industrial, and Military Sources*, 3rd edn, Academic Press, New York.

RADON CAS No. 10043-92-2

Rn

Radon, a radioactive member of Group 18 (noble gases) of the Periodic Table, has an atomic number of 86, an atomic weight of ~ 222, a specific gravity of the gas of $9.73\,g\,l^{-1}$ at STP, a melting point of $-71°$, and a boiling point of $-61.8°$. It is an inert gas with an oxidation state of zero. Over 30 isotopes are known. The three naturally occurring radon isotopes, with atomic masses 219, 220 and 222, are found in the decay series U-235, Th-232 and U-238, respectively. Radon-219 has a very short half-life of about 4 s. Radon-222 from α-decay of ^{226}Ra has a half-life of 3.83 days and is an α-emitter. Radon-220 emanating from thorium has a half-life of 54.5 s, and is also an α-emitter. Every square mile of soil to a depth of 6 in contains about 1 g of radium, which releases small amounts of radon to the atmosphere. Found in trace amounts in uranium ores. Also minute amounts in human systems.

Uses. Anti-cancer agent.

The main hazard is from the inhalation of the element and its solid daughters, which collect on the dust in the air. Good ventilation is required where radium, thorium or actinium is stored to prevent build-up of radon. The air concentration of radon is fairly low, $\sim 0.003\text{--}2.6\,pCi\,l^{-1}$. Indoor radon air levels are generally about $1.5\,pCi\,l^{-1}$. Indoor radon has been widely recognized as a problem in Europe and the Scandinavian countries since the 1970s. It is particularly acute in granite areas as in parts of Devon and Cornwall, UK. There is considerable regional and local variation in the amount of radon emanating from the earth and concentrating inside homes; 10% of homes have radon levels above the acceptable limit of $\sim 4\,pCi\,l^{-1}$ of up to $200\,pCi\,l^{-1}$. The average level of radon in groundwater is $\sim 350\,pCi\,l^{-1}$.

The incidence of lung cancer among miners was linked epidemiologically to radon exposure in mines. Iron ore, potash, tin, fluorspar, gold, zinc and lead mines also have been found to have significant levels of radon. Underground uranium mines pose the greatest risk to miners because of their high concentration of radon. Long-term exposure to radon in air $(50\text{--}150\,pCi\,l^{-1})$ increases the risk of getting lung cancer. When exposures are high, other diseases of the lungs may occur, such as thickening of certain lung tissues. A number of studies have shown that smokers exposed to radon are at greater risk of lung cancers than others (~ 20 times). Further, children are at greater risk than adults.

Futher reading

Cohen, B. L. (1993) Relationship between exposure to radon and various types of cancer, *Health Phys.*, **65**, 529–531.

Cole, L. A. (1993) *Elements of Risk: The Politics of Radon*, AAAS Press, Washington, DC.
ToxFAQS (1990) *Radon*, ATSDR, Atlanta, GA.

RHENIUM CAS No. 7440-15-5

Re

Rhenium, a member of Group 7 of the Periodic Table, has an atomic number of 75, an atomic weight of 186.21, a specific gravity of 21.02 at 20°, a melting point of 3186°, and a boiling point of 5596°. The element is silvery white with a metallic lustre and forms compounds with oxidation states of -1, $+2$, $+3$, $+4$, $+5$, $+6$ and $+7$. Natural rhenium is a mixture of two stable isotopes, ^{185}Re (37.40%) and ^{187}Re (62.60%). Rhenium-187 is radioactive with a half-life of 4.4×10^{10} years.

The average concentration in the earth's crust is 0.7 ppm. The element is found in *gadolinite* and *molybdenite* (typically 0.002–0.2% rhenium). Commercially, rhenium is obtained from molybdenite roaster flue dusts. World production is about $40\,t\,year^{-1}$.

Uses. Additive to tungsten and molybdenum alloys. Filaments for mass spectrometers. Thermocouples. Rhenium compounds are used in various areas of homogeneous and heterogeneous catalysis in petrochemistry, the pharmaceutical industry and organic synthesis (e.g. alkylation, hydrogenation/dehydrogenation, oxidation). ^{186}Re ($t_{1/2} = 89\,h$) and ^{188}Re ($t_{1/2} = 17\,h$) are used in radio(immuno)therapy.

No reported cases of human toxicity. For experimental animals, Re^{3+} species are more toxic than the perrhenate ion, $Re^{VII}O_4^-$. Until more data are available, rhenium and its compounds should be handled with care.

Occupational exposure limits. N.a. Rhenium powder is flammable.

Further reading
Kirk–Othmer ECT, 4th edn, Wiley, New York, 1997, Vol. 21, pp. 315–346.

Rhenium(III) chloride CAS No. 13569-63-6

$ReCl_3$

Uses. Preparation of rhenium complexes.

Toxicity. LD_{50} intraperitoneal mouse $280\,mg\,kg^{-1}$. Symptoms of rhenium poisoning include drowsiness and elevation of blood pressure.

RHODIUM CAS No. 7440-16-6

Rh

Rhodium, a member of Group 9 of the Periodic Table, has an atomic number of 45, an atomic weight of of 102.9055, a specific gravity of 12.41 at 20°, a melting point of 1966° and a boiling point of $\sim 3700°$. Rhodium compounds usually contain Rh in the oxidation states $+2$, $+3$, $+4$, $+5$ and $+6$. Natural rhodium exists as only one isotope, ^{103}Rh (100%).

Found in the minerals *rhodite, sperrylite* and *iridosmine* and some nickel-copper ores. Constitutes ~ 1 ppb of the earth's crust. The pure metal is obtained by procedures designed to separate silver, gold and all the platinum metals.

Uses. Alloy agent to harden platinum and palladium. Electrical contact material. Jewellery. Catalyst in oxidation and hydrogenation reactions. Catalytic converters for car exhaust systems.

Little is known about the toxicity of rhodium or its compounds, therefore they should be handled with care.

Occupational exposure limits. OSHA TLV (TWA) 1 mg Rh m^{-3} (insoluble compounds) ; 0.01 mg Rh m^{-3} (soluble compounds). Rhodium powder is flammable.

Further reading

ECETOX (1991) *Technical Report 30(4)*, European Chemical Industry Technology and Toxicology Centre, Brussels.

King, R. B. (ed.) (1994) *Encyclopedia of Inorganic Chemistry*, Wiley, Chichester, Vol. 7, pp. 3469–3488.

Kirk–Othmer ECT, 4th edn, Wiley, New York, 1997, Vol. 21, pp. 335–346.

SI 1991 No. 472 The Environmental Protection (Prescribed Processes and Substances) Regulations 1991, HMSO, London, 1991.

Rhodium(III) chloride, CAS No. 10049-07-7; hydrate 20765- 98-4

RhCl$_3$

Uses. Preparation of rhodium compounds and complexes. Alloys.

Occupational exposure limits. US TLV (TWA) 1 mg m^{-3}. UK short-term limit 0.003 mg m^{-3}.

Toxicity. Harmful by all routes. LD$_{50}$ oral rat 1300 mg kg^{-1}; LD$_{50}$ intravenous rabbit 215 mg kg^{-1} . Affects respiration, lungs, and blood (leukaemia).

RUBIDIUM CAS No. 7440-17-7

Rb

Rubidium, a member of Group 1 (alkali metals) of the Periodic Table, has an atomic number of 37, an atomic weight of 85.468, a specific gravity of 1.532 at 20°, a melting point of 39.31° and a boiling point of 688°. It is a soft, silvery white metallic element and forms compounds principally in the +1 oxidation state. Naturally occurring rubidium is made up of two isotopes, ^{85}Rb (72.17%) and ^{87}Rb (27.83%). Rubidium-87 is a radioactive β-emitter with a half-life of 5×10^{11} years. Ordinary rubidium is sufficiently radioactive to expose a photographic film after 30–60 days with a specific activity of 0.02 pCi g^{-1}. The ^{87}Rb content of ocean water has been reported to be ~ 2.8 pCi l^{-1}. It is estimated that the whole body dose from ^{87}Rb is 0.6 mrem year^{-1}.

It is the 16th most abundant element in the earth's crust with a concentration of 80 ppm. Found with potassium and caesium minerals in granitoid rocks and pegmatites. Rubidium occurs in *pollucite, carnallite, leucite* and *zinnwaldite* (up to 1% as the oxide). It is also found in *lepidolite*, K$_2$Li$_3$Al$_4$Si$_7$O$_{21}$(OH, F)$_3$ ($\sim 1.5\%$), which is the commercial source of rubidium. The pure metal can be prepared by reducing rubidium chloride with calcium. World annual tonnage is less than 5 t year^{-1}. Typical concentrations in the environment are sea water 0.12 mg l^{-1}; river water 0.6–1.1 µg l^{-1}; coal 14.4 mg kg^{-1}; corn 3 mg kg^{-1}; daily intake for adult human 1.5–6 mg; average total body level for 70 kg adult 1.1 g (32% in bones, 25% in muscles); human red blood cells 4.18 µg ml^{-1}; plasma 0.16 µg ml^{-1}; urine 1.52 µg ml^{-1}.

Uses. Rubidium is used as a getter in vacuum tubes. Photocell component.

Rubidium metal is a fire hazard since it ignites spontaneously in air and reacts violently with water. In the human body, rubidium substitutes for potassium as intracellular ion. It is said to stimulate metabolism. The ratio of the Rb/K intake is important in the toxicology of rubidium. A ratio above 40% is dangerous. Workers exposed to rubidium compounds for between 2 and 10 years complained of increased excitability and sweating, headaches, numbness in fingers and impaired kidney and gastrointestinal functions.

Further reading
Kirk–Othmer ECT, 4th edn, Wiley, New York, 1997, Vol. 21, pp. 591–600.

Rubidium chloride CAS No. 7791-11-9

RbCl

Uses. Rubidium-activated catalyst for production of organic chemicals. Catalyst for petrol octane number improvement. Antidepressant.

Toxicity. Moderately toxic. Symptoms of poisoning are hyperactivity, aggressiveness, anxiety and death. LD_{50} oral rat $4440 \, mg \, kg^{-1}$. LD_{50} intravenous mouse $233 \, mg \, kg^{-1}$.

Rubidium hydroxide CAS No. 7440-17-7

RbOH

Uses. Photoelectric cells. Preparation of rubidium salts. Catalyst.

Toxicity. Solution is irritating to the eyes. Solid is corrosive. LD_{50} oral rat $586 \, mg \, kg^{-1}$, mouse $900 \, mg \, kg^{-1}$.

RUTHENIUM CAS No. 7440-18-8

Ru

Ruthenium, a member of Group 8 of the Periodic Table, has an atomic number of 44, an atomic weight of 101.07, a specific gravity of 12.41 at 20°, a melting point of 2334°, and a boiling point of 4150°. Ruthenium is a hard, white metal and has four crystal modifications. It forms compounds with oxidation states of +1, +2, +3, +4, +5, +6 and +7. Natural ruthenium comprises of seven stable isotopes: ^{96}Ru (5.52%), ^{98}Ru (1.88%), ^{99}Ru (12.70%), ^{100}Ru (12.60%), ^{101}Ru (17.0%), ^{102}Ru (31.6%) and ^{104}Ru (18.7%).

It is found native with other members of the platinum group. It is also found in *pentlandite* (nickel ore) and in *pyroxinite* deposits. Constitutes about 0.4 ppb of the earth's crust. The metal is produced in powder form by hydrogen reduction of ammonium ruthenium chloride.

Uses. Hardener for platinum and palladium. Manufacture of corrosion-resistant alloys. Catalyst in synthesis of long-chain hydrocarbons. Substitute for platinum in jewellery. ^{97}Ru ($t_{1/2} = 69 \, h$) is used in tumour and liver diagnosis.

Very similar chemistry to osmium. Most ruthenium compounds are poisonous. Risk of poisoning from ruthenium carbonyl complexes and organometallic compounds of ruthenium. Reported anticancer and immunosuppressive properties.

Ruthenium-103 ($t_{1/2} = 39.5$ days), ruthenium-105 ($t_{1/2} = 1.50$ days) and ruthenium-106 ($t_{1/2} = 366$ days) are produced in the core of nuclear reactors but are not considered too dangerous. The two longer lived isotopes have been used as radioactive labels. ^{106}Ru has been detected in the Irish Sea following discharges from the nuclear processing plant at Sellafield, UK.

Toxicity. LD$_{50}$ (RuO$_2$) oral rat 4580 mg kg^{-1}.

Further reading
Kirk–Othmer ECT, 4th edn, Wiley, New York, 1996, Vol. 19, pp. 347–406.
King, R. B. (ed.) (1994) *Encyclopedia of Inorganic Chemistry*, Wiley, Chichester, Vol. 7, pp. 3514–3533.

Ruthenium(III) chloride CAS No. 10049-08-8: hydrate 148-98-67-0

RuCl$_3$

Uses. Catalyst for oxidation reactions.

Toxicity. Harmful by all routes. LD$_{50}$ intraperitoneal rat 360 mg kg^{-1}.

Ruthenium(VIII) oxide CAS No. 20427 56 9

RuO$_4$

Uses. Powerful oxidant.

Toxicity. Highly toxic.

SAMARIUM CAS No. 7440-19-9

Sm

Samarium, a member of the lanthanide or rare earth series of elements, has an atomic number of 62, an atomic weight of 150.4, a specific gravity of 7.52 at 20°, a melting point of 1077°, and a boiling point of 1791°. Natural samarium is a mixture of seven isotopes, three of which are unstable with long half-lives: ^{144}Sm (3.1%), ^{147}Sm (15.00% , $t_{1/2} = 1.06 \times 10^{11}$ years), ^{148}Sm (11.3%, $t_{1/2} = 1.2 \times 10^{13}$ years), ^{149}Sm (13.8%, $t_{1/2} = 4 \times 10^{14}$ years), ^{150}Sm (7.4%), ^{152}Sm (26.7%) and ^{154}Sm (22.7%). Samarium forms compounds in the +2 and +3 oxidation states.

Average concentration in the earth's crust is ~ 7 ppm. Samarium is found with many other minerals of the rare earths, including *bastnasite* and *monazite* ($\sim 2.8\%$ Sm$_2$O$_3$). Samarium metal can be produced by reducing the oxide with barium. Levels in the human body include blood 0.008 mg dm^{-3}, bone very low, liver very low. Total body mass for an adult (70 kg) 0.05 mg.

Uses. Misch metal ($\sim 1\%$ Sm) used as lighter flints. Carbon-arc lighting. Alloyed with cobalt to produce permanent magnets. Dopant for salts used in optical lasers and masers. Samarium metal is used as a chemical intermediate as a reactant with organic iodo compounds. Samarium(II) iodide is used in organic synthesis. ^{153}Sm ($t_{1/2} = 46$ h) is used in radio(immuno)therapy.

No biological role. Little is known about the toxicity of samarium or its compounds. Therefore, they should be handled with care. The metal dust is a fire and explosion hazard.

Further reading
Gschneidner, K. A., Jr, and LeRoy, K.S. (eds) (1988) *Handbook on the Physics and Chemistry of the Rare Earths*, North-Holland, Amsterdam.
Kirk–Othmer ECT, 4th edn, Wiley, New York, 1995, Vol. 14, pp. 1091–1115.
King, R. B. (ed.) (1994) *Encyclopedia of Inorganic Chemistry*, Wiley, Chichester, Vol. 7, pp. 3595–3618.

Samarium(III) chloride hexahydrate CAS No. 14365-55-9

$SmCl_3 \cdot 6H_2O$

Uses. Preparation of samarium compounds and complexes.

Toxicity. LD_{50} intraperitoneal mouse $565\,mg\,kg^{-1}$.

Samarium(III) oxide CAS No. 12060-58-1

Sm_2O_3

Uses. Preparation of samarium salts. The oxide has been used in optical glass to absorb infrared radiation. The oxide is also used as a catalyst in the hydrogenation/ dehydrogenation of ethanol.

Toxicity. May be harmful by inhalation, ingestion, skin absorption and eye contact.

SARIN CAS No. 107-44-8
LD_{50} rat $550\,\mu g\,kg^{-1}$; LC_{50} (10 min) inhalation rat $150\,mg\,m^{-3}$

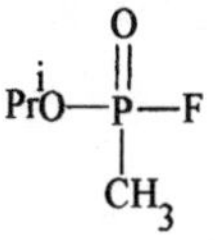

$C_4H_{10}FO_2P \quad M = 140.1$

Isopropyl methylphosphonofluoridate, GB; m.p. $-57°$; b.p. $56°/16\,Torr$; dens. 1.10. Miscible with and hydrolysed by water.

Sarin is a G-agent (see **Tabun**), a highly toxic anticholinesterase inhibitor similar in its action on the nervous system to the phosphorus group of pesticides, but more severe in its effect (see **Parathion**).

Sarin was used by the Aum Supreme Truth sect in their terrorist attack on the Tokyo underground railway on 20 March 1995, when 12 passengers died and 5000 were overcome by fumes (*Daily Telegraph*, 1995).

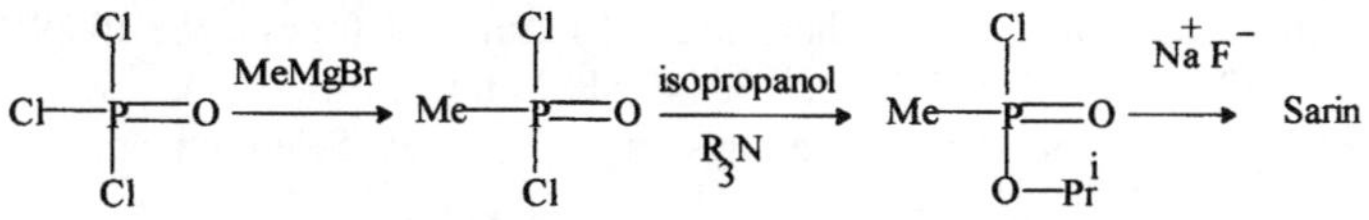

It is synthesized from phosphorus oxychloride by the following sequence:
Sarin is the most toxic of the first group of G-agents with an estimated human LD_{50} of 5–20 mg (see **Soman** and **Tabun**)

Reference
Daily Telegraph, 22 April 1995.

Further reading
Chem. Eng. News 1953, **31**, 4676.
Marrs, T. C., Maynard, R. L. and Sidell, F. R. (1996) *Chemical Warfare Agents*, Wiley, New York.

SCANDIUM CAS No. 7440-20-2

Sc

Scandium, a member of Group 3 of the Periodic Table, has an atomic number of 21, an atomic weight of 44.956, a specific gravity of 2.99 at 25°, a melting point of 1541°, and a boiling point of 2831°. Scandium is a silvery white metal forming compounds in the +3 oxidation state. Sc exists as only one isotope , ^{45}Sc (100%). The pure metal is obtained by reducing the trifluoride with calcium.

Average concentration in the earth's crust is $\sim$ 10–20 ppm. It occurs as the principal component of *thortveitite*, $Sc_2Si_2O_7$, and is found in *wolframite* and *davidite* ($\sim$ 0.02% Sc_2O_3). It is obtained as a by-product in the extraction of uranium from *davidite*.

Uses. Catalysts. Phosphors for display tubes.

Little is known about the toxicity of scandium or its salts. Therefore, the metal and its compounds should be handled with care.

Further reading
Gschneidner, K. A., Jr, and LeRoy, K.S. (eds) (1988) *Handbook on the Physics and Chemistry of the Rare Earths*, North-Holland, Amsterdam.
King, R. B. (ed.) (1994) *Encyclopedia of Inorganic Chemistry*, Wiley, Chichester, Vol. 7, pp. 3595–3618.
Kirk–Othmer ECT, 4th edn, Wiley, New York, 1995, Vol. 14, pp. 1091–1115.

Scandium(III) chloride hydrate CAS No. 25813-71-2

$ScCl_3.H_2O$

Uses. Preparation of other scandium compounds.

Toxicity. LD$_{50}$ oral mouse 3980 mg kg^{-1}; intraperitoneal mouse 314 mg kg^{-1}.

SELENIUM CAS No. 7782-49-2

Se

Selenium, a member of Group 16 of the Periodic Table, has an atomic number of 34, an atomic weight of 78.96, a specific gravity of 4.79 at 20°, a melting point of 217° and a boiling point of $\sim$ 685°. There are six natural isotopes: ^{74}Se (0.89%), ^{76}Se (9.36%), ^{77}Se (7.63%), ^{78}Se (23.78%), ^{80}Se (49.61%) and ^{82}Se (8.73 %). ^{77}Se, ^{77m}Se and ^{82m}Se are used in neutron activation analysis and radiology. Selenium can exist in five oxidation states, -2, 0, $+2$, $+4$ and $+6$.

The earth's crust contains between 0.05 and 0.09 ppm Se. Since the chemistry of selenium and sulphur is very similar, selenium is invariably found with sulphide ores. The principal source of selenium is as a by-product in the treatment of copper sulphide ores. World production is $\sim$ 2100–2300 t year^{-1}. An important part of the world's production is derived from the anode mud deposited during the electrolytic refining of copper. Environmental selenium compounds originate from metal smelting, phosphate production, coal combustion (1–10 ppm Se) and disposal of wastes. The predominant Se compounds in the atmosphere are dimethyl selenide, Me_2Se, selenium dioxide, SeO_2, and selenium. Typical selenium concentrations are air 1–10 $ng\,m^{-3}$, soil 0.1–2 $\mu g\,g^{-1}$ (seleniferous environment 1–200 $mg\,kg^{-1}$), freshwater 0.02–10 $\mu g\,l^{-1}$, sea water 0.03–0.2 $\mu g\,l^{-1}$, food crops 0.05–1 $\mu g\,g^{-1}$, cigarettes 0.001–0.063 μg/cigarette, human (tissues) 0.05–5 $\mu g\,g^{-1}$, (blood) 0.05–0.5 $\mu g\,ml^{-1}$, (blood serum) 85 $\mu g\,l^{-1}$, (urine) 5–30 $\mu g\,l^{-1}$, (ingestion) 10–300 $\mu g\,day^{-1}$ (recommended level $\sim$ 100 $\mu g\,day^{-1}$), (inhalation) 0.0002–0.7 $\mu g\,day^{-1}$ rural to urban, (whole body content, 70 kg adult) $\sim$ 20 mg.

Uses. Rectifiers. Decolorization of glass. Semiconductors. Photoelectric cells. Xerography. Dehydrogenation catalyst. Stainless steel manufacture. Pigments. In Sweden, selenium is used to control atmospheric mercury pollution, some 360 $kg\,year^{-1}$, from cremations. Addition of selenium produces non-volatile mercury selenide, HgSe.

Selenium has a similar chemistry to sulphur and is found as organic compounds such as amino acids in plants. Selenomethionine is one predominant form of selenium found in wheat, soybeans and Se-enriched yeast. Although selenium is an essential trace element in animal nutrition, it is not essential for plant growth. Concentrations of > 4 $mg\,Se\,kg^{-1}$ over a long period of time are toxic to most plants. In these non-tolerant plant species selenium compounds may impair germination and growth and lead to chlorosis. Some plants are selenium tolerant and are able to accumulate up to 10 000 times normal (0.05–1 ppm) Most plants convert selenium into the volatile methyl and dimethyl selenides.

The principal source of selenium for animals including humans is the diet. The uptake within the body is dependent on the chemical form of selenium. Marine creatures tend to bioaccumulate selenium, e.g. fish up to a factor of 400 times. Selenium is detoxified by conversion of dimethyl selenide to $(CH_3)_3Se^+$, the trimethylselenonium ion, a urinary selenium metabolite.

Selenium is antagonistic to mercury, cadmium, arsenic, silver, lead and copper and reduces their metal toxicity. Zinc and tellurium are antagonistic to selenium and interfere with its action. Metal selenates are more toxic than selenites, SeO_3^{2-}, or selenides, Se^{2-}. There is increasing medical evidence that selenium posseses antineoplastic properties and that selenium dietary supplements can inhibit chemically induced tumours, particularly of the skin, liver, colon, and mammary glands.

Sheep, cattle, pigs and poultry are all sensitive to a deficiency or excess of selenium in their feed. For example, 'white muscle disease' in horses, sheep and calves, hepatosis dietetica (liver degeneration) in pigs and exudative diathesis (pancreatic degeneration) in chickens are all symptoms of selenium-deficient diets. Addition of sodium selenide/selenate at a level of 0.1–0.2 $mg\,Se\,kg^{-1}$ of feed is usually curative for all symptoms. Selenium deficiency has been linked to a variety of diseases: cancer; sickle cell anaemia, ischaemic heart disorder, pancreatitis, cystic fibrosis, etc. Keshan disease is potentially fatal cardiomyopathy, and is found in certain areas of China with soil containing low levels of available selenium, and hence low levels of selenium in staple foods.

Many cases of selenium toxicity have been reported for laboratory and domestic animals. The level of toxicity is dependent on several parameters: chemical form of selenium and concentration, intake route, animal species, age, sex, etc. Excess selenium given to laboratory animals usually results in varying degrees of damage to internal organs and haemorrhages. Two diseases are associated with selenium poisoning. 'Blind staggers' is a form of acute selenium poisoning. Common symptoms in livestock and laboratory animals are garlic odour of breath, dyspnea, pulmonary oedema, tachycardia, emesis, diarrhoea, depression, ataxia, paralysis and excessive salivation. 'Alkali disease', is a form of chronic Se poisoning in livestock fed with grain containing >5 ppm Se for several months. Symptoms are loss of vitality, lameness, degeneration of the internal organs and disfigured hooves. In laboratory animals, symptoms include poor appetite, reduced growth rate, decreased blood haemoglobin and liver atrophy, necrosis and cirrhosis. As little as 7.2 ppm Se in grain was found to be fatal to dogs.

Humans are less sensitive than animals to overdoses of selenium. The acute cases have been accidental due to the inhalation of Se fumes and dust with exposure arising from the manufacture of Se and various industries using it. Symptoms include irritation of ear, nose, throat, nausea and vomiting, indigestion and other intestinal disturbances. Five cases of industrial Se poisoning (selenosis) have been reported due to the presence of < 0.2 ppm H_2Se in the air of a metal etching operation. In a study of 62 workers in a selenium rectifier plant, many showed symptoms of Se poisoning. Although Se poisoning is normally seen in industrial environments, cases of chronic toxicity in seleniferous areas of China have shown that daily Se intakes up to $5\,mg\,day^{-1}$ give mean blood selenium levels of up to $3.2\,mg\,l^{-1}$ (normal values 0.05–$0.15\,mg\,l^{-1}$). The symptoms most often found were garlic odour of the breath, bad teeth, dermatitis, gastrointestinal disturbances, hair loss and abnormalities of the nervous system.

Copper refinery workers exposed to selenium showed various symptoms, nose and eye irritation, indigestion, stomach pains, fatigue and garlic-like breath. A number of selenium compounds and selenium containing anions have important commercial uses.

Occupational exposure limit.   OSHA TLV (TWA) $0.2\,mg\,Se\,m^{-3}$.

Further reading
Environmental Health Criteria 58. Selenium, WHO/IPCS, Geneva, 1987.
Fishbein, L. (1988) in Merian, E., Frei, R. W., Härdi, W. and Schlatter, C. (eds), *Carcinogenic and Mutagenic Metal Compounds*, Gordon & Breach, New York, pp. 96–112.
Fowler, B. A., Yamauchi, H., Conner, E. A. and Akkerman, M. (1993) Cancer risk for humans from exposure to the semiconductor metals, *Scand. J. Work Environ. Health*, **19**, 101–103.
Högberg, J. and Alexander, J. (1988) Selenium, in Friberg, L., Nordberg, G. F. and Vouk, V. B. (eds), *Handbook on the Toxicology of Metals*, 2nd edn, Elsevier Amsterdam, Vol. II, Ch. 19, pp. 482–520.
King, R. B. (ed.) (1994) *Encyclopedia of Inorganic Chemistry*, Wiley, Chichester, Vol. 7, pp. 3667–3708.
Kirk–Othmer ECT, 4th edn, Wiley, New York, 1997, Vol. 21, pp. 686–719.
Newland, L. W. (1982) in Hutzinger, O. (ed.), *The Handbook of Environmental Chemistry, Vol. 3, Part B, Anthropogenic Compounds*, Springer, Heidelberg, pp. 45–57.
Oldfield, J. E. (1992) Selenium in agriculture, *Environ. Geochem. Health*, **14**, 81–86.
Olson. O. E. (1986) Selenium toxicity in animals with emphasis on man, *J. Am. Coll. Toxicol.*, **5**, 45–70.

Selenium(VI) fluoride CAS No. 7783-79-1

SeF_6

Uses. Gaseous electric insulator.

Occupational exposure limit. USA OEL (TWA) 0.05 ppm (0.16 mg m^{-3}).

Toxicity. LC$_{Lo}$ (1 h) inhalation rat 10 ppm.

Selenium(IV) oxide CAS No. 7446-08-4

SeO$_2$

Uses. Catalyst for oxidation, dehydrogenation and hydrogenation of organic compounds.

Toxicity. Highly toxic. Possible mutagen. LD$_{50}$ subcutaneous rabbit 4 mg kg^{-1}.

Selenium(IV) oxychloride CAS No. 7791-23-3

SeOCl$_2$

Uses. Solvent for sulphur, selenium, tellurium, rubber, Bakelite, etc.

Toxicity. LD$_{50}$ subcutaneous rabbit 7 mg kg^{-1}.

Selenium(IV) sulphide CAS No. 7488-56-4

SeS$_2$

Uses. Dandruff removal.

Toxicity. Toxic. Severe irritant. Possible carcinogen. Causes dermatitis. Affects liver, spleen and teeth. LD$_{50}$ oral rat 138 mg kg^{-1}.

SILICON CAS No. 7440-21-3

Si

Silicon, a member of Group 14 of the Periodic Table, has an atomic number of 14, an atomic weight of 28.09, a specific gravity of 2.33 at 20°, a melting point of 1410°, and a boiling point of 3265°. Natural silicon exists as three stable isotopes: ^{26}Si (92.23%), ^{29}Si (4.67%) and ^{30}Si (3.10%). The principal oxidation state is +4.

Silicon makes up 25.7% (257 000 ppm) by weight of the earth's crust, and is the second most abundant element. Silicon occurs mainly as the oxide and as silicates. *Sand, quartz, amethyst, agate, flint* and *jasper* are some of the forms in which the oxide appears. *Granite, hornblende, asbestos, feldspar, clay, mica,* etc., are just some of the numerous minerals. Silicon is prepared commercially by heating silica and carbon in an electric furnace, using carbon electrodes. Typical concentrations in the environment are: sea water 0.03–4 ppm, human blood 3.9 mg l^{-1}, human bone 18 ppm, daily intake (70 kg adult) 20–1200 mg, average total human body content (adult, 70 kg) 1.4 g.

Uses. Extremely pure silicon, doped with boron, gallium, phosphorus, arsenic, etc., can be used as semiconductor material in transistors, integrated circuits, solar cells, rectifiers and other solid-state devices. Steel manufacture.

Silicon is important in plant and animal life. Silicon is thought to be essential for many living species, including humans. It is found in the body up to 200 ppm. It is a relatively inert element and is of low toxicity. However, silicon-containing dust, deposited in airways, tends to cause slowly developing pathological changes, catarrh, chronic bronchitis and pneumoconiosis.

Occupational exposure limit.   OSHA PEL (TWA) $5\,mg\,Si\,m^{-3}$

Further reading
King, R. B. (1994) (ed.) *Encyclopedia of Inorganic Chemistry*, Wiley, Chichester, Vol.
     7, pp. 3770–3821.
Ullmann's Encyclopedia of Industrial Chemistry, 5th edn, VCH, Weinheim, 1993, Vol.
     A23, pp. 721–748.

Silane, methyltrichloro- CAS No. 75-79-6

CH_3SiCl_3

Uses.   Preparation of silicones.

Toxicity.   Harmful by all routes. Inhalation may cause spasm, inflammation and
oedema of the larynx and bronchi with the risk of chemical pneumonitis.

Silicates CAS No. 12627-13-3
Widely occurring compounds containing silicon, oxygen and one or more metals.
Many different structural types of silicates and aluminosilicates with a wide variety of
industrial and commercial applications. Certain silicates, such as asbestos (hydrated
magnesium silicate) and talc, can produce fibrotic changes in the lungs and may be
carcinogenic to experimental animals.

Further reading
Ullmann's Encyclopedia of Industrial Chemistry, 5th edn, VCH, Weinheim, 1993, Vol.
     A23, pp. 661–719.

Silicides CAS No. 12651-10-4
Silicides are prepared by direct fusion of the elements. Comparable with borides and
phosphides.

Further reading
Ullmann's Encyclopedia of Industrial Chemistry, 5th edn, VCH, Weinheim, 1993, Vol.
     A24, pp. 57–93.

Silicon carbide CAS No. 409-21-2

SiC

See **Carbides**.

Silicones
(Organopolysiloxanes)
Silicones are important products of silicon. They are obtained by hydrolysing organic
silicon chlorides. Silicone oils, elastomers and resins have become major industrial
products. They are largely non-toxic.

Further reading
Ullmann's Encyclopedia of Industrial Chemistry, 5th edn, VCH, Weinheim, 1993, Vol.
     A24, pp. 57–93.

Silicon hydride CAS No. 7803-62-5
(Silane)

SiH_4

Uses. Preparation of ultrapure silicon.

Toxicity. Irritant to mucous membranes and upper respiratory tract. LC_{50} (4 h) inhalation rat 9600 ppm; ignites in air.

Silicon(IV) oxide CAS No. 7631-86-9
(Silica)

SiO_2 (as the minerals *cristobalite, quartz, tridymite*)

Uses. Manufacture of concrete and brick. Refactory material for high-temperature work. Glass manufacture.

Occupational exposure limits. OSHA PEL (TWA) $0.05 \, mg \, Si \, m^{-3}$.

Toxicity. The low solubility of silica in water makes it non-toxic to most species. Diatoms in both fresh and salt water can extract silica from water to build up their cell walls. Miners and others engaged in work where siliceous dust is breathed in large quantities often develop a serious lung disease known as silicosis. It may also be carcinogenic to this group of workers.

Further reading
Koskela, R.-S., Klockars, M., Laurent, H. and Holopainen, M. (1994) Silica dust exposure and lung cancer, *Scand. J. Work Environ. Health*, **20**, 407–416.

Silicon tetrachloride CAS No. 10026-04-7

$SiCl_4$

Uses. Preparation of other silicon compounds. Smoke screens. Manufacture of silicones and organosilicates. Source of pure silicon and silicon dioxide. Laboratory agent.

Toxicity. Toxic and corrosive. Inhalation may be fatal as a result of spasm, inflammation and oedema of the larynx and bronchi with a risk of chemical pneumonitis. LC_{50} (4 h) inhalation rat 8000 ppm.

SILVER CAS No. 7440-22-4

Ag

Silver, a member of Group 11 of the Periodic Table, has an atomic number of 47, an atomic weight of 107.87, a specific gravity of 10.50 at 20°, a melting point of 961.93° and a boiling point of 2162°. Natural silver consists of only two isotopes: ^{107}Ag (51.84%) and ^{109}Ag (48.16%). It is a soft, ductile, malleable, lustrous, white metal which forms compounds mainly in the +1 oxidation state.

The abundance of silver in the earth's crust is ~ 0.1 ppm, 66th in order of relative abundance. Found native or in ores such as *argentite*, Ag_2S, or *horn silver*, AgCl. Principal sources are copper, nickel, lead, zinc or gold ores. Silver has not been monitored in detail, the following concentrations show the low level of silver availability: sea water $0.04 \, \mu g \, l^{-1}$, probably existing as $[AgCl_2]^-$; freshwater $0.01–3.5 \, \mu g \, l^{-1}$; soils $0.03–4.0 \, \mu g \, g^{-1}$; plants $< 1 \, \mu g \, g^{-1}$. Coal burning introduces small amounts into the atmosphere.

Uses. Jewellery. Tableware. Coinage. Silver solder. Electronics. Catalyst. Dental alloys. High-capacity Ag–Zn and Ag–Cd batteries. Mirrors. Silver sulphodiazine is used in the treatment of burns.

Not essential for humans or animals. Silver and some silver compounds are germicidal to many lower organisms, e.g. soil microbes. Silver itself is probably not toxic; silver salts usually are poisonous only when certain anions are present. Salts generally are not toxic to plants up to $2\,\mu g\,g^{-1}$ but are toxic at concentrations $> 5–10\,\mu g\,g^{-1}$. For humans, silver salts can be absorbed into the circulatory system and the subsequent deposition of reduced silver in various tissues of the body may lead to the production of grey pigmentation of the skin and mucous membranes, a condition known as argyria.

The ingestion of 1–30 g of soluble silver salts or the long-term inhalation of relatively high levels of silver dusts or silver compounds can lead to the disease argyrosis. This disease, which can arise after 2–25 years of exposure, causes pigmentation changes to the conjuctivae, in the mucous membranes of the mouth and gums and in the skin. Persons exhibiting this disease show a permanent silver pigmentation. Despite the increased presence of Ag in the blood, faeces and hair, there is no evidence that chronic silver exposure adversely affects the health of workers.

Occupational exposure limit.   OSHA TLV (TWA) $0.1\,mg\,Ag\,m^{-3}$.

Further reading
ECETOX (1991) *Technical Report 30(5)* European Chemical Industry Ecology and Toxicology Centre, Brussels.
King, R. B. (ed.) (1994) *Encyclopedia of Inorganic Chemistry*, Wiley, Chichester, Vol. 7, pp. 3822–3834.
Ullmann's Encyclopedia of Industrial Chemistry, 5th edn, VCH, Weinheim, 1993, Vol. A24, pp. 107–163.

Silver bromide CAS No. 7785-23-1

AgBr

Uses.   Photography.

Toxicity.   Very low.

Silver chloride CAS No. 7783-90-6

AgCl

Uses.   Photography.

Toxicity.   Very low.

Silver iodide CAS No. 7785-23-1

AgI

Uses.   Weather modification. Photography.

Toxicity.   Very low.

Silver nitrate CAS No. 7761-88-8

AgNO$_3$

Uses.   Photography. Disinfectant.

Toxicity.   Toxic. LD_{50} oral mouse $50\,mg\,kg^{-1}$; LD_{Lo} oral rabbit $800\,mg\,kg^{-1}$. LD_{50} human (unrecorded weight) $29\,mg\,kg^{-1}$.

Silver oxide CAS No. 20667-12-3

Ag$_2$O

Uses. Manufacture of silver oxide–zinc alkaline batteries (> 200 t year^{-1}). In organic chemistry as 'AgOH.'

Toxicity. Highly irritating to skin, eyes, mucous membranes and respiratory tract.

SIMAZINE CAS No. 122-34-9
(Caliber, Simadex)
Selective systemic herbicide. Toxicity class IV. LD$_{50}$ rat > 5 g kg^{-1}; LC$_{50}$ rainbow trout > 100 mg l^{-1}

$C_7H_{12}ClN_5$ $M = 201.7$

2-Chloro-4,6-bis(ethylamino)-1,3,5-triazine; m.p. 227° (dec.); v.p. 2.94 μPa (25°); water solubility 6.2 mg l^{-1}

First reported by Gast *et al.* in 1956 and introduced by Ciba-Geigy. It has systemic action against annual grasses and broad-leaved weeds in fruit, vines, olives, some vegetables, maize, tea, coffee, cocoa, etc. Simazine is phytotoxic to some plants, e.g. lettuce, spinach, carrots and sugar beet.

It is decomposed by UV radiation and has a half-life in soil of about 75 days. The lowest application rate is 400×10^3 kg year^{-1} on corn crops in the USA.

In the US cornbelt run-off post planting can raise river levels to 7 μg l^{-1}; a typical median level in the Mississippi is 60 ng l^{-1}. It has been estimated that simazine transport to the Gulf of Mexico was 6 t in 1987 and 68 t in 1989.

Despite its low toxicity, simazine has been banned in some EC countries owing to its penetration of water supplies. In the USA the EPA has set a limit for drinking water of 1 μg l^{-1}.

Further reading
Dinham, B. (1993) *The Pesticide Hazard*, The Pesticide Trust, ZED Books, London.

SODIUM CAS No. 7440-23-5

Na

Sodium, a member of Group 1 (alkali metals) of the Periodic Table, has an atomic number of 11, an atomic weight of 22.99, a specific gravity of 0.971 at 20°, a melting point of 97.81° and a boiling point of 882.9°. Natural sodium exists as only one isotope, ^{23}Na (100%). In solution, sodium is stable only in the +1 oxidation state.

Sodium is the sixth most abundant element, comprising about 2.6% (26 000 ppm) of the earth's crust, and occurring in over 200 minerals. The more important minerals include *halite*, NaCl, *chile saltpeter*, NaNO$_3$, *thenardite*, Na$_2$SO$_4$, and *mirabilite*, Na$_2$SO$_4$. Elemental sodium is obtained commercially by the electrolysis of dry fused sodium chloride. Typical environmental concentrations are sea water ~ 11–

$290\,\text{mg}\,\text{g}^{-1}$, freshwater $6.3\,\text{mg}\,\text{g}^{-1}$, coal $2.0\,\text{mg}\,\text{g}^{-1}$, petroleum $2\,\mu\text{g}\,\text{g}^{-1}$. In the human body, most of the sodium is found in extracellular fluid. Typical concentrations for an average (70 kg) person are bone $10\,000\,\text{ppm}$, blood $1970\,\text{mg}\,\text{l}^{-1}$, dietary intake 2–15 g day^{-1}, total body content 100 g.

Uses. Sodium–potassium alloy is used as a heat transfer agent. Reducing agent in organic and inorganic chemistry.

Toxicity. Sodium metal should be handled with great care since it is a very reactive metal. The lethal dose toxicities of sodium salts are a function of the anion but are usually in the range 350–$6500\,\text{mg}\,\text{kg}^{-1}$.

Further reading
King, R. B. (ed.) (1994) *Encyclopedia of Inorganic Chemistry*, Wiley, Chichester, 1994, Vol. 1, pp. 35–54.
Ullmann's Encyclopedia of Industrial Chemistry, 5th edn, VCH, Weinheim, 1993, Vol. A24, pp. 277–368.

Disodium methylarsenate(V) (DSMA) CAS No. 144-21-8
(Disodium methanearsonate)

$CH_3AsO(ONa)_2$

Uses. In cotton fields to control weeds/crabgrass in crop and non-crop areas.

Toxicity. Confirmed carcinogen. Moderately toxic to humans.

Monosodium methylarsenate(V) (MSMA) CAS No. 2163-80-6
(Monosodium methanearsonate)

$CH_3AsO(OH)ONa$.

Uses. In cotton fields to control weeds/crabgrass in crop and non-crop areas.

Toxicity. Poison. Skin and eye irritant.

Sodium antimonate(V) CAS No. 12208-13-8

$Na[Sb(OH)_6]$

Uses. High-temperature oxidizing agent. Ingredient of acid-resistant sheet steel enamels.

Toxicity. LD_{Lo} oral human $2\,\text{mg}\,\text{kg}^{-1}$; LD_{50} oral rat $115\,\text{mg}\,\text{kg}^{-1}$.

Sodium arsenate CAS No.7778-43-0

Na_2AsHO_4

Uses. Wood preservative.

Toxicity. Poison.

Sodium arsenite CAS No. 7784-46-5

$NaAsO_2$

Uses. Control of baits. Non-selective herbicide. Rodenticide. Insecticide (termites).

Occupational exposure limit. US TLV (TWA) $0.01\,\text{mg}\,\text{m}^{-3}$.

Toxicity. Extremely toxic to humans. Causes DNA damage. LD_{50} oral rat $41\,mg\,kg^{-1}$.

Sodium azide CAS No. 26628-22-8

NaN_3

Uses. Bactericide.

Toxicity. Poison. A severe irritant to skin, eyes and mucous membranes. Inhalation can cause CNS problems and renal damage. High concentrations cause hypotension, ataxia, weakness and possibly death. LD_{Lo} oral human $29–143\,mg\,kg^{-1}$. LD_{50} oral rat $27\,mg\,kg^{-1}$.

Sodium bromide CAS No. 7647-15-6

NaBr

Uses. Medicine. Photography.

Toxicity. LD_{50} oral rat $3500\,mg\,kg^{-1}$; intraperitoneal mouse $5\,g\,kg^{-1}$.

Sodium carbonate CAS No. 497-19-8

Na_2CO_3

Uses. Manufacture of glass, aluminium, soap, detergents. In petroleum refining.

Toxicity. Acute and chronic doses cause respiratory tract problems. Fine dust causes burns, eczema of the skin and nasal defects. LD_{50} subcutaneous mouse $2210\,mg\,kg^{-1}$; oral rat $4090\,mg\,kg^{-1}$.

Sodium chlorate CAS No. 7775-09-9

$NaClO_3$

Uses. Non-selective herbicide. Defoliant. Manufacture of chlorine dioxide (ClO_2).

Toxicity. LD_{50} oral rat $1200\,mg\,kg^{-1}$.

Sodium chloride CAS No. 7647-14-5

NaCl

Uses. Widely used chemical. Manufacture of sodium hydroxide, sodium carbonate, chlorine, hydrochloric acid, etc.

Toxicity. Long-term exposure leads to hypertension, chronic bronchitis, ephysema, liver disorders. Sodium chloride dust causes chronic bronchitis. Oral doses 2000–5000 $mg\,kg^{-1}$ to laboratory animals causes twitching and gross cyanosis of paws, tail and ears and possibly death. LD_{50} subcutaneous mouse $3150\,mg\,kg^{-1}$; oral rat $3\,g\,kg^{-1}$.

Sodium chlorite CAS No. 7758-19-2

$NaClO_2$

Uses. Production of ClO_2 and as bleaching agent in textile, paper and pulp industry.

Toxicity. LD_{50} oral rat $165\,mg\,kg^{-1}$. Affects liver, kidney and reproduction.

Sodium cyanide CAS No. 143-33-9

NaCN

Uses. Poison for rats and rabbits.

Toxicity. IARC group 3 (not classifiable). LD_{50} oral rat $165\,mg\,kg^{-1}$.

Sodium fluoride CAS No. 7681-49-4

NaF

Uses. Fluorinating agent. Manufacture of glass, aluminium and beryllium. Fluoridation of drinking water.

Toxicity. LD_{Lo} oral human $71\,mg\,kg^{-1}$; LD_{50} oral rat $52\,mg\,kg^{-1}$.

Sodium hydrogen carbonate CAS No. 144-55-8

$NaHCO_3$

Uses. Food industries.

Toxicity. LD_{50} inhalation rat $77\,mg\,m^{-3}$; oral rat $4220\,mg\,kg^{-1}$. Affects lungs and thorax.

Sodium hydroxide CAS No. 1310-73-2

NaOH

Uses. Manufacture of artificial fibres, soaps, paints, paper and pulp, textiles, etc.

Toxicity. Causes severe burns. Scarring of tissues. Conjunctivitis. Swallowing gives severe pain in mouth and stomach. Collapse and death may occur. LD_{50} intraperitoneal mouse $40\,mg\,kg^{-1}$.

Sodium iodide CAS No. 7681-82-5

NaI

Uses. Medicines. Source of iodine.

Toxicity. LD_{50} oral rat $4340\,mg\,kg^{-1}$; intraperitoneal mouse $430\,mg\,kg^{-1}$.

Sodium molybdate(VI) dihydrate CAS No. 7631-95-0

$Na_2MoO_4.2H_2O$

Uses. Analytical reagent. Paint pigments. Metal finishes. Catalyst in dye production. Additive to animal feeds and fertilizers. Micronutrient.

Toxicity. LD_{50} intraperitoneal mouse $257\,mg\,kg^{-1}$; oral rat $3120\,mg\,kg^{-1}$.

Sodium nitrate CAS No. 7631-99-4

$NaNO_3$

Uses. Manufacture of nitric acid and sodium nitrite. Glass enamelling. Matches. Fertilizer.

Toxicity. LD_{Lo} oral human $114\,mg\,kg^{-1}$. LD_{50} oral rat $1267\,mg\,kg^{-1}$.

Sodium nitrite CAS No. 7632-00-0

$NaNO_2$

Uses. Food industry. Dyes. Meat curing.

Toxicity. LD_{Lo} oral human 71 mg kg^{-1}. Affects blood and cardiovascular system.

Sodium nitroprusside dihydrate CAS No. 13755-38-9

$Na_2[Fe(CN)_5NO].2H_2O$

Sodium nitroprusside is a potent vascular smooth-muscle relaxant. It reacts with haemoglobin to release cyanide.

Sodium perborate tetrahydrate CAS No. 10486-00-7

$NaBO_4.4H_2O$

Uses. Strong bleach in detergents. Local antiseptic.

Toxicity. LD_{Lo} oral human 214 mg kg^{-1}. LD_{50} oral rat 1200 mg kg^{-1}.

Sodium peroxide CAS No. 1313-60-6

Na_2O_2

Uses. Bleaching wood pulp and linen. Manufacture of sodium peroxoborate hexahydrate and sodium carbonate peroxohydrate.

Toxicity. Strong oxidizing agent. Irritant. Corrosive.

Sodium selenate CAS No. 13410-01-0; decahydrate 10102-23-5

Na_2SeO_4

Uses. Insecticide. Glass manufacture. Veterinary medicine.

Toxicity. Highly toxic. Possible carcinogen and mutagen. LD_{50} oral rat 1000 μg kg^{-1}.

Sodium selenite CAS No. 10102-18-8; pentahydrate 26970-82-1

Na_2SeO_3

Uses. Glass manufacture. Soil additive in Se-deficient areas.

Toxicity. Very toxic. Possible mutagen.

Sodium sulphate decahydrate CAS No. 7727-73-3

$Na_2SO_4.10H_2O$

Uses. Manufacture of glass, soap, textiles. Medicines.

Toxicity. LD_{50} oral mouse 5989 mg kg^{-1}. Mutation and tumorigenic data.

Sodium sulphite CAS No. 7757-83-7

Na_2SO_3

Uses. Photography. Pharmaceutical industry. Textiles.

Toxicity. Possible mutagen. LD_{50} oral mouse 820 mg kg^{-1}.

Sodium tellurite(IV) CAS No. 10102-20-2

Na_2TeO_3

Uses.   Corrosion resistance of electroplated nickel.

Toxicity. LD_{50} oral rat $83\,mg\,Te\,kg^{-1}$. LD_{Lo} oral human $30\,mg\,kg^{-1}$.

Sodium tetraborate decahydrate CAS No. 1303-96-4

$Na_2B_4O_7.10H_2O$

Uses.   Herbicide available only as mixtures.

Toxicity. LD_{Lo} oral human $709\,mg\,kg^{-1}$; LD_{50} oral rat $2660\,mg\,kg^{-1}$.

Sodium thiocyanate CAS No. 540-72-7

NaSCN

Uses.   Chemical agent.

Toxicity.   Irritant. Teratogen. LD_{50} oral rat $764\,mg\,kg^{-1}$; intraperitoneal rat $540\,mg\,kg^{-1}$.

Sodium thiosulphate CAS No. 7772-98-7

$Na_2S_2O_3$

Uses.   Photography. Analytical chemistry.

Toxicity. LD_{50} intraperitoneal mouse $5200\,mg\,kg^{-1}$.

Sodium tripolyphosphate CAS No. 7758-29-4

$Na_5P_3O_{10}$

Uses.   Builder for synthetic detergents. Curing hams and bacon.

Toxicity. LD_{50} oral rat $3800\,mg\,kg^{-1}$; intravenous mouse $71\,mg\,kg^{-1}$.

Sodium tungstate(VI) dihydrate CAS No. 10213-10-2

$Na_2WO_4.2H_2O$

Uses.   Growth promoter in certain plants. Fireproofing and waterproofing fabrics. Preparation of complex tungstates.

Toxicity. LD_{50} (66 day) oral rat $250\,\mu g\,kg^{-1}$. LD_{50} intraperitoneal rat $204\,mg\,kg^{-1}$. The most soluble tungsten salt. Acute symptoms – diarrhoea and respiratory arrest and death.

Sodium uranate(VI) CAS No. 13721-34-1

$Na_2U_2O_7$

Uses.   Manufacture of yellow/green fluorescent glass. Painting porcelain and enamels.

Toxicity.   Moderately toxic.

SOMAN CAS No. 96-64-0

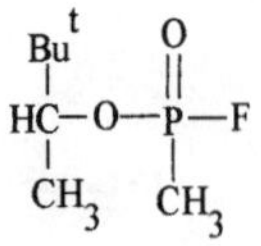

$$\begin{array}{cc} Bu^t & O \\ | & || \\ HC\!-\!O\!-\!\!P\!-\!F \\ | & | \\ CH_3 & CH_3 \end{array}$$
$C_7H_{16}FO_2P$ $M = 182.2$

Pinacolyl methylphosphonofluoridate.
A G-agent to be compared with **Sarin** and **Tabun**.

STRONTIUM CAS No. 7440-24-6

Sr

Strontium, a member of Group 2 of the Periodic Table, has an atomic number of 38, an atomic weight of 87.62, a specific gravity of 2.54 at 20°, a melting point of 777° and a boiling point of 1384°. Natural strontium is a mixture of four stable isotopes : ^{84}Sr (0.56%), ^{86}Sr (9.86%), ^{87}Sr (7.00%) and ^{88}Sr (82.58%). Twelve other unstable isotopes are known to exist. Principal oxidation state is $+2$.

Strontium is found mainly as *celestite*, $SrSO_4$, and *strontianite*, $SrCO_3$. Strontium (384 ppm) is the 15th element in order of crustal abundance. The metal can be produced by reduction of the oxide SrO with aluminium. Finely divided Sr is a fire hazard. Average daily intake for humans $0.8–5\,mg\,kg^{-1}$. Average total human body content is $\sim 200\,mg$, with the bones containing 40–140 ppm.

Uses. Pyrotechnics. Special glass manufacture. Phosphors.

Strontium salts are of low toxicity although this is sometimes dependent on the anion type; they resemble calcium salts in their chemical and biological behaviour. Normal entry to the human body is via ingestion. The gastrointestinal absorption of soluble Sr salts ranges from 5 to 25%. Symptoms of acute toxicity are excessive salivation, vomiting, colic, diarrhoea, and possibly respiratory failure. Workers in a strontium salt plant showed reduced activity of cholinesterase and acetylcholine. Drinking water with a concentration $13\,mg\,l^{-1}$ caused impaired tooth development in a 1-year-old child.

Of great environmental interest is ^{90}Sr, with a half-life of 28 years, which results from nuclear fallout and presents health problems. The principal ecological pathway is as for calcium: grass $\rightarrow$ cow's milk $\rightarrow$ human. Tendency to concentrate in the bone. The Chernobyl accident released to the atmosphere $\sim 0.22\,MCi$ of strontium-90. A safe skeletal burden for humans is $\sim 1.0\,\mu Ci$.

Further reading
Eisenbud, M. (1987) *Environmental Radiochemistry. From Natural, Industrial, and Military Sources*, 3rd edn, Academic Press, Orlando, FL.
King, R. B. (ed.) (1994) *Encyclopedia of Inorganic Chemistry*, Wiley, New York, Vol. 1, pp. 67–87.
Thompson, R. (ed.) (1995) *Industrial Inorganic Chemicals. Production and Uses.* Royal Society of Chemistry, Cambridge.
Ullmann's Encyclopedia of Industrial Chemistry, 5th edn, VCH, Weinheim, 1994, Vol. A25, pp. 321–327.

Strontium carbonate CAS No. 1633-05-2

$SrCO_3$

Uses. As a component of glass which absorbs X-rays from CRTs. Ceramic permanent magnets (strontium ferrite). Manufacture of strontium metal and strontium steels.

Toxicity. N.a.

Strontium chloride hexahydrate CAS No. 10025-70-4

$SrCl_2.6H_2O$

Uses. Preparation of other strontium salts. Toothpaste.

Toxicity. LD_{50} intraperitoneal mouse 1253 mg kg^{-1}.

Strontium nitrate CAS No. 10042-76-9

$Sr(NO_3)_2$

Uses. Pyrotechnics and signalling devices (red flame).

Toxicity. LD_{50} oral rat > 2000 mg kg^{-1}.

STYRENE CAS No. 100-42-5
LD_{50} oral rat 5 g kg^{-1}

C_8H_8 $M = 104.1$

Ethenylbenzene, phenylethylene; b.p. 145°, 33° at 1.33 kPa (10 Torr); v.p. 0.87 kPa; v. dens. 3.6; dens. (4°, 20°) 0.91; λ_{max} 245 nm. Water solubility 300 mg l^{-1} (20°), soluble in most organic solvents.

Produced from ethylbenzene by either catalytic dehydrogenation at 600° with the formation of minor amounts of benzene (0.7%) and toluene (1.0%), or oxidation to the hydroperoxide and reaction with propylene:

$$Ph-\underset{O-OH}{CH}-CH_3 \;+\; CH_3-CH{=}CH_2 \longrightarrow Ph-\underset{OH}{CH}-CH_3 \;+\; CH_3-CH-CH_2$$

Styrene

Styrene rapidly polymerizes owing to aerial oxidation and it requires a stabilizer such as *t*-butylcatechol. In 1976 production in the USA was 2864×10^3 t and in Japan 1090×10^3 t. It is used principally for the production of polystyrene and also for styrene–butadienes as synthetic rubbers and for acrylonitrile–butadiene–styrene (ABS) resins.

In addition to release during manufacture and handling, styrene is a component of vehicle exhaust and cigarette smoke (ca 40 µg/cigarette). It forms 2.6% of the emission from rotatory engines and airborne levels in California have averaged 5 ppb.

It is not a serious contaminant of water but occurs in foodstuffs owing to leakage from polystyrene and ABS packaging. The concentration of monomer in these materials lies in the range 700–3300 ppm.

There have been numerous studies in the workplace which showed that styrene enters the bloodstream and lipid-rich tissues, where it is metabolized. There is evidence of CNS depression and of liver damage but only at high rates, e.g. 2500 ppm as vapour in a single dose or 300 ppm in repeated doses.

Styrene acts as a mutagen through its metabolite, the 7,8-oxide. There is evidence of chromosomal abnormalities in workers in the reinforced plastics industry and this oxide induces squamous cell carcinomas in rats. No causal relationship has been established with cancer in humans. The IARC classes it as group 2B – carcinogenic in animals. The World Health Organization guideline for drinking water is $20\,\text{mg}\,\text{l}^{-1}$ and the ACGIH has set a TLV (TWA) of $50\,\text{ppm}$ ($213\,\text{mg}\,\text{m}^{-3}$).

Further reading

Berlin, A., Draper, M., Krug, E., Rui, R. and Van der Venne, M.Th. (1990) *The Toxicology of Chemicals*, Commission of the European Community, Brussels, Vol. 2, pp. 93–102.

Environ. Health Criteria, World Health Organization, Geneva, 1987, Vol. 26.

SULPHUR CAS No. 7704-34-9

S

Sulphur, a member of Group 18 of the Periodic Table, has an atomic number of 16, an atomic weight of 32.06, a specific gravity of 1.96 at $20°$ (monoclinic form), a melting point of $115.2°$ (monoclinic form) and a boiling point of $444.67°$. Sulphur is a pale yellow, odourless, brittle solid forming compounds in the $+2$, $+4$ and $+6$ oxidation states. Natural sulphur consists of four stable isotopes: ^{32}S (95.0%), ^{33}S (0.76%), ^{34}S (4.22%) and ^{36}S (0.02%).

It is widely distributed in nature as the minerals *pyrites, cinnabar, stibnite, gypsum, barite*, etc. It is the sixteenth element in order of abundance, to the extent of about $340\,\text{ppm}$ in the earth's crust. Sulphur is recovered from deep wells using the Frasch process, or from natural gas and petroleum refining. Sea water contains up to $900\,\text{ppm}$ (as SO_4^{2-}) but fresh water normally contains between 1–$10\,\text{ppm}$. Sulphur is present in the human body ($70\,\text{kg}$), total mass 140–$170\,\text{g}$. Other levels are bone 500–$2400\,\text{ppm}$, blood $1800\,\text{mg}\,\text{l}^{-1}$, dietary intake $\sim 900\,\text{mg}\,\text{day}^{-1}$.

Uses. Vulcanization of natural rubber. Manufacture of sulphur dioxide. Organic sulphides such as carbon disulphide. Fungicide, insecticide and fumigant. Sulphide chemicals.

Sulphur is an essential element for all life forms. It is present in the two amino acids cysteine and methionine which are constituents of peptides.

Inhalation of sulphur dust by humans may irritate the eyes, respiratory tract and lungs.

Further reading

ECETOX (1991) *Technical Report 30(4)*, European Chemical Industry Technology and Toxicology Centre, Brussels.

King, R. B. (ed.) (1994) *Encyclopedia of Inorganic Chemistry*, Wiley, Chichester, Vol. 7, pp. 3954–3986.

SI 1991 No. 472 The Environmental Protection (Prescribed Processes and Substances) Regulations 1991, HMSO, London, 1991.

Ullmann's Encyclopedia of Industrial Chemistry, 5th edn, VCH, Weinheim, 1994, Vol. A25, pp. 507–567.

Sulphates CAS No. 14808-79-8

SO_4^{2-}

Uses. Widely used salts.

Toxicity. Variable, dependent on cation and solubility.

Sulphides CAS No. 18496-25-8

S^{2-}

Inorganic sulphides react with acids to give the very toxic hydrogen sulphide. When burnt in air they are oxidized to the noxious sulphur dioxide. Organic sulphides are very toxic but fortunately have very strong smells. These compounds are volatile and are easily inhaled.

Sulphites CAS No. 18496 25 8

SO_3^{2-}

Uses.   Reducing agent. Paper manufacture.

Toxicity. LD_{50} oral rat 6379 mg kg^{-1}. IARC group 3 (not classifiable).

Sulphur Dioxide CAS No. 7446-09-5

SO_2

STEL 5 ppm (13 mg m^{-3}).
   Air pollutant, contributes to acid rain in the atmosphere.

Uses.   Bleaching agent for textiles and paper. Disinfectant. Preservative in cooked meats, fruit and vegetables. Preparation of sulphur trioxide. Nonaqueous solvent.
   Pollution from sulphur dioxide arises from combustion of fuels containing sulphur. The average level in coal is 1.3%, although exceptionally this may rise to 15%. Of this sulphur, 40% is as iron pyrites with the remaining 60% combined organic material. In 1989 production of sulphur dioxide from coal worldwide was 39×10^6 t. The percentage sulphur levels in petroleum fractions vary:

	%	Amount consumed in UK (1990) (10^6 t)
Motor spirit	0.1	24
Diesel fuel	0.3	10.6
Gas oil	0.7	8
Fuel oil	2.0	6.5
Heavy fuel oil	>3.5	11

   In Germany, whose forests have been badly affected by acid rain, the annual emission in 1990 of SO_2 was 3.5×10^6 t year^{-1}, and of NO_x 3.0×10^6 t year^{-1}. At this time the worldwide emission of SO_2 from coal usage (2.7×10^9 t oil equivalent) was 3.9×10^6 t year^{-1}. Emissions from oil usage (3.0×10^9 t) were 28×10^6 t year^{-1}.
   Released SO_2 is oxidised by ozone and $^{\bullet}OH$ to SO_3 and then to sulphuric acid.

$$^{\bullet}OH + SO_2 + (O_2, H_2O) \longrightarrow H_2SO_4 + {^{\bullet}OOH}$$

   By 1987, heavy damage by acids to both coniferous and deciduous trees was detected with 50% of species such as silver firs suffering between 20% and 60% needle loss; damage in all species of up to 25% was found all across Europe. In spite of the undoubted damage done to trees close to industrialized areas, the tendency to attribute all forest decline to air pollution post 1970 should be resisted; actual improvements in the quality of some European forests have been recorded in the 1980s. There have been a number of serious historical episodes of pollution by smoke and sulphur dioxide. An early example occurred in 1930 in the Meuse valley during a temperature inversion, when levels built up to over 10 ppm, resulting in 60 deaths from heart failure linked to respiratory disorders.

In the London smog of December 1952, SO_2 levels exceeded 100 ppm and in the following months deaths from lung and heart disease rose by 4000 above the normal expectancy. In 1993 in major cities the position was:

City	SO_2 emitted (10^3 t)	Peak hourly maximum in air (ppb)
Los Angeles	50	30
London	49	283
Bombay	157	40
Madrid	38	113

Sulphur dioxide has an odour threshold of 1.5 ppm. The critical concentration for toxicity at 0.0005% v/v is lower than that for H_2S and CO; the trigger concentration for a potential hazard is 0.000125% v/v.

Breathing is affected at levels above 1.5 ppm and 1 min at 200 ppm causes great discomfort. The limit for prolonged exposure is 5 ppm. Levels indoors are typically about 20 ppb, which is less than those outdoors. On 20 occasions hourly maxima (1995) in Liverpool and Barnsley, UK, exceeded 300 ppb and Leeds was also badly affected. The DOE ratings for SO_2 levels in air (ppb) are as follows:

very good	< 60
good	60–124
poor	125–399
very poor	> 400

The World Health Organization guideline for the 1 h maximum is 122 ppb. In the USA the National Air Quality level for the annual arithmetic mean is 80 mg kg^{-1} (0.03 ppm).

Occupational exposure limit.   US TLV (TWA) 2 ppm (5.2 mg m^{-3}).

Toxicity.   LC$_{50}$ (1 h) inhalation rat 2520 ppm; TC$_{Lo}$ (5 days) inhalation human 3 ppm (pulmonary effects).

Further reading
ECETOX (1994) *Technical Report 30*(5), European Chemical Industry Ecology and Toxicology Centre, Brussels.
IARC (1992) *Monograph. Sulfur Dioxide, and Some Sulfites, Bisulfites and Metabisulfites*, IARC, Lyon, Vol. 54, pp. 131–188.
Kandler, O. and Innes, J. L. (1995) Air pollution and forest decline in Central Europe, *Environ. Pollut.*, **90**, 171–180.

Sulphuric acid CAS No. 7664-93-9

H_2SO_4 $M = 98.1$

Uses.   Manufacture of phosphate fertilizers, paints and pigments. Metal pickling. Manufacture of phosphoric acid. Manufacture of detergents and explosives. Many other uses.

Occupational exposure limit.   US (TLV) TWA 1 mg m^{-3}; US (TLV) STEL 3 mg m^{-3}.

Toxicity. Irritant. Corrosive to all body tissues. Toxic by all routes. Ingestion can cause severe internal injuries and death. LD_{50} oral rat $2140 \, mg \, kg^{-1}$.

Further reading
IARC (1992) *Monograph. Sulfur Dioxide, and Some Sulfites, Bisulfites and Metabisulfites*, IARC, Lyon, Vol. 54, pp. 41–119.

Sulphurous acid CAS No. 7782-99-2

H_2SO_3 $M - 82.1$

Uses. Bleach. Preservation of food, fruits, nuts, wines, etc.

Toxicity. Similar to sulphur dioxide. TD_{Lo} oral human $0.50 \, mg \, kg^{-1}$.

Sulphur trioxide CAS No. 7446-11-9

SO_3 $M = 80.1$

Uses. Manufacture of sulphuric acid and oleum. Sulphonation of organic compounds.

Toxicity. Irritant. Skin and eye damage. TC_{Lo} inhalation human $30 \, mg \, m^{-3}$.

Further reading
IARC (1992) *Monograph. Sulfur Dioxide, and Some Sulfites, Bisulfites and Metabisulfites*, IARC, Lyon, Vol. 54, pp. 121–130.

2,4,5,-T CAS No. 93-76-5
(Silvex)
Selective systemic herbicide. LD_{50} rat $300 \, mg \, kg^{-1}$; LC_{50} rainbow trout 1.0–$8.0 \, mg \, l^{-1}$ depending on conditions.

O—CH₂—CO₂H ... (structure) $C_8H_5Cl_3O_3$ $M = 255.5$

2,4,5-Trichlorophenoxyacetic acid; m.p. 156°; v.p. $6.4 \times 10^{-6} \, mmHg$ (25°); water solubility $150 \, mg \, l^{-1}$ (25°)

Anaerobic degradation leads to dechlorination and side-chain cleavage.

Introduced in 1948 by Dow, who later spent $10 million in the courts defending its use before voluntarily withdrawing it. 2,4,5-T was formerly used in rice fields and orchards and on sugar cane and rangelands.

Its use was first questioned in 1970 when a teratogenic effect was discovered, and following a preliminary ban in 1979 by the USEPA all registered use ceased in March 1985 on the grounds of oncogenic, foetotoxic and teratogenic risks.

The butyl ester was a component of Agent Orange, which was used to defoliate supply trails during the Vietnam War. The technical material then used contained traces of 2,3,7,8- tetrachlorodibenzo-*p*-dioxin [see **Polychlorodibenzodioxins (PCDD) and Polychlorodibenzofurans (PCDF)**] in the range $2–50 \, \mu g \, g^{-1}$. US veterans who had been exposed to risk initiated claims against the government; an out-of-court claim by seven manufacturers was settled in 1984.

The EPA has set a limit in drinking water of $70\,\mu g\,l^{-1}$ and the WHO recommends a total daily intake of $0.03\,mg\,kg^{-1}$. In the USA and the UK the limit set for the workplace is a TWA of $10\,mg\,m^{-3}$.

TABUN CAS No. 77-81-6
LD_{50} dog $200\,\mu g\,kg^{-1}$; LD_{50} i.p. mouse $0.6\,mg\,kg^{-1}$

$$\underset{\displaystyle NMe_2}{\overset{\displaystyle O}{EtO-\overset{\displaystyle \|}{P}-CN}}$$

$C_5H_{11}N_2O_2P$ $M = 162.1$

Ethyl N-dimethylphosphonamidocyanidate, GA; b.p. 240°; v.p. (25°) 0.07 Torr; v.dens. 5.63. Miscible with water but rapidly hydrolysed.

Discovered by Schrader of I.G. Farben and used as the first military nerve gas. Prepared from phosphorus oxychloride by the following sequence:

$$\underset{\displaystyle Cl}{\overset{\displaystyle Cl}{Cl-\overset{\displaystyle |}{\underset{\displaystyle |}{P}}=O}} \xrightarrow{Me_2NH} \underset{\displaystyle Cl}{\overset{\displaystyle Me_2N}{Cl-\overset{\displaystyle |}{\underset{\displaystyle |}{P}}=O}} \xrightarrow[\overset{+}{Na}\,\overset{-}{CN}]{EtOH} EtO\underset{\displaystyle CN}{\overset{\displaystyle Me_2N}{-\overset{\displaystyle |}{\underset{\displaystyle |}{P}}=O}}$$

Tabun has an odour of bitter almonds. The signs of exposure are stomach cramps, constricted pupils, abnormal vision, nausea, vomiting, dizziness and convulsions. It is taken in through the skin and especially the eyes. The lethal dose in humans is 1–7 mg, which should be compared with the lethal dose of hydrogen cyanide of ca 50 mg. Tabun is rapidly detoxified in alkaline solution.

See also **Sarin**.

TANTALUM CAS No. 7440-25-7

Ta

Tantalum, a member of Group 5 of the Periodic Table, has an atomic number of 73, an atomic weight of 180.95, a specific gravity of 16.65 at 20°, a melting point of 3017° and a boiling point of 5458°. Natural tantalum contains only two isotopes, ^{181}Ta (99.988%) and the radioactive ^{180}Ta (0.0123%) with half-life $> 10^{13}$ years. It is a grey, very hard, malleable and ductile metal. The principal oxidation state is $+5$.

The abundance of Ta in the earth's crust is ca 1 ppm. Invariably found with niobium, e.g. in the mineral *columbite–tantalite,* $[(Fe, Mn)(Ta, Nb)_2O_6]$. Present in sea water at ~0.004 ppb. The sea squirts (class Ascidians) can bioaccumulate up to 400 ppb.

Uses. In surgery as wire, clips and mesh. In pen nibs. Acid proof alloys (Ta–W–Co and Ta–W–Mo). Filaments. Electrolytic capacitors. Diagnostic radiology for cancer.

Not essential for humans or animals. Despite high affinity for proteins, it appears to be biologically inert. Some industrial skin injuries have been reported.

Occupational exposure limit. OSHA TLV (TWA) $5\,mg\,Ta\,m^{-3}$; UK short-term limit $10\,mg\,m^{-3}$.

Further reading
De Meester, C., (1988) Tantalum, in: Seiler, H. G., Sigel, H. and Sigel, A. (eds), *Handbook on Toxicity of Inorganic Compounds*, Marcel Dekker, New York, pp. 661–663.
King, R.B. (ed.) (1994) *Encyclopedia of Inorganic Chemistry*, Wiley, Chichester, Vol. 5, pp. 2444–2462.
Kirk–Othmer ECT, 4th edn, Wiley, New York, 1996, Vol. 17, pp. 43–68.

Tantalum(V) chloride CAS No. 7721-01-9

$TaCl_5$

Uses.   Chemical agent.

Toxicity.   Poison by intraperitoneal route. Moderately toxic by ingestion. LD_{50} oral rat 1900 mg kg^{-1}; LD_{50} intravenous rat 75 mg kg^{-1}.

Tantalum(V) fluoride CAS No. 7783-71-3

TaF_5

Uses.   Friedel–Craft catalyst.

Toxicity.   Poisonous and corrosive. LD_{50} intravenous rat 110 mg kg^{-1}

Tantalum(V) oxide CAS No. 1314-61-0

Ta_2O_5

Uses.   Preparation of Ta metal. Optical glass. Preparation of TaC. Piezoelectric, maser and laser applications.

Toxicity.   Mildly toxic by ingestion. LD_{50} oral rat 8000 mg kg^{-1}

TECHNETIUM CAS No. 7440-26-8

Tc

Technetium, a member of Group 7 of the Periodic Table, has an atomic number of 43, an atomic weight of 98.91, a specific gravity of 11.50 at 20°, a melting point of 2157° and a boiling point of 4265°. All isotopes of technetium are radioactive, with atomic masses ranging from 86 to 113. The element forms compounds in the +2, +4, +5, +6 and +7 oxidation states. The isotopes $^{97}Tc(t_{1/2} = 2.6 \times 10^6$ years), $^{98}Tc(t_{1/2} = 4.2 \times 10^6$ years) and $^{99}Tc(t_{1/2} = 2.12 \times 10^5$ years) all have long half-lives. The last isotope has a specific activity of 6.2×10^8 disintegrations s^{-1}g^{-1}, and is therefore a contamination hazard and should be handled in a glove-box.

Technetium was the first element to be artificially produced in 1937 by bombarding molybdenum with deuterons in a cyclotron. Nowadays it can be obtained in tonne quantities as a by-product in the separation of the products of uranium fission.

Uses.   Technetium-95m, with a half-life of 61 days, is useful for tracer work. $^{99m}Tc(t_{1/2} = 6$ h) is used in radiodiagnostics.
One of the most stable forms of technetium is the very soluble pertechnetate ion, TcO_4^-. It has a high mobility rate in soils with rapid plant root uptake. For humans, the biological half-life, whole body, is 2 days.

Further reading

Eisenbud, M., (1987) *Environmental Radioactivity. From Natural, Industrial, and Military Sources*, 3rd edn, Academic Press, New York.

Holm, E. (1993) Radioanalytical studies of Tc in the environment – progress and problems, *Radiochim. Acta*, **63**, 57–62.

King, R.B. (ed.) (1994) *Encyclopedia of Inorganic Chemistry*, Wiley, Chichester, Vol. 5, pp. 4094–4099.

Ullmann's Encyclopedia of Industrial Chemistry, 5th edn, VCH, Weinheim, 1993, Vol. A22, pp. 499–591.

TELLURIUM CAS No. 13494-80-9

Te

Tellurium, a member of Group 16 of the Periodic Table, has an atomic number of 52, an atomic weight of 127.60, a specific gravity of 6.24 at 20°, a melting point of 450° and a boiling point of 990°. Natural tellurium consists of eight isotopes; ^{130}Te (33.80 %), ^{128}Te (31.69%), ^{126}Te (18.95%), ^{125}Te (7.14%), ^{124}Te (4.82%), ^{122}Te (2.60%), ^{120}Te (0.089%) and ^{123}Te (0.89%, $t_{1/2} = 1.3 \times 10^{13}$ years). Tellurium exists in two allotrophic forms, as an amorphous black powder and the crystalline hexagonal form with a metallic lustre. Te forms compounds in the $-2, +2, +4$ and $+6$ oxidation states.

The abundance of tellurium in the earth's crust is about 0.01 ppm, 71st in order of relative abundance. Tellurium is the main component in over 40 minerals, and it is usually associated with sulphur and selenium. Commercially, it is obtained from the anode sludge of the electrorefining of copper. World production of tellurium is about 350 t. The main source of tellurium in the air originates from coal burning. Little or no information is available about concentrations of tellurium in the environment; sea water/fresh water, ng quantities; mean levels in food samples, ppb quantities. Human (70 kg) daily intake $\sim 100\,\mu$g. Human blood contains about $0.25\,\mu$g Te l^{-1}.

Uses. Improving the characteristics of alloys of copper, steel, lead and bronze. Catalyst in petrol refining and the organic chemicals industry. Glass manufacture. Vulcanization agent in natural rubber and in styrene–butadiene rubbers.

Unlike selenium, tellurium has no essential biochemical function in plants or mammals. Inorganic compounds are moderately toxic and tellurites are more toxic than tellurates. Hydrogen telluride and tellurium hexafluoride are the most toxic tellurium compounds. Little is known about the toxicity of organotellurium compounds.

In animals, acute tellurium poisoning results in tremor, reduced reflexes, paralysis, convulsions, coma and death. The liver, kidney and central nervous system are the organs and system most affected by exposure to tellurium and its compounds. Only a few tellurium compounds (e.g. tellurium dioxide, hydrogen telluride) are industrially significant to be health hazards to workers. Tellurium poisoning is easily identified by the strong garlic odour of the breath of workers in contact with tellurium. Other symptoms include suppression of sweat and dryness of mouth. To date, there have been no reports of carcinogenic, mutagenic or teratogenic effects in humans. There is only one reported case of tellurium poisoning, caused by mistaken administration of sodium tellurite in a military hospital to three patients, two of whom died.

Occupational exposure limits.   OSHA TLV (TWA) 0.1 mg Te m^{-3}.

Toxicity.   LD$_{50}$ inhaled rat 10 mg m^{-3} (Te aerosol).

Further reading
Alexander, J., Thomassen, Y. and Assethy, J. (1988) Tellurium, in Seiler, H. G., Sigel, H. and Sigel, A. (eds) *Handbook of Toxicology of Inorganic Compounds*, Marcel Dekker, New York, Ch. 6, pp. 669–674.
Fishbein, L. (1977) Toxicology of selenium and tellurium, *Adv. Mod. Toxicol.*, **2**, 191–240.
Keall, J. H. H., Martin, N. H. and Tunbridge, R. E. (1946) Report of three cases of accidental poisoning by sodium tellurite, *Br. J. Ind. Med.*, **3**, 175–176.
King, R.B. (ed.) (1994) *Encyclopedia of Inorganic Chemistry*, Wiley Chichester, Vol. 8, pp. 4105–4117.

Tellurides of Ag, Cu, Ge, Mn, Pb and Sb

Uses.   In thermoelectric materials and as semiconductors.

Tellurides of Bi and Cd

Uses.   As pigments, semiconductors and solar cells in pocket calulators and cameras.

Tellurides of Mo, Tc, W and Zr

Uses.   In solid self-lubricating composites in electronics, instrumentation and aerospace industries.

Tellurium(IV) bromide CAS No. 10031-27-3

$TeBr_4$

Uses.   Catalyst in organic chemistry.

Toxicity.   Irritant.

Tellurium(IV) chloride CAS No. 10026-07-0

$TeCl_4$

Uses.   Catalyst in organic chemistry.

Toxicity.   Harmful. Possible mutagen.

Tellurium(IV) oxide CAS No. 7446-07-3

TeO_2

Uses.   Catalyst in organic preparations. IR-transparent glass. Additive in bright copper electroplating. Rubber vulcanizing agent.

Toxicity.   Ingestion or inhalation may cause irreversable damage to the lungs and the gastrointestinal system.

TERBIUM CAS No. 7440-27-9

Tb

Terbium, a member of the lanthanide or rare earth group of elements, has an atomic number of 65, an atomic weight of 158.93, a specific gravity of 8.23 at 20°, a melting point of 1359° and a boiling point of 3221°. Nineteen isotopes with atomic masses ranging from 147 to 164 are known. Ordinary terbium consists of only one isotope,

^{159}Tb (100%). It is a silvery grey, malleable, ductile and soft metal and forms compounds in the +3 and +4 oxidation states.

Average concentration in the earth's crust is ~ 1 ppm. It is found in *gadolinite, cerite, monazite ($\sim 0.03\%$ Tb), euxenite and xenotime.* The pure metal is produced by reducing the anhydrous fluoride/chloride with calcium metal in a tantalum crucible. Concentrations of terbium compounds in human organs are very low.

Uses. Sodium terbium borate is used as laser material. As a dopant in calcium and strontium compounds, used in solid-state devices.

No biological role. Little is known about the toxicity of terbium. It should be handled with care, as with other lanthanide elements. It is believed to act as an anticoagulant. Terbium compounds are skin and eye irritants. The metal dust is a fire and explosion hazard.

Further reading
Gschneidner, K.A., Jr, and LeRoy, K.S. (eds), (1988) *Handbook on the Physics and Chemistry of the Rare Earths*, North-Holland, Amsterdam.
Kirk–Othmer ECT, 4th edn, Wiley, New York, 1995, Vol. 14, pp. 1091–1115.

Terbium(III) chloride hexahydrate CAS No. 13798-24-8

$TbCl_3.6H_2O$

Uses. Chemical agent.

Occupational exposure limits. N.a.

Toxicity. Eye and skin irritant. Moderately toxic to humans depending on intake route. LD$_{50}$ oral mouse 5100 mg kg^{-1}.

TETRACHLOROETHYLENE CAS No. 127-18-4
LC$_{50}$ (96 h) fish 5–20 mg l^{-1}

$$\begin{array}{c} Cl \quad\quad Cl \\ \diagdown\quad\quad\diagup \\ C{=}C \\ \diagup\quad\quad\diagdown \\ Cl \quad\quad Cl \end{array} \quad\quad\quad C_2Cl_4 \quad M = 165.8$$

1,1,2,2-Tetrachloroethylene, ethylene tetrachloride, perchloroethylene; b.p. 121°; density (20°) 1.62 g ml^{-1}; v. dens. 5.8× air; v.p. 1.9 kPa (14 Torr, 20°).

Many trade names, e.g. Ankilostin, Antisol, Perclene, Tetralex. Produced by perchlorination or dehydrochlorination of compounds such as 1,2-dichloroethane, ethane, propylene, acetylene. It is mainly used for dry cleaning and metal cleaning, also as an extractant, grain fumigant and in CFC manufacture.

Sampling by absorption on Porapak N followed by GC–ECD gave a detection limit of 200 ng m^{-3} or an order of magnitude better by direct GC–MS detection (1 ppm = 6.8 mg m^{-3}).

Tetrachloroethylene is widespread in the oceans, surface water and soils. Industrial spillage has produced levels up to 22 µg l^{-1} in groundwater. In bays along the coast of the UK sea water levels lay in the range 0.02–2.6 µg l^{-1}. The half-life in rivers is 30–300 days.

Over 80% of that in use is lost by evaporation and worldwide emission (1975) was 450×10^3 t year^{-1}. In mammals the proportion exhaled rises with the level of exposure; following epoxidation it is largely excreted by mammals as trichloroacetic acid.

The odour threshold is $32 \, \text{ng m}^{-3}$ and eye irritation and headache follow 30 min exposure at $1.4 \, \text{g m}^{-3}$. In humans pulmonary oedema follows prolonged (3–6 h) exposure at levels in the range of 2–$7 \, \text{g m}^{-3}$. Workers in dry cleaning were exposed to levels in the range 700–2700 mg m^{-3} (100–400 ppm) and were found to have over $40 \, \text{mg l}^{-1}$ of trichloroacetic acid in their urine – there was evidence in this group of CNS depression; neurotoxic effects were also observed in those so employed.

TCE is carcinogenic in mice but not in rats; there is no evidence of carcinogenicity in humans and it is placed in the IARC group 2B. However, workers in dry cleaning are held to be at risk from pancreatic and liver cancer, also from cancer of the larynx and bladder (Richardson and Gangolli, 1994). The short-term occupational exposure limit in the USA is 200 ppm and the WHO guideline for drinking water is $10 \, \text{mg l}^{-1}$. This substance is no longer permitted in cosmetics in the EC.

Reference
Richardson, M.L. and Gangolli, S. (1994) *Dictionary of Substances and Their Effects*, Royal Society of Chemistry, Cambridge, pp. 310–314.

Further reading
Berlin, A., Draper, M., Krug, E., Rui, R. and Van der Venne, M. Th. (1990) *The Toxicology of Chemicals*, Commission of the European Community, Brussels, Vol. 2.
Environ. Health Criteria, World Health Organization, Geneva, 1986, Vol. 31.

THALLIUM CAS No. 7440-28-0

Tl

Thallium, a member of Group 13 of the Periodic Table, has an atomic number of 81, an atomic weight of 204.38, a specific gravity $11.85 \, \text{g cm}^{-3}$ at $20°$, a melting point of $303°$ and boiling point $1473°$. It is a grey-white, soft, ductile metal with stable oxidation states of $+1$ and $+3$. In addition to two stable isotopes, ^{203}Tl (29.52%) and ^{205}Tl (70.48%), over 26 radioactive isotopes are known including ^{208}Tl($t_{1/2} = 4.19 \, \text{min}$), ^{207}Tl($t_{1/2} = 4.79 \, \text{min}$), ^{208}Tl($t_{1/2} = 3.1 \, \text{min}$), ^{210}Tl($t_{1/2} = 1.30 \, \text{min}$) and ^{204}Tl($t_{1/2} = 3.56 \, \text{years}$), which is used as a source of β-radiation.

Widely distributed in nature, probably due to its volatility. Thallium is a rare element, $\sim 1.0 \, \text{ppm}(0.5 \, \mu\text{g g}^{-1})$ of the earth's crust. It occurs with other heavy metals and is found in pyrites, blendes and the mineral *crookosito*, $(\text{Cu, Tl, Ag})_2\text{Se}$. Commercial source is as a by-product of zinc/lead/copper smelting.

Thallium as a pollutant arises from the burning of coal and the smelting of Cu, Pb and Zn ores which produces thallium-containing dusts and aerosols. A typical coal contains 1 ppm thallium. Typical concentrations in the environment are air $< 1 \, \text{ng m}^{-3}$, sea water $0.01 \, \mu\text{g l}^{-1}$; freshwater 0.01–$14 \, \mu\text{g l}^{-1}$; soils 0.02–$2.8 \, \text{mg kg}^{-1}$; herbage 0.02–$1.0 \, \text{mg kg}^{-1}$. For plants concentrations of > 7 ppm Tl cause injury or distress. Thallium accumulates in cabbage, lettuce, spinach and leeks. Thallium levels in some Rubiaceae have been found to be as high as $\sim 17\,000 \, \text{mg kg}^{-1}$ ash wt in the flower. Daily human intake $\sim 2 \, \mu\text{g}$. Human body concentrations: urine 0.05–$1.5 \, \mu\text{g l}^{-1}$, bone 2 ppb, blood 0.5–$2 \, \mu\text{g l}^{-1}$, liver/kidney 0.5–$4 \, \mu\text{g l}^{-1}$; amount in body 0.1 mg with ng quantities in hair, nails, etc.

Uses. electrical, semiconductor and chemical industries. Thallium metal is used in lead-, silver- or gold-bearing alloys which have very low friction coefficients, high endurance limits and high resistance to acids. Manufacture of some optical lenses (thallium–arsenic–selenium). Low-temperature thermometers. Other uses for thallium salts continue, especially in their increasing demand as an ingredient of fibreglass in communications systems. ^{201}Tl($t_{1/2} = 73$ h) is used in heart diagnosis.

Thallium is considered to be one of the most toxic heavy metals to animals and humans. Plants and fish are also sensitive to Tl^+ ion concentration. For animals, including humans, both Tl(I) and Tl(III) salts are readily absorbed by the gastrointestinal tract and the skin. Excretion is slow, with a half life of several weeks. Thallium concentrates in the brain, kidney and testes. The Tl^+ ion can substitute for the similarly sized K^+ ion and interfere in K^+-dependent processes. Recovery from thallotoxicosis takes months and may be incomplete as nervous system damage may be irreversible.

A number of mechanisms have been proposed to explain the toxicity of thallium including ligand formation with sulphydryl groups of enzymes such as flavoenzymes, pyridoxal phosphate- dependent enzymes and thiol proteases, alteration of K^+ ion-dependent processes and disruption of intracellular calcium ion homeostasis. For industrial workers, thallium exposures have occurred by inhalation, ingestion or skin absorption. NIOSH has recommended that $15 \, \text{mg m}^{-3}$ of thallium be considered immediately dangerous to life and health. The clinical aspects of thallium toxicity (thallotoxicosis) include polyneuropathy, gastrointestinal changes, increased liver activity and kidney impairment. Acute poisoning in humans is accompanied by nausea, vomiting, diarrhoea, coma, convulsions and possibly death. Chronic exposure has been associated with leg weakness and pain, hair loss and psychological disturbances. Exposure to more than 100 mg thallium causes the above symptoms and a dose of 500–800 mg for adults is usually fatal. Thallium has been known to cause eye disorders such as retrobulbar neuritis. Cardiac abnormalities including irregular pulse, hypertension and angina-like pains frequently occur during the second week after exposure. Standard treatment of thallium poisoning usually consists of giving iron(III) hexacyanoferrate(II) (Prussian blue) or activated charcoal to enhance faecal elimination of thallium or inducing diuresis by K^+ ion loading to increase renal thallium clearance.

Median lethal doses for humans and rats are of the order 8–12 and 15.8–72 kg^{-1}, respectively. The lethal dose for humans is thus less than 1 g of a thallium compound in a single ingestion. Thallotoxicosis from ingestion of thallium-based rodenticides remains one of the most common acute toxic diseases caused by metals. Fatal acute thallium poisoning is rare but a recent case has been reported associated with thrombocytopenia. A 43-year-old man was admitted with a 12 h history of backache radiating to the chest, painful paresthesia of the legs, tingling in the mouth and a weak feeling particularly in the feet. On the basis of information supplied by the patient, a screen was run for thallium, which showed raised blood and urine levels. Despite the standard treatment for thallium poisoning the patient died on the eleventh day of hospitalization. Thrombocytopenia resulted from the direct toxic effect of thallium on the megakaryocytes.

During 1960–77, 189 cases of chronic thallium poisoning were reported in a rural area of Guizhou Provence, China. Environmental and epidemiological data indicated that the soil was polluted by thallium in the waste slag from metal and coal mines. Consumption of cabbage in containing soils was the route of entry into the body. Thallium has been shown to be teratogenic to chick embryos but similar investigations in laboratory animals have produced conflicting results.

Occupational exposure limits. OSHA (TLV) TWA 0.1 mg m^{-3} (skin)

Further reading

Environmental Health Criteria 182. *Thallium*, WHO/IPCS, Geneva, 1996.

Kazantzis, G. (1986) Thallium, in Friberg, L., Nordberg, G.F. and Vouk, V.B. (eds), *Handbook on the Toxicology of Metals*, 2nd edn, Elsevier, Amsterdam, pp. 549–567.

King, R.B. (ed.) (1994) *Encyclopedia of Inorganic Chemistry*, Wiley, Chichester, Vol. 8, pp. 4134–4151.

Manzo, L. and Sabbioni, E. (1988) Thallium, in Seiler, H.G., Sigel, H. and Sigel, A. (eds), *Handbook on Toxicology of Inorganic Compounds*, Marcel Dekker, New York, pp. 677–688.

Mulkey, J.P. and Oehme, F.W. (1993) A review of thallium toxicity, *Vet. Hum. Toxicol.* **35**, pp. 445–453.

Sager, M. (1994) Thallium, *Toxicol. Environ. Chem.*, **45**, 11.

ToxFAQs (1995) *Thallium*, ATSDR, Atlanta, GA. Georgia.

Ullmann's Encyclopedia of Industrial Chemistry, 5th edn, VCH, Weinheim, 1995, Vol. A26, pp. 607–619.

Thalllum(I) acetate CAS No. 563-68-8

CH_3CO_2Tl

Uses. Ore flotation.

Toxicity. Very toxic if inhaled or swallowed. LC_{50} (96 h) Atlantic salmon 0.03 ppm Tl. LD_{Lo} oral human 12 mg kg^{-1}; LD_{50} oral rat 41300 μg kg^{-1}.

Thallium(I) carbonate CAS No. 6533-73-9

Tl_2CO_3

Uses. Manufacture of imitation diamonds.

Toxicity. Very toxic if inhaled or swallowed. LD_{50} oral mouse 23 mg kg^{-1}.

Thallium(I) chloride CAS No. 7791-12-0

TlCl

Uses. Catalyst in chlorinations.

Toxicity. Very toxic if inhaled or swallowed. Mutation and teratogenic data reported. LD_{50} oral mouse 24 mg kg^{-1}.

Thallium Fluoride CAS No. 7789-27-7

TlF

Uses. Preparations of fluoro esters.

Toxicity. Very toxic if swallowed.

Thallium nitrate CAS No. 10102-45-1

$TlNO_3$

Uses. Reagent in analytical chemistry. Sea flares.

Toxicity. Highly toxic. LD_{50} intraperitoneal rat 21 μg kg^{-1}; oral mouse 15 mg kg^{-1}.

Thallium(I) sulphate CAS No. 7446-18-6

Tl_2SO_4

Uses. Rodenticide.

Toxicity. LD_{Lo} oral mouse $3\,mg\,kg^{-1}$; LD_{50} oral rat $16\,mg\,kg^{-1}$.

THIABENDAZOLE CAS No. 148-79-8
(Mertect, Tecto)
Systemic fungicide with preventative and curative action. Toxicity class III. LD_{50} rat $3100\,mg\,kg^{-1}$; Not toxic to fish or bees.

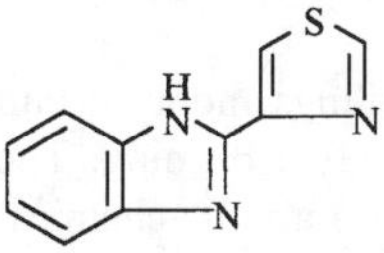

$C_{10}H_7N_3S \quad M = 201.2$

2-(Thiazol-4-yl)benzimidazole; m.p.305°; v.p. negligible at 20°; water solubility $< 50\,mg\,l^{-1}$ (pH5).

First reported by Robinson *et al.* in 1964 and introduced originally by Merck as an anthelmintic.

It is a systemic fungicide effective against pests in citrus fruits, cereals, rice, potatoes, tobacco, vines, tomatoes, etc. Also used to control diseases of fruit and vegetables during storage.

Thiabendazole is rapidly excreted by mammals as its 5- hydroxy derivative.

Codex Alimentarius acceptable daily intake $0.3\,mg\,kg^{-1}$(1977). Maximum residue levels: cereal grains 0.2, milk 0.1, potatoes 5.0, citrus fruits $10.0\,mg\,kg^{-1}$

THIOCYANIC ACID AND ITS SALTS, THIOCYANATES CAS No. 463–56;–9

$HSCN, SCN^-$

The acid is unstable. Sodium, potassium and ammonium thiocyanate are commonly obtained from water gas or coke oven gas.

Uses. Analytical chemistry, disinfectants, dyeing and printing textiles.

Toxicity. Low compared with cyanides. LD_{50} mammals $0.5\text{-}10.0\,g\,kg^{-1}$.

Further reading
Ullmann's Encyclopedia of Industrial Chemistry, 5th edn, VCH, Weinheim, 1995, Vol. A26, pp. 759–766.

THORIUM CAS No. 7440-29-1

Th

Thorium, the second member of the actinide series of elements, has an atomic number of 90, an atomic weight of 232.04, a specific gravity of $11.72\,g\,cm^{-3}$ at 20°, a melting point of 1750°, and a boiling point of $\sim 4790°$. Pure metal is obtained by reduction of thorium tetrachloride with sodium. It is a silvery-white air-stable metal and the principal oxidation state is +4. ^{232}Th (100%) occurs naturally and has a half-life of 1.41×10^{10} years. Thorium decays with the production of ^{220}Rn, which is an α-emitter and presents a radiation hazard.

Thorium, a ubiquitous element in the upper layers of the earth (~ 13 ppm), occurs in *thorite*, $ThSiO_4$ and *thorianite*, $ThO_2 + UO_2$. Commercially, thorium is recovered from the mineral *monazite* (3–9% ThO_2). The pure metal can be obtained by reducing thorium dioxide with calcium.

Typical environmental concentrations are soils 6 ppm; normal 70 kg human intake $\sim 3\,\mu g/0.03$ Bq day^{-1}; bone 0.002–0.12 ppm Total body content $40\,\mu g/0.45$ Bq. Residence time: bones 8×10^3 days liver/tissues 700 days.

Uses. The principal use of thorium has been in the preparation of incandescent gas mantles. Refractory materials. Alloy with magnesium. Thorium-232 is used as a nuclear fuel in nuclear reactors.

Thorium is found in elevated concentrations in uranium mill tailings and in certain phosphate fertilizers. It is also present in the smoke from fossil fuel burning. The biological effects of thorium are due to both the ionizing radiation and its chemical toxicity. Thorium may be a carcinogen. In a thorium processing plant, male workers showed increased risk of all cancers, including lung cancers. Thorium toxicity has been thoroughly investigated by analysing the effect on $> 10\,000$ patients of Thorotrast, a colloidal suspension of thorium dioxide used as a contrast agent in radiography. Radiation damage increased cases of leukaemia, caused changes in the nervous system, damaged lung and bone tissues (sarcomas) and caused chromosome mutations. Liver diseases in this group were attributed to the chemical toxicity of thorium. The poor solubility of thorium compounds in water prevents an enrichment along the food chain. Thorium is poorly absorbed from the lung and digestive tract.

Occupational exposure limit. USA TLV (TWA) 0.05 mg Th m^{-3}.

Toxicity. LD$_{50}$ oral rat 9160 mg kg^{-1}.

Radiotoxicity exposure limits. Inhalation 1×10^2 Bq; ingestion 3×10^4 Bq.

Further reading

Buckart, W. (1988) Radiotoxicity in Seiler, H. G., Sigel, H. and Sigel, A. (eds), *Handbook on Toxicity of Inorganic Compounds*, Marcel Dekker, New York, pp. 805–827.

Eisenbud, M. (1987) *Environmental Radiochemistry. From Natural, Industrial, and Military Sources*, 3rd edn, Academic Press, New York.

Métivier, H.J. (1988) Thorium in Seiler, H. G., Sigel, H. and Sigel, A. (eds), *Handbook on Toxicity of Inorganic Compounds*, Marcel Dekker, New York, pp. 689–694.

ToxProfiles (1990) *Thorium*. ATSDR, Atlanta. GA.

Ullmann's Encyclopedia of Industrial Chemistry, 5th edn, VCH, Weinheim, 1996, Vol. A27, pp. 1–37.

Thorium(IV) oxide CAS No. 1314-20-1

ThO_2

Uses. Gas mantles used in portable lamps. High-temperature crucibles. High-quality lenses. Catalyst (ammonia to nitric acid). Petroleum cracking. Manufacture of sulphuric acid.

Occupational exposure limit. N.a., but avoid dust.

Toxicity. Carcinogen. LD_{50} intravenous rat $>1140\,mg\,kg^{-1}$. Affects liver, kidneys and blood.

THRESHOLD LIMIT VALUE (TLV)
The TLV is that concentration in air to which healthy workers can be exposed without adverse effect. It is normally reported as a time-weighted average (TWA) for an 8 h day or 40 h week. This average allows for varying exposure levels; for example exposure to a level of $0.5\,mg\,m^{-3}$ for 4 h and to one of $1.0\,mg\,m^{-3}$ for 4 h is equivalent to an 8 h TWA of $(0.5 \times 4 + 1.0 \times 4)/8 = 0.75\,mg\,m^{-3}$

THULIUM CAS No. 7440-30-4

Tm

Thulium, a member of the lanthanide series of elements, has an atomic number of 69, an atomic weight of 168.93, a specific gravity of 9.321 at 20°, a melting point of 1545°, and a boiling point of 1947°. The pure metal has a bright, silvery lustre, reasonably stable in air and forms compounds in the +2 and +3 oxidation states. Ordinary thulium exists as one stable isotope, ^{169}Tm (100%).

It is obtained commercially from *monazite*, which contains about 0.007% of the element. Average concentration in the earth's crust is $\sim0.48\,ppm$. Separated from the other rare earths using ion-exchange and solvent extraction techniques.

Uses. Radiation source in portable X-ray equipment. Blue phosphor, $ZnS.Tm^{3+}$.

No biological role. As with the other lanthanides, thulium and its salts have a low to moderate toxicity. They should be handled with care. Harmful for all routes. High concentrations of salts are extremely destructive to soft tissues. The metal dust is a fire and explosion hazard. Concentrations of thulium compounds are very low in human organs.

Further reading
Gschneidner, K.A., Jr, and LeRoy, K.A. (eds) (1988) *Handbook on the Physics and Chemistry of the Rare Earths*, North-Holland, Amsterdam.
Kirk–Othmer ECT, 4th edn, Wiley, New York, 1995, Vol. 14, pp. 1091–1115.
Ullmann's Encyclopedia of Industrial Chemistry, 5th edn, VCH, Weinheim, 1993, Vol. A22, pp. 607–649.

Thulium(III) chloride hexahydrate CAS No. 1331-74-4

$TmCl_3.6H_2O$

Uses. Chemical agent.

Toxicity. Irritant to skin and eyes. Poison by intraperitoneal route. LD_{50} oral mouse $4299\,mg\,kg^{-1}$.

TIN CAS No. 7440-31-5

Sn

Tin, a member of Group 14 of the Periodic Table, has an atomic number of 50, an atomic weight of 118.71, a specific gravity of 7.31 at 20°, a melting point of 231.9° and a boiling point of 2602°. Metallic tin exists in two different forms. The stable form is grey cubic or α-tin, which changes at 18° to the tetragonal β-tin. Natural tin is composed of ten stable isotopes: ^{112}Sn (0.97%), ^{114}Sn (0.65%), ^{115}Sn (0.34%), ^{116}Sn (14.54%), ^{117}Sn (7.68%), ^{118}Sn (24.22%), ^{119}Sn (8.59%), ^{122}Sn (4.63%), ^{120}Sn (32.4%) and ^{124}Sn (5.94%). There are two oxidation states for Sn compounds, +2 and +4.

The earth's crust contain 2–3 ppm of tin, and it is ranked 49th in relative abundance. The main tin ore is *cassiterite*, SnO_2. Typical concentrations in the environment: air < 4.4 µg m^{-3}; sea water < 1 µg l^{-1}; soil < 10 mg kg^{-1}; canned food 1–500 ppm; daily intake (adult male) 0.2–1 mg; stored tin (bones, liver, lungs) 0.1–1.4 mg kg^{-1}, the average adult body (70 kg) contains about 30 mg of tin.

Uses.   Protection of steel–tin plating. Solder (tin alloys containing lead, antimony, silver and zinc). Brasses. Bronzes. Pewter.

Tin is an essential element and is found in the hormone gastrine, produced by the stomach.

Inorganic tin salts are generally considered to be of low toxicity since they are poorly absorbed into the body. Tin salts that have gained access to the blood stream are highly toxic. Occupational exposure of workers to tin/tin oxide fumes or dusts by inhalation can lead to stannosis – a benign form of pneumoconiosis. Examples of acute intoxication have mainly resulted from ingestion of tin in food containing > 200 mg kg^{-1} with symptoms of headache, gastroenteritis, etc. In the case of fatal poisoning death occurred in a state of coma or from cardiac and pulmonary abnormalities. Nutrient solutions containing more than 40 mg l^{-1} of tin are fatal to plant seeds. The much higher toxicity of organotin compounds (compared with inorganic tin) is determined by the type and number of organic groups. The most toxic organotin compounds are those with the formula R_3SnX, where R = alkyl or aryl group and X = halide, acetate, etc. Some are used commercially as biocides, fungicides, molluscides, etc. Their use in anti-fouling paints for boats is now discouraged or forbidden since they are highly toxic to most aquatic species (at the ppm to ppt level). The sequence of biological activity is $R_3SnX > R_2SnX_2 > RSnX_3$. The basic cause of acute organotin toxicity arises from the disruption of ion transport and inhibition of ATP synthesis. Organotins usually degrade to SnO_2 but some compounds are stable to water and oxidation. Bioaccumulation is concerned mainly with Bu_3Sn species and their degradation products. Some examples are, sheepshead minnow (up to 2600×), oysters (up to 10 000×), algae (up to 2000×). Methylated organotins are detectable in the environment from methylation of Sn^{2+}/Sn^{4+} with yeast or methylcobalamin. Widespread poisoning occurred in France as a result of taking capsules containing diethyltin diiodide for the treatment of staphylococcal skin infection; the active poison was probably trace amounts of triethyltin iodide. There is a restriction to 1 µg Sn g^{-1} organotin compound allowed for food wrapping by PVC. There is little evidence for migration of organotins from PVC bottles to liquid food inside.

Tin compounds tend to be corrosive and toxic. Tin salts are usually poorly absorbed into the body but those that enter the blood stream are highly toxic and may cause neurological damage and paralysis.

Occupational exposure limits.   OSHA PEL (TWA) metal, oxide, inorganic compounds 2 mg m^{-3}; organotin compounds 0.1 mg m^{-3}.

Further reading
King, R.B. (ed.) (1994) *Encyclopedia of Inorganic Chemistry*, Wiley, Chichester, Vol. 8, pp. 4159–4197.

Magos, L. (1986) Tin, in Friberg, L., Nordberg, G.F. and Vouk, V.R. (eds), *Handbook on Toxicology of Metals*, 2nd edn, Elsevier, Amsterdam, Vol II, pp. 568–593.

Environmental Health Criteria 15. Tin and Organotin Compounds, WHO/IPCS, Geneva, 1980.

ECETOX (1991) *Technical Report 30(5)*, European Chemical Industry Technology and Toxicology Centre, Brussels.

SI 1991 No. 472 The Environmental Protection (Prescribed Processes and Substances) Regulations 1991, HMSO, London, 1991.

ToxFAQs (1995) *Tin*, ATSDR, Atlanta, GA.

Ullmann's Encyclopedia of Industrial Chemistry, 5th edn, 1996, Vol. A27, pp. 49–81.

Bis[tributyltin(IV)] oxide CAS No. 56-35-9

$(Bu_3Sn)_2O$

Uses. Wood preservative. Banned for small boats, UK, May 1987 (see **Tin**).

Toxicity. Toxic to commercial oysters. LD_{50} oral rat 150–234 mg kg^{-1}.

Tetrabutyltin(IV) CAS No. 1461-25-2

$Sn(CH_2CH_2CH_2CH_3)_4$

Uses. Polymerization catalyst. Lubricant and fuel additive. Stabilizing agent for silicones.

Toxicity. Attacks tissues of the mucous membranes and upper respiratory tract, eyes and skin. Inhalation can be fatal. LC_{50} (96 h) fathead minnow 45 μg l^{-1}; LD_{Lo} oral mouse 1400 mg kg^{-1}; LD_{50} intravenous mouse 56 mg kg^{-1}.

Tetraethyltin(IV) CAS No. 597-64-8

$Sn(C_2H_5)_4$

Uses. Preparation of other organotin compounds.

Toxicity. Highly toxic. Affects lungs, muscles, liver and brain cells. LD_{50} oral mouse 40 mg kg^{-1}; LD_{50} oral rabbit 7 mg kg^{-1}.

Tin(II) chloride CAS No. 7772-99-8

$SnCl_2$

Uses. Plating steel. Reducing agent. Mordant in dyeing. Tanning agent. Analytical reagent. Manufacture of tin chemicals, pharmaceuticals.

Toxicity. LD_{50} oral rat 700 mg kg^{-1}; LD_{50} intravenous mouse 17.8 mg kg^{-1}.

Tin(IV) chloride CAS No. 7646-78-8

$SnCl_4$

Uses. Preparation of electrically conductive SnO_2 films. Ceramics. Preparation of organotins.

Toxicity. Highly toxic by all routes. Corrosive. LC_{50} (10 min) inhalation rat 2300 mg kg^{-1}. LD_{50} intraperitoneal mouse 9 mg kg^{-1}.

Tin(II) fluoride CAS No. 7783-47-3

SnF_2

Uses.   Dental preparations. Toothpaste.

Toxicity. LD_{50} oral rat 377 mg kg^{-1}; intraperitoneal mouse 16.15 mg kg^{-1}.

Tin(II) oxalate CAS No. 814-94-8

$Sn(C_2O_4)_2$

Uses.   Transesterification catalyst.

Toxicity.   Harmful if swallowed, inhaled or absorbed through the skin. Affects nerves, kidneys and gastrointestinal systems. LD_{50} oral rat 3620 mg kg^{-1}.

Tin(IV) oxide CAS No. 18282-10-5

SnO_2

Uses.   Ceramics. Pigments.

Toxicity.   Tin oxide inhaled as dust or fumes leads to a benign pneumoconiosis.

Tin(II) sulphate CAS No. 7488-55-3

$SnSO_4$

Uses.   Plating steel.

Toxicity. LD_{50} oral rat 2207 mg kg^{-1}; oral mouse 2152 mg kg^{-1}.

Triphenyltin(IV) acetate CAS No. 9000-95-8

$(C_6H_5)_3SnOOCH_3$

Uses.   Fungicide. Algicide. Molluscide.

Toxicity.   Irritant to skin. LD_{50} oral rat 4.0 mg kg^{-1}.

Triphenyltin(IV) hydroxide CAS No. 76-87-9

$(C_6H_5)_3SnOH$

Uses.   Antifungicide for crops.

Toxicity. LD_{50} oral rat 46 mg kg^{-1}.

TITANIUM CAS No. 7440-32-6

Ti

Titanium, a member of Group 4 of the Periodic Table has an atomic number of 22, an atomic weight of 47.88, a specific gravity of 4.54 at 20°, a melting point of 1667° and a boiling point of 3285°. It is a lustrous white metal with compounds in the +2, +3 and +4 oxidation states; the last is the most stable. Natural isotopes of titanium are ^{46}Ti (8.00%), ^{47}Ti (7.44%), ^{48}Ti (73.72%), ^{49}Ti (5.41%) and ^{50}Ti (5.48%).

Widely distributed in nature (6000 ppm), in the earth's crust, it is the ninth most abundant element. Commercially useful ores are *ilmenite* [iron titanate ($FeOTiO_2$, 35–60% TiO_2)] and *rutile* [titanium dioxide (~ 95% TiO_2)]. Typical concentrations are sea water 1.2 µg l^{-1} and plants 1 mg kg^{-1} dry weight. Daily human intake

0.3–2.0 mg. The adult body store is ~ 15 mg, principally in the lungs. Urine levels $< 1.5 \mu g l^{-1}$. In humans, Ti levels in different organs have been found to vary widely.

Uses. Various alloys in aeronautics. Surgical implants. Because of its low density and strength, it is used in the manufacture of aircraft and missiles.

Titanium is not essential for humans but may be important in N_2 fixation of legumes. At concentrations $\gtrsim 5$ ppm Ti is toxic to some plants. Toxicity symptoms include chlorosis, necrosis and stunted growth. The bulk manufacture of titanium dioxide, titanium tetrachloride and their derivatives presents the greatest occupational hazard; workers exposed to these chemicals were found to have pathological changes in the respiratory tract. Similarly, accidentally inhaled fumes of metatitanic acid, $TiO_2.H_2O$, and titanium oxychloride, $TiOCl_2$, caused marked congestion of mucous membranes in the pharynx, vocal chords and trachea. There are reports that titanium metal placed in the body as a component part of a surgical device can cause malignant carcinomas. Slight fibrosis in animal studies using titanium hydride, TiH_4, and titanium carbide, nitride and boride (TiC, TiN and TiB).

Occupational exposure limits. N.a. Finely divided titanium is flammable.

Further reading
ECETOX (1991) *Technical Report 30(5)*, European Chemical Industry Technology and Toxicology Centre, Brussels.
Environmental Health Criteria 24. Titanium, WHO/IPCS, Geneva, 1982.
King, R.B. (ed.) (1994) *Encyclopedia of Inorganic Chemistry*, Wiley, Chichester, Vol. 8, pp. 4197–4206.
Ullmann's Encyclopedia of Industrial Chemistry, 5th edn, VCH, Weinheim, 1996, Vol. A27, pp. 95–122.
Wennig, R. and Kirsch, N. (1988) Titanium, in Seiler, H. G., Sigel, H. and Sigel, A. (eds), *Handbook on Toxicity of Inorganic Compounds*, Marcel Dekker, New York, pp. 705–714.

Titanium(IV) oxide CAS No. 1317-80-2

TiO_2

Uses. White pigment in paints, plastics, rubber, paper, ceramics, printing inks, cosmetics. Rutile is used for making Ti metal, alloys, carbides, ceramics and fibreglass. Titanium dioxide pigments account for the largest use of the element (4×10^6 t, 1992); the pigment is extensively used for both house paints and artists' paints.

Occupational exposure limit. 10 mg TiO_2 m^{-3} total dust; 5 mg TiO_2 m^{-3} respiratory dust.

Toxicity. Possible carcinogen. May be harmful by all routes. Dermatitis from exposure to 300 μg for 3 days. LC_{50} (28 days) rainbow trout 7.31 mg l^{-1}.

Titanium(IV) tetrachloride CAS No. 7550-45-0

$TiCl_4$

Uses. Preparation of other titanium compounds including organotitanium compounds. Catalyst. Manufacture of artificial pearls.

Toxicity. Highly toxic. Harmful vapour. Lachrymator. LC_{50} inhalation rat 400 mg m^{-3}; inhalation mouse 100 mg m^{-3} (1 h)

TOLUENE CAS No. 108-88-3

LD_{50} rat $7.5\,g\,kg^{-1}$; LC_{50} rat $45\,mg\,m^{-3}$; LC_{50} fish $5-60\,mg\,l^{-1}$ (96 h).

$$C_7H_8 \quad M = 92.1$$

Methylbenzene, methoxide, toluol. b.p. 111°; v.p. (25°) 28 Torr; v.dens 3.2× air. Water solubility $535\,mg\,l^{-1}$ (20°); soluble in most organic solvents. Flash point +4.4° 1 ppm = $3.75\,mg\,m^{-3}$.

When pure, toluene contains <0.1% of benzene but in the industrial grade this level may reach 20%. Toluene was originally produced by the carbonization of coal; today it comes largely from the reforming of petroleum. Worldwide production (1985) 10^7 t; USA production (1994) 6.75×10^9 lb.

Up to 90% of production is added to gasoline to raise the octane number; it replaces benzene as a solvent in paints, thinners and coatings and for degreasing. Toluene is a precursor for chemicals such as the diisocyanate, benzaldehyde, xylene, for dyestuffs and as a resin modifier. It enters the environment from its use as a solvent, from vehicle exhaust and from cigarette smoke.

It is easily degraded in sewage and by soil microorganisms. Toluene is there metabolized by mixed function oxidases to form benzoic acid and smaller amounts of the isomeric cresols. In humans toluene is metabolized to benzoic acid and excreted as its conjugate with glycine, hippuric acid, which is an indicator of exposure to toluene:

$$Ph-\overset{O}{\overset{\|}{C}}-OH \;+\; \overset{NH_2}{\underset{}{CH_2}}-CO_2H \longrightarrow Ph-\overset{O}{\overset{\|}{C}}-NH-CH_2-CO_2H$$
hippuric acid

Detection by GC allows a limit of $1\,\mu g\,m^{-3}$ (3 ppb). The odour threshold is $9\,mg\,m^{-3}$ and workers experience eye irritation and light headiness at $700\,mg\,m^{-3}$. In humans levels in the range of 200–1500 ppm affect the CNS and produce nausea and lack of coordination; substantially higher levels, e.g. $30\,g\,m^{-3}$, arising from industrial accidents, lead to loss of consciousness and coma. Toluene is suspected to be neurotoxic, causing encephalopathy in children. It is responsible for renal damage in glue sniffers and has been detected at 2.2 ppm in their urine and 3.9 ppm in their blood (Uehori et al., 1987); they also suffer CNS depression, headache and blurred vision. Toluene is a major component in these solvent mixtures, which include benzene.

The occupational exposure limit TLV (TWA) in the USA is 50 ppm ($188\,mg\,m^{-3}$). The WHO guidline for drinking water is $700\,\mu g\,l^{-1}$.

Reference

Uehori, T., Nagata, T., Kimura, K., Kuido, K. and Noda, M. (1987) Screening of volatile compounds present in human blood using retention indices in GLC, *J. Chromatogr.*, **411**, 251–257.

TOLUENE DIISOCYANATES CAS No. 584-84-9 (2,4-); 91-08-7 (2,6-); 26471-62-5 (commercial 80:20 mixture).

LD_{50} rat $3\,g\,kg^{-1}$; LC_{50} inhalation rat $60\,mg\,kg^{-1}$; LC_{50} (24 h) fathead minnow $194\,mg\,l^{-1}$

2,4-TDI   2,6-TDI

These are obtained by the action of phosgene ($COCl_2$) on diaminotoluenes and the most commonly used product is an 80:20 mixture of 2,4- and 2,6-isomers. Pure 2,4-TDI has some special applications and its properties are summarized: m.p. 22°; b.p. 251°/760 Torr, 120°/10 Torr; v.p. 1 Torr (80°); v.dens. 6.0; odour threshold 0.4–0.9 mg m^{-3}. Both isomers react with water; they are soluble in most common organic solvents and dimerize slowly at ambient temperature.

The TDIs are important as co-reactants with polyether and polyester polyols in the manufacture of polyurethane foams, paints and coatings. These are widely used in the car and furniture industries. In 1983 production in the USA was 300×10^6 kg and in the EC it was 350×10^6 kg.

Their reaction with water yields polymers of ureas; in body fluids they have a half-life of 20 min or less. They can react with OH, NH_2, CO_2H and SH groups and so can inactivate proteins.

In rats long-term inhalation (6 h for 6 days per week) at a level of 3.5 mg m^{-3} led to 45% fatalities in those below 120 g bodyweight. At a level of 7 mg m^{-3} there were 75% fatalities in all animals.

TDIs induce asthmatic responses in humans and irritate the eyes at a level of 0.35 mg m^{-3}. Their decomposition products released from polyurethane foam fires affected firefighters, leading to dizziness, loss of consciousness and personality changes. They are proven animal carcinogens and potential human carcinogens. The parent amine, p-toluidine, is carcinogenic in rats.

In the USA the TLV (TWA) is 0.005 ppm (0.036 mg m^{-3}). In the UK the long-term limit (MEL) is 0.02 mg m^{-3}. The EC directive on drinking water quality (80/778/ EEC) is 50 µg l^{-1} (as cyanide).

Further reading
Environ. Health Criteria, World Health Organization, Geneva, 1987, Vol. 75.

TOXAPHENES CAS No. 8001-35-2 Toxaphene A 102489-36-1
Toxaphene B 51775-36-1

M (empirical) = 414

8 Cl subst.

Chlorinated camphene, Campheclor; an amber waxy solid, m.p. 65–90°; dens.1.65 (20°); log P_{ow} 5.28. Solubility in water 3 mg l^{-1}, soluble in most organic solvents. A priority pollutant.

The technical product is a mixture of compounds of chlorine number 4–12, containing overall 67–69% of chlorine in which 177 individuals have been identified (Holmstead et al., 1974).

It was first used in 1948 for control of grasshoppers and cotton pests and also flies, lice and ticks on livestock.

In 1974 US production was 40×10^6 lb year^{-1} and it was estimated that 10^9 lb had been produced in the previous 25 years.

Toxaphene A has the composition $C_{10}H_{10}Cl_8$ but its structure has not been established. It was separated from the technical mixture by GLC using 3% SE30 on Gas Chrom Q at 180° upward with EC detection (Holmstead *et al.*, 1974). Its half-life on a sandy loam soil is 20 years (1951–71).

Toxaphene B($C_{10}H_{11}Cl_7$) is 2,2,5 endo 6 exo 8,9,10 heptachlorobornane and forms 2–6% of the technical mixture. LD$_{50}$ oral pheasant 40 mg kg^{-1}; LD$_{50}$ i.p. mouse 6.6 mg kg^{-1}; very toxic to fish, LC$_{50}$ trout 3 ppb. The bioaccumulation factor water to fish is 10×10^4 and the half-life in soil is up to 10 years. Toxaphene B is carcinogenic in rats and mice.

The human acute lethal dose of toxaphenes has been estimated at 2–7 g/adult and five fatal cases of child poisoning have been recorded.

When in use the US maximum level in drinking water was 5 ng l^{-1}, very similar to those for endrin and lindane. At the time of its peak usage toxaphene levels in Bermuda ranged from 0.02 to 3.3 ng m^{-3}, but in cotton growing areas they range from zero in winter to a maximum of 1540 ng m^{-3} in late summer. The TLV (TWA) for uptake through the skin was 0.5 mg m^{-3}.

In view of its toxicity, toxaphene is on the United Nations list of banned substances.

Reference

Holmstead, R.L., Khalifa, S. and Casida, J.E., (1974), *J. Agric. Food Chem.*, 22, 939–947.

Further reading

IARC (1979) *Some Halogenated Hydrocarbons*, IARC, Lyons, Vol. 20.
Kirk–Othmer ECT, 3rd edn, Wiley, New York, 1984, Vol. 24, pp. 386, 421.

TRIAZOPHOS CAS No. 24017-47-8
(Hostathion)

Broad-spectrum insecticide, acaricide, nematocide. Toxicity class II. LD$_{50}$ rat 60 mg kg^{-1}; LC$_{50}$ trout 10 µg l^{-1}; Toxic to bees

$$C_{12}H_{16}N_3O_3PS \quad M = 313.3$$

O,O-Diethyl-*O*-1-phenyl-1*H*-1,2,4-triazol-3-yl phosphorothioate; m.p. 5° v.p. 0.39 mPa (30°); water solubility 40 mg l^{-1} (20°)

First reported by Vulic *et al.* in 1970 and introduced by Hoechst.

Deeply penetrative but non-systemic in its action. Used to control aphids, beetles and soil insects in ornamental plants, fruit trees, vines, vegetables, cereals, cotton, rice, coffee, etc. Degraded by complete hydrolysis of the phosphate ester and also by partial hydrolysis of the *O*-ethyl groups.

In Egypt during a period of 20 years diarrhoeal infections have occurred from eating fruit contaminated with inappropriate pesticides, including triazophos.

The No Effect Limit in rats is $1\,mg\,kg^{-1}$ of diet fed over 2 years.

Further reading
Dinham, B. (1993) *The Pesticide Hazard*, Pesticide Trust, ZED Books, London.

TRICHLOROACETIC ACID CAS No. 76-03-9
Herbicide. LD_{50} rat $400\,mg\,kg^{-1}$; Sodium salt: LD_{50} rat $3200\,mg\,kg^{-1}$

$$Cl_3CCO_2H \qquad\qquad C_2HCl_3O_2 \quad M = 163.4$$

1,1,1-trichloroethanoic acid; m.p. 56°; b.p. 197°; v.p. 1 mmHg (50°) very soluble in water ($939\,g\,l^{-1}$, 20°)

Trichloroacetic acid irritates the eyes, burns the skin and damages the respiratory system. The sodium salt is less toxic but the ammonium salt is a carcinogen. It is also used as a fixative in microscopy and to precipitate protein. It is a mammalian metabolite of trichloroethylene.

The UK exposure limit is $5\,mg\,m^{-3}$ and in the USA the TLV (TWA) is $7\,mg\,m^{-3}$. The WHO guideline for drinking water is $0.1\,mg\,l^{-1}$

TRICHLORFON CAS No. 52-68-6
(Dipterex, Cekufon)
Non-systemic insecticide. Toxicity class II. LD_{50} rat $250\,mg\,kg^{-1}$; LC_{50} rainbow trout $0.7\,mg\,l^{-1}$. Not toxic to bees

$$\underset{\underset{MeO}{|}}{\overset{\overset{O}{\|}}{MeO-P}}-\underset{\underset{OH}{|}}{\overset{\overset{H}{|}}{C}}-CCl_3 \qquad\qquad C_4H_8Cl_3O_4P \quad M = 257.4$$

Dimethyl 2,2,2-trichloro-1-hydroxyethylphosphonate; m.p. 84°; v.p. 0.21 mPa (20°); water solubility $120\,g\,l^{-1}$.

First reported by Unterstenhofer and introduced by Bayer in 1957.

It is mobile in soils but is rapidly metabolized. It is excreted by mammals as mono- and dimethylphosphates.

Trichlorfon is manufactured in Brazil and is in common use in Paraguay and elsewhere for the control of insect pests such as flies, cockroaches, fleas, bed bugs, ants etc., and also as fly bait in farm buildings and for domestic animal parasites.

China has set a workplace limit(TWA) of $1.0\,mg\,m^{-3}$

Codex Alimentarius: acceptable daily intake $0.01\,mg\,kg^{-1}$ (1978). Maximum residue levels: soya bean 0.1, spinach 0.5, sugar beet $0.05\,mg\,kg^{-1}$. The EC has an MRL of $0.1\,mg\,kg^{-1}$ for cereals only.

1,1,1-TRICHLOROETHANE CAS No. 71-55-6
LD_{50} female rat $11\,g\,kg^{-1}$; LC_{50} fish $33–110\,mg\,l^{-1}$

$$CH_3—CCl_3 \qquad\qquad C_2H_3Cl_3 \quad M = 133$$

Methylchloroform; m.p. −30°; b.p. 74°; dens. 1.339; water solubility $0.95\,g\,l^{-1}$. Very soluble in acetone, benzene and $CHCl_3$. 1 ppm $= 5.40\,mg\,m^{-3}$. Non-flammable except for the vapour above 500°.

It is produced either by the addition of HCl to vinyl chloride and further chlorination, or by the addition of HCl to 1,1-dichloroethylene:

$$CH_2\!\!=\!\!CH\text{-}Cl \ + \ HCl \longrightarrow CH_3\!\!-\!\!CHCl_2 \xrightarrow[\text{UV}]{Cl_2} CH_3\!\!-\!\!CCl_3$$

or

$$CH_2\!\!=\!\!CCl_2 \ + \ HCl \xrightarrow[300^\circ]{FeCl_3} CH_3\!\!-\!\!CCl_3$$

In 1988 world production was 680 000 t; production in Western Europe in 1990 was 229 000 t. Trichloroethane is used for degreasing metals and as a solvent for adhesives and spot removers and in aerosol cans. During welding of degreased metal phosgene can be formed from residual solvent.

It leaches readily into groundwater where it is slowly hydrolysed to 1,1-dichloro-ethene, hydrogen chloride and acetic acid. The half-life in water at 20° is 1.7–2.8 years, at 80° it is about 4 h. 1,1,1-Trichloroethane may be detected in drinking water by GC–MS at a level of $0.1\,\text{mg}\,l^{-1}$ and in blood by GC–ECD at $0.05\,\text{mg}\,l^{-1}$

Some levels found in water (1975) were rain water 0.005–0.009, river Rhine 0.01–3300 and Liverpool Bay 0.25–3.3 ppb. A 1981 review of a 100 German cities found levels in the range of <0.1–$1.7\,\mu\text{g}\,l^{-1}$.

Some 13% of anthropogenic chlorine is due to 1,1,1-trichloroethane and it has been detected in air far from areas of its manufacture and use, some typical levels were: Runcorn, UK (1973) 16 ppb; urban Los Angeles (1972) 0.1 2.3 ppb; Barbados (1985) 90–144 ppt; South Pole (1980) 102 ppt. A typical indoor median level is $17\,\mu\text{g}\,\text{m}^{-3}$. It has been found in butter at a maximum level of $7500\,\mu\text{g}\,\text{kg}^{-1}$ but at a mean of $16\,\mu\text{g}\,\text{kg}^{-1}$; in margarine the mean level was $45\,\mu\text{g}\,\text{kg}^{-1}$.

Inhalation by humans has the following consequences: at 70 000 ppm, death within a few minutes; above 5000 ppm, death within 10 min; at 1000 ppm, mild eye discomfort and lack of coordination on brief exposure.

Abuse of 1,1,1-trichloroethane as a solvent in spot remover caused 29 deaths in the 1960s. During 1971–81 20 deaths in the UK were attributed to TCE with an overall total of 140 deaths.

The ACGIH has set a TLV (TWA) of 350 ppm ($1910\,\text{mg}\,\text{m}^{-3}$ 1976).

Further reading
Environ. Health Criteria, World Health Organization, Geneva, 1992, Vol. 136.

TRIFLURALIN CAS No. 1582-09-8
(Treflan, Tri-4)
Selective soil herbicide. Toxicity class III or IV. LD_{50} rat $> 5\,\text{g}\,\text{kg}^{-1}$; LC_{50} rainbow trout $10\,\mu\text{g}\,l^{-1}$

$C_{13}H_{16}F_3N_3O_4$ $M = 335.3$

α,α,α-Trifluoro-2,6-dinitro-N,N-dipropyl-p-toluidine; m.p. 49°; v.p. 9.5 mPa (25°); water solubility $0.2\,\text{mg}\,l^{-1}$(25°).

First reported by Alder *et al.* in 1960 and introduced in 1961 by Eli Lilly.

Trifluralin acts by disrupting cell division and by inhibiting root growth. It is used for the pre-emergence control of weeds and annual grasses in brassicas, root vege-tables, vines, citrus fruits, strawberries, cotton, sugar cane, etc.

The activity persists for 6–8 months but trifluralin is not mobile in soil; metabolism results in N-dealkylation, reduction of NO_2 and partial hydrolysis of the CF_3 group to CO_2H.

The USEPA has set a limit in drinking water of $5\,\mu g\,l^{-1}$, the Canadian limit is $45\,\mu g\,l^{-1}$.

TUNGSTEN CAS No. 7740-33-7

W

Tungsten, a member of Group 6 of the Periodic Table, has an atomic number of 74, an atomic weight of 183.86, a specific gravity of 19.3 at 20°, a melting point of 3422° and a boiling point of 5555°. Natural tungsten contains five stable isotopes: ^{180}W (0.12%), ^{182}W (26.50%), ^{183}W (14.31%), ^{184}W (30.642%) and ^{186}W (28.426%). Many unstable radionuclides are known. It is a shiny white metal and forms compounds in oxidation states +2 to +6.

The abundance of tungsten in the earth's crust is $\sim 1.2\,ppm$. Tungsten occurs in *wolframite*, $(Fe, Mn)WO_4$, *scheelite*, $CaWO_4$, *huebnerite*, $MnWO_4$, and *ferberite*, $FeWO_4$. Typical concentrations in the environment are scarce but include: air $< 1.5\,\mu g\,m^{-3}$; sea water $\sim 0.1\,\mu g\,l^{-1}$; intake for adult humans up to $13.0\,\mu g\,day^{-1}$, human blood 0.001 ppm, bone 0.00025 ppm.

Uses. High-speed tool steels. Special alloys. Filament in electric tubes and bulbs. X-ray targets. Heating elements for electric furnaces. Space/rocket material.

Not essential for humans or animals although several tungsten-containing enzymes have been identified. Main occupational hazard is the exposure during manufacture of tungsten carbide, WC. Exposure leads to two types of respiratory diseases, bronchial asthma and progressive interstitial fibrosis. Also dust from the crushing and milling of ores. Tungsten dust is flammable.

Occupational exposure limit. ACGIH TLV (TWA) $5\,mg\,W\,m^{-3}$ insoluble compounds, $1\,mg\,m^{-3}$ soluble compounds.

Further reading
King, R.B. (ed.) (1994) *Encyclopedia of Inorganic Chemistry*, Wiley, Chichester, 1994, Vol. 8, pp. 4240-4268.
Thompson, R. (ed.) (1995) *Industrial Inorganic Chemicals. Production and Uses*, Royal Society of Chemistry, Cambridge.
Wennig, R. and Kirsch, N. (1988) Tungsten, in Seiler, H. G., Sigel, H. and Sigel, A. (eds), *Handbook on Toxicity of Inorganic Compounds*, Marcel Dekker, New York, pp. 731–738.
Ullmann's Encyclopedia of Industrial Chemistry, 5th edn, VCH, Weinheim, 1996, Vol. A27, pp. 229–266.

Tungsten carbide CAS No. 12070-12-1

WC

Uses. Metalworking, mining and petroleum industries.

Toxicity. Irritant as dust. May be harmful by inhalation, ingestion and skin absorption.

Tungsten(IV) disulphide CAS No. 12138-09-9

WS$_2$

Uses. High-temperature dry lubricant.

Toxicity. Irritant to mucous membranes. May be harmful by all routes.

Tungsten(VI) oxide CAS No. 1314-35-8

WO$_3$

Uses. Catalytic converters. Manufacture of tungstates used in X-ray machines.

Toxicity. LD$_{50}$ oral rat 1059 mg kg^{-1}.

URANIUM CAS No. 7440-61-1

U

Uranium, a member of the actinide group of elements, has an atomic number of 92, an atomic weight of 238.03, a specific gravity of 18.95 at 20°, a melting point of 1135° and a boiling point of 4131°. All uranium isotopes are radioactive. Naturally occurring uranium contains only three isotopes: ^{238}U (99.275%, $t_{1/2} = 4.51 \times 10^9$ years), ^{235}U (0.720%, $t_{1/2} = 7.1 \times 10^8$ years) and ^{234}U (0.0055%, $t_{1/2} = 2.45 \times 10^5$ years). Uranium is a heavy, silvery white metal forming compounds in the +2, +3, +4, +5 and +6 oxidation states.

The earth's crust contains about 2.4 ppm of uranium. It is found in well over 100 minerals, such as *pitchblende, uraninite* (U$_3$O$_8$), *carnotite, davidite and tobernite*. It is also found in *phosphate rock, lignite* and *monazite sands*, which are the present commercial sources. Uranium can be prepared by reducing uranium halides with sodium or calcium. Typical concentrations in the environment are: sea water 1–3 ppb, river water 0.5 µg l^{-1}, drinking water 0.4–1.4 µg l^{-1}, soil 3.4–10.5 mg kg^{-1} dry weight, food 0.08–70 µg l^{-1}, granite rock ~15–80 ppm, daily intake for humans ~1.9 µg/ 0.023 Bq, blood 5×10^{-4} mg l^{-1}. Total body content 90 µg/1.12 Bq.

Uses. Uranium is of great importance as a nuclear fuel. Uranium-238 together with a small percentage of uranium-235 is used to fuel nuclear power reactors for the generation of electricity. The use of nuclear fuels to make isotopes for peaceful purposes and to make explosives (uranium-235) is well known. Starting material for the synthesis of transuranium elements. In X-ray targets for production of high energy X-rays. Pottery and tile glazes. Colouring glass. Manufacture of armour-piercing weapons.

Since uranium is present over all the earth, everyone normally eats or drinks a small amount of uranium daily. About 99% of it leaves the body within a few days. The remaining 1% enters the bloodstream, and most of it is eliminated in the urine in a few days. A very small amount of the original 1% is stored in bones. Uranium and its compounds are highly toxic, from both chemical and radiological standpoints. The mining and milling of uranium ores produce large quantities of low-level radioactive wastes, including the emission of uranium in dusts, with potential effects on workers of damage to any organ and tissue. Uranium dust is also pyrophoric and presents a dangerous fire hazard. The handling of uranium or its compounds can cause dermatitis. The critical organ for chemical toxicity is the kidney, causing structural damage to tubules. Constant exposure to U, an emitter of low specific activity, leads to chronic radiation sickness and bone sarcomas. Exposure limits are inhalation 3×10^4 and ingestion 5×10^4 Bq. Chronic exposure to insoluble U compounds trapped in the lungs leads to blood disorders, pulmonary fibrosis and pneumoconiosis. A case of acute poisoning was caused by accidental exposure to uranium hexafluoride, UF_6. Rupture of a tank containing the hexafluoride led to two deaths. Survivors had damaged lungs and the gastrointestinal tract was also affected.

Depleted uranium, a form of recycled radioactive waste, was used for the first time in the Gulf war. It makes the tips of shells 2.5 times denser than the hardest steel and capable of piercing all known tank armour. According to international risk estimates, the quantity of depleted uranium used (50 000 lb), if inhaled, could cause 500 000 deaths from cancer.

Occupational exposure limits. OSHA PEL (TWA) 0.05 mg U m^{-3} soluble compounds, 0.2 mg U m^{-3} insoluble compounds. Finely divided uranium is flammable.

Further reading
Buckart, W. (1988) Radiotoxicity, in: Seiler, H. G., Sigel, H. and Sigel, A. (eds), *Handbook on Toxicity of Inorganic Compounds*, Marcel Dekker, New York, pp. 805–827.
ECETOX (1991) *Technical Report 30(4)* European Chemical Industry Technology and Toxicology Centre. Brussels.
Eisenbud, M. (1987a) *Environmental Radioactivity.From Natural, Industrial and Military Sources*, 3rd edn, Academic Press, New York.
Eisenbud, M. (1987b) *Environmental Radiochemistry*, Academic Press, Orlando, FL.
Fisher, D. R. (1988) Uranium, in Seiler, H. G., Sigel, H. and Sigel, A. (eds), *Handbook on Toxicity of Inorganic Compounds*, Marcel Dekker, New York, pp. 739–748.
SI 1991 No. 472 The Environmental Protection (Prescribed Processes and Substances) Regulations 1991, HMSO, London, 1991.
ToxProfiles (1990) *Uranium*, ATSDR, Atlanta, GA.
Ullmann's Encyclopedia of Industrial Chemistry, 5th edn, 1996, VCH, Weinheim, Vol. A27, pp. 281–332.

Uranium hexafluoride CAS No. 7783-81-5

UF_6

Uses. Separation of uranium-235 and uranium-238.

Toxicity. Radioactive poison. Corrosive. Irritant to skin, eyes and mucous membranes.

Uranyl acetate CAS No. 541-09-3

$(CH_3CO_2)_2UO_2$

Uses. Dry copying inks.

Occupational exposure limits. US TLV (TWA) 0.2 mg U m^{-3}.

Toxicity. Very toxic by inhalation or if swallowed.

VANADIUM CAS No. 7447-62-2

V

Vanadium, a member of Group 5 of the Periodic Table, has an atomic number of 23, an atomic weight of 50.94, a specific gravity of 6.11 at 20°, a melting point of 1910° and a boiling point of 3407°. Only two isotopes are naturally occurring: ^{50}V (0.25%) and ^{51}V (99.75%). Several radionuclides, $^{46-49}$V, $^{52-54}$V, have been obtained artificially. Vanadium is a bright, white, soft, ductile metal, slightly radioactive with valencies from +2 to +5.

Widespread in nature with over 70 minerals, including *patronite*, VS$_4$, *carnotite*, K(UO$_2$)(VO$_4$).1.5H$_2$O, *roscoelite*, Mg–Fe–V, *vanadinite*, Pb–V, and *mottramite*, Pb–Cu–V, ores. Vanadium is also found in some crude oils up to 0.1%, particularly those from Venezuela and Canada. Abundance in the earth's crust about 136 ppm, 19th element in order of relative abundance. The pure metal is obtained by reduction of the pentoxide with calcium.

Vanadium is released into the atmosphere from metal works and the burning of fossil fuels. In the production of vanadium, dust contains 0.1–30 mg m^{-3}. Vanadium compounds are common air contaminants: rural < 1 ng m^{-3}, urban–industrial 400–1300 ng m^{-3}. Other mean V concentrations are sea water 2 μg l^{-1} (as H$_2$VO$_4^{2-}$), drinking water 0.3–200 μg l^{-1}, vegetables ~ 1 ppb (dry weight). Main sources of vanadium as food intake: sea food 2.4–44 ppb, saltwater fish 0.03 mg kg^{-1}, milk 0.13 ng g^{-1}, probable human daily intake 10–30 μg. Vanadium concentration is generally low in muscle, blood, fat, < 1 ng g^{-1} (ng ml^{-1}), with higher values in liver, kidney and lungs (1.1–30 ng g^{-1}). Human blood serum up to 0.03–0.05 ng ml^{-1}. Average total human body content (adult, 70 kg) is 20 mg.

Widely distributed in most biological systems (0.5–2 ppm dry weight). Bioaccumulation of vanadium occurs in various species; sea squirts accumulate up to 10 000 times the concentration in the surrounding sea water; the mushroom *Amanita muscaria* can concentrate up to 100 times the level of other mushrooms. Vanadium is known to be an essential element for animal nutrition; it is generally non-toxic at low levels. The daily requirement for humans is around 0.1 ppm. Although esssential for growth of certain bacteria and algae, it is not essential for higher plants. It may enhance chlorophyll formation. Of biochemical interest is the ability of certain organisms to bioaccumulate vanadium in their blood. For example, the ascidian seaworm *Phallusia mammilia* has a blood concentration of V up to 1900 ppm. Vanadium also occurs in the protein haemovanadin.

Uses.   Its main use is in the manufacture of stainless steel and together with its compounds as an industrial catalyst.

The toxicity of vanadium on both animals and humans is well documented. Vanadium in high concentrations can cause toxic symptoms in humans. On injection, the toxicity of vanadium compounds is dependent on the nature of the compound administered, but it is poisonous in all but minute doses. On ingestion, vanadium is only slightly less toxic than selenium and more so than tellurium, molybdenum and arsenic. Inhalation of vanadium pentoxide, V$_2$O$_5$, dust at a concentration of 70 mg m^{-3} is fatal within a few hours. The levels of vanadium in coal conversion and petroleum refining can cause particular problems. No matter by which route the metal is absorbed, it acts by depressing the respiratory centre, constricting periphereal arteries of the viscera and causing hyperperistalsis and enteritis. It has been shown to be one the few metal ions or their compounds that can modify amino acid secretion. The effects of the vanadate(V) ion on the cardiovascular system have been studied. Numerous enzymes are known to be inhibited by vanadium complexes.

Vanadium poisoning gives rise to a number of systemic effects, consisting of polycythaemia, followed by red blood cell destruction and anaemia, loss of appetite, pallor, emaciation, albuminuria, and haematuria, gastrointestinal disorders, nervous complaint and cough, sometimes severe enough to cause haemocoptysis. More recent reports describe symptoms that, for the most part, are restricted to the conjunctivae and respiratory systems, no evidence being found of disturbances of the blood or the central nervous system. However, analysis of the blood of metal workers handling V showed levels of 2.5–$54.5\,\mu g\,l^{-1}$ (normal $<2.5\,\mu g\,l^{-1}$). Vanadate (VO_3^-) is a potent inhibitor of the sodium pump, that is, of an enzyme universally present in eukaryotic organisms. The absorption of V_2O_5 by inhalation is nearly 100%. Chimney sweeps who inhaled 0.7–$14\,mg\ V_2O_5$ per day in soot from the cleaning operation had urine levels between 0.15 and $13\,\mu g\,l^{-1}$ (normal $< 3\,\mu g\,l^{-1}$). Dermatitis was observed in workers exposed to levels of $6.5\,\mu g\,m^{-3}$. High exposure levels 5–$150\,mg\,m^{-3}$ cause atrophic rhinitis and bronchitis. Although certain workers believe that it is only the pentoxide that is harmful, other investigators have found that patronite dust (chiefly vanadium sulphide) is toxic to animals, causing acute pulmonary oedema. Acute poisoning in animals by ingestion of vanadium compounds causes nervous disturbances, paralysis of legs, respiratory failure, convulsions, bloody diarrhoea and death. Poisoning by inhalation causes bleeding of the nose and acute bronchitis. Some compounds have reported mutation effects.

Vanadium is a natural component of fuel oil and exposed workers have developed vanadium poisoning, which is characterized by chronic respiratory effects and eczematous skin lesions. Vanadium has been reported to have insulin-like properties and has been demonstrated to be beneficial in the treatment of diabetic animals. A relationship exists between the vanadium concentration in drinking water and incidence of caries in children; the higher the level, the fewer caries that develop.

Occupational exposure limits.   OSHA (TLV) TWA $0.05\,mg\,V\,m^{-3}$ (respirable dust and fume). OSHA PEL-Cl $0.1\,mg\,(V_2O_5)\,m^{-3}$.

Toxicity. LD_{50} subcutaneous rabbit $59\,mg\,kg^{-1}$.

Further reading
ECETOX (1991) *Technical Report 30(4)*, Euopean Chemical Industry Ecology and Toxicology Centre, Brussels.
Environmental Health Criteria 81. Vanadium, WHO/IPCS, Geneva, 1988.
King, R. B. (ed.) (1994) *Encyclopedia of Inorganic Chemistry*, Wiley, Chichester, Vol. 8, pp. 4304–4321.
Lagerkvist, B. (1986) Vanadium, in Friberg, L., Nordberg, G. F. and Voux, V. B. (eds), *Handbook on the Toxicology of Metals*, 2nd edn, Elsevier, Amsterdam, Vol. II, pp. 638–663.
Leonard, A. and Gerber, G. B. (1994) Mutagenicity, carcinogenicity and teratogenicity of vanadium compounds, *Mutat. Res.*, **317**, 81–88.
ToxFAQs (1995) *Vanadium*, ATSDR, Atlanta, GA.
Wennig, R. and Kirsch, N. (1988) Vanadium, in Seiler, H. G., Sigel, H. and Sigel, A. (eds), *Handbook on Toxicity of Inorganic Compounds*, Marcel Dekker, New York, pp. 749–765.

Ullmann's Encyclopedia of Industrial Chemistry, 5th edn, VCH, Weinheim, 1996, Vol. A27, pp. 367–386.

Vanadium (V) oxytrichloride CAS No. 7727-18-6

$VOCl_3$

Uses. Catalyst.

Toxicity. Corrosive. LD_{50} oral rat $140 \, mg \, kg^{-1}$.

Vanadium (V) pentoxide CAS No. 1314-62-1

V_2O_5

Uses. Catalyst in organic and inorganic chemistry. Mordant in dyeing.

Toxicity. Poison by all routes. Possible mutagen. A respiratory irritant affecting the lungs. LD_{50} oral rat $10 \, mg \, kg^{-1}$; oral mouse $5 \, mg \, kg^{-1}$

VINYL ACETATE CAS No. 108-05-4
LD_{50} rat $2.5 \, g \, kg^{-1}$

$$CH_3CO_2CH = CH_2 \qquad\qquad C_4H_6O_2 \quad M = 86.1$$

Ethenyl acetate, VAC: b.p. 72°; v.p. 100 Torr (23°); dens. 0.932 (20°); log P_{ow} 0.73. Water solubility $20 \, g \, l^{-1}$ (20°); soluble in acetone, benzene, diethyl ether, etc. Polymerizes in the light. Ethereal odour, threshold 0.05–0.5 ppm ($1.7 \, mg \, m^{-3}$).

Originally prepared by Klatte in 1912 by the direct addition of acetic acid to acetylene, catalysed by Hg(II) salts. In 1994 production in the USA was 3.0×10^6 t; the industrial route is:

$$CH_2{=}CH_2 \ + \ HO{-}\overset{\overset{\displaystyle O}{\|}}{C}{-}CH_3 \ \xrightarrow[\substack{Pd/Cu \\ salts}]{O_2} \ CH_2{=}CH{-}O{-}\overset{\overset{\displaystyle O}{\|}}{C}{-}CH_3$$

It is also obtained industrially by the reaction of vinyl chloride with sodium acetate, catalysed by $PdCl_2$.

Vinyl acetate is used to prepare polymers and copolymers. Poly(vinyl acetate) – adhesives for paper, wood, glass and metals; also in latex water paints, a base for inks and lacquers, an emulsifier in cosmetics and food additives. Ethylene–VAC – hot melt and pressure sensitive adhesives, food packaging and medical tubing. PV chloride/acetate – cable and wire coverings, protective garments.

It may be determined in air by adsorption on Carbowax, desorption with CH_2Cl_2– MeOH followed by GC with FID, when a $1 \, \mu g$ sample is detectable. It can be quantified in groundwater at a concentration of $50 \, \mu g \, l^{-1}$. It has been found at a level of $50 \, mg \, l^{-1}$ in the wastewater from a manufacturing plant.

It has been estimated that in 1983 129 000 employees were exposed to vinyl acetate in the USA. In 1992 levels of 0.07–0.57 ppm (0.25–$2.0 \, mg \, m^{-3}$) were detected in the vicinity of industrial units. The release to air in the whole of the USA in 1987 was 4000 t (8.8×10^6 lb). Emissions indoors come from carpets and cigarette smoke; one cigarette emits 400 ng.

The threshold for irritation of the nose and throat is 10–22 ppm (35–$77\,\mathrm{mg\,m^{-3}}$). Rats and mice suffered no toxic effects from drinking water containing VAC at 5000 ppm. It is rapidly metabolized by esterases in human blood to acetaldehyde and acetic acid.

The ACGIH has set a TLV (TWA) of 10 ppm ($35\,\mathrm{mg\,m^{-3}}$) and have classed it as a group A3 carcinogen – carcinogenic in animals, but at relatively high doses not related to human industrial exposure.

Further reading
Carraher, C.E. (1996) *Polymer Chemistry*, 4th Edn, Marcel Dekker, New York.

VINYL CHLORIDE CAS No. 75-01-4

$CH_2=CHCl$ $\qquad\qquad\qquad C_2H_3Cl \quad M = 62.5$

Chloroethene, VCM; b.p. $-14°$; v.p. 240 Torr ($-40°$); dens. 0.91; v. dens 2.15. Solubility in water $1.1\,\mathrm{g\,l^{-1}}$; soluble in most organic solvents.

Vinyl chloride was originally made by the addition of hydrogen chloride to acetylene. The modern process is one of oxychlorination. Ethylene is passed over a $CuCl_2$ catalyst supported on alumina at $225°$ and 3 atm pressure; potassium chloride is also present as an activator:

$$CH_2{=}CH_2 \;+\; 2\,CuCl_2 \;\longrightarrow\; \underset{\underset{Cl}{|}}{CH_2}{-}\underset{\underset{Cl}{|}}{CH_2} \;+\; 2\,CuCl$$

Oxygen and hydrogen chloride are used to regenerate the $CuCl_2$ via an intermediate double salt:

$$4\,CuCl \;+\; O_2 \longrightarrow 2\,CuCl_2{\cdot}CuO \xrightarrow{\;4\,HCl\;} 4\,CuCl_2 \;+\; 2\,H_2O$$

Vinyl chloride is finally obtained from the dichloroethane by cracking at $500°$ with the elimination of HCl, which is recycled.

In 1994 in the USA, 24% of the total chlorine production was used to make 14.8×10^9 lb of vinyl chloride, making it the sixteenth in order of bulk chemicals. It is used for the production of polyvinyl chloride (PVC), when it is polymerized in glass-lined steel kettles at $50°$ with a peroxide catalyst. Growth proceeds to give a granular material of about 1000 monomer units with a molecular weight of about 63 000. Vinyl chloride is also consumed in the manufacture of copolymers and for the synthesis of chloroethylenes; the use of these substances is decreasing because of their persistence and tendency to accumulate in body fat.

Vinyl chloride is formed from trichloro- and tetrachloroethylene in water bodies by the action of methanogenic bacteria; this may account for an incident in Los Angeles in 1993 when leakage of VCM from a waste tip was detected indoors. A survey of UK rivers at 70 different sites revealed no vinyl chloride above the detection limit of $0.005\,\mathrm{\mu g\,l^{-1}}$. In air, vinyl chloride reacts with the OH radical with a half-life of 1.2 days; the reaction with ozone has a half-life of 4.0 days.

Vinyl chloride is a primary carcinogen, inducing tumours of the liver and blood in those occupationally exposed, e.g. workers engaged in PVC production. It is also mutagenic in rats and mice and is a suspected cause of human mutations. Large cohorts of exposed workers have been studied long-term in the USA, UK, Canada and elsewhere, leading to the conclusion that there exists a significant risk of angio-sarcoma of the liver. This was first reported in 1974.

The activity of vinyl chloride is consequent upon metabolic change induced by the cytochrome P450 system, which converts it into the epoxide and then by rearrangement to chloroacetaldehyde:

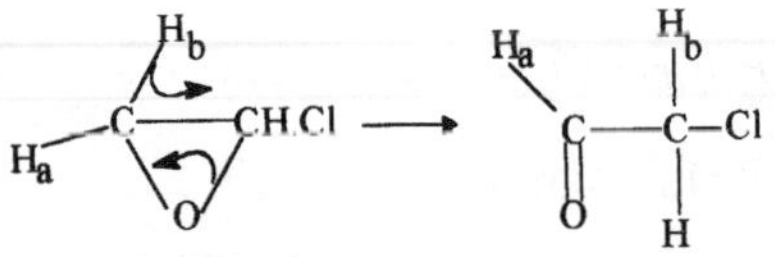

In view of its toxicity, vinyl chloride is no longer available as a laboratory reagent, however, economic demands are paramount as its manufacture currently accounts for 13% of total ethylene consumption in the USA. The applications of PVC are so extensive that its production must continue and there are at present no substitutes.

The ACGIH and the IARC rate vinyl chloride as a group 1 carcinogen. The TLV (TWA) is 5 ppm ($13\,mg\,m^{-3}$), which is higher than the OSHA Permissible Exposure Limit (1 ppm).

Further reading
Berlin, A., Draper, M., Krug, E., Rui, R. and Van der Venne, M. Th. (eds) (1990) *The Toxicology of Chemicals – Carcinogenicity*, Commission of the European Communities, Brussels, Vol. 2, pp. 127–136.

WARFARIN CAS No. 81-81-2
5543-58-8 (*R*-isomer)
Rodenticide. Toxicity class 1. LD_{50} rat $186\,mg\,kg^{-1}$; LC_{50} harlequin fish $12\,mg\,l^{-1}$

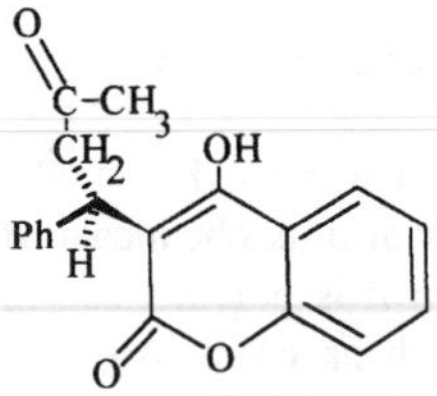

$C_{19}H_{16}O_4$ $M = 308.3$

(*R*) (+) isomer

3-(α-Acetonylbenzyl)-4-hydroxycoumarin; m.p. 162°; v.p. 9.0 Pa (21°); water solubility $17\,mg\,l^{-1}$

Reported to be an anticoagulant by Link in 1944 and manufactured by All-India Medical.

Warfarin controls rats and mice by inducing internal bleeding following repeated ingestion. The (*S*)(−)-isomer has seven times the activity of the (*R*)(+)- form.

In the UK and USA the workplace limit is TWA $0.1\,mg\,m^{-3}$; no standards have been set for water.

WARFARE AGENTS
See **Mustard Gas, Phosgene, Sarin, Soman, Tabun.**

XENON CAS No. 7440-63-3

Xe

Xenon, a member of Group 18 (noble gases) of the Periodic Table, has an atomic number of 54, an atomic weight of 131.30, a specific gravity (gas) of 5.897 at STP, a melting point of $-111.9°$ and a boiling point of -108 . The stable oxidation state is 0. Compounds with $+2$, $+4$, $+6$ and $+8$ oxidation states are strongly oxidizing, very reactive to explosive. Natural xenon is composed of nine stable isotopes with atomic masses in the range 124–136: ^{124}Xe (0.10%), ^{126}Xe (0.09%), ^{128}Xe (1.91%), ^{129}Xe (26.4%), ^{130}Xe (4.1%), ^{131}Xe (21.2%), ^{132}Xe (26.9%), ^{134}Xe (10.4%) and ^{136}Xe (8.9%).

Xenon is present in the atmosphere to the extent of about one part in 2×10^7. It is obtained commercially by extraction from liquid air.

Uses. Electron tubes. Electroluminescent plasma display screens. Stroboscopic lamps. Lasers. Bubble chambers in atomic energy.

^{133}Xe, with a half-life of 5.2 days, has useful applications as a radioactive tracer. It was one of the isotopes liberated as a result of the Chernobyl accident.

Xenon can act as an asphyxiant. Xenon is not toxic but its compounds are highly toxic because of their strong oxidizing characteristics. They also liberate the noxious hydrogen fluoride on reaction with water. Examples of xenon compounds are xenon difluoride, XeF_2, xenon tetrafluoride, XeF_4, xenon hexafluoride, XeF_6, xenon trioxide, XeO_3, and and sodium perxenate, Na_4XeO_6.

Further reading
Kirk–Othmer ECT, 4th edn, Wiley, New York, 1995, Vol. 13, pp. 1–53.
King, R. B. (ed.) (1994) *Encyclopedia of Inorganic Chemistry*, Wiley, Chichester, Vol. 5, pp. 2660–2680.

XYLENES

Me

C_8H_{10} $M = 106.2$

Me

Dimethylbenzenes. Technical xylene is a mixture of the three isomers whose properties are summarized below. All three are classified as dangerous under the EC directive (76/464/EEC).

Xylene	CAS No.	B.p. (°)	V.p. (20°) (Torr)	Dens.	Water solubility (20°) (mg l^{-1})	Log P_{ow}
ortho	95-47-6	144	5	0.88	175	2.77
meta	108-38-3	139	6	0.86	147	3.20
para	106-42-3	138	6.5	0.86	198	3.15

The xylenes are flammable with a sweet odour. They are produced industrially by the reforming of crude oil (b.p. 75–140°): after separation of other products, such as benzene and toluene, the *o*-xylene can be separated from its isomers by further fractionation. *p*-Xylene (m.p. 13°) may be separated from *m*-xylene by cooling at −65°.

The mixture of xylenes is used as a solvent and for blending in gasoline; they also constitute 10% of the composition of paints (primers, fillers and topcoats) – production in the UK (1991) was 380 × 10^3 t. *o*-Xylene is the precursor for phthalic anhydride and *p*-xylene is the precursor for terephthalic acid.

The LD_{50} values for the rat are mixture 4300, *ortho* 3567, *meta* 4988 and *para* 3910 mg kg^{-1}. The lowest LC_{50} values are *o*-xylene, rainbow trout 7.6 mg l^{-1}, and *p*-xylene, striped bass 1.7 mg l^{-1}.

The xylenes cause irritation of the eyes, nose and throat at a level of 200 ppm. Higher levels lead to headache, nausea, anorexia and flatulence; there may be injury to the liver and kidneys and nervous system (Riihimaki and Hanninen, 1990).

In air, high levels of 77 mg m^{-3} have been found in samples from landfill sites. In 10 US cities typical levels were 4–20 μg m^{-3} and alongside a motorway in the UK 1060 μg m^{-3} was recorded. Vehicle emissions (UK) in 1986 were cars (gasoline) 34 × 10^3, cars (diesel) 210 and trucks (gas and diesel) 3.2 × 10^3 t year^{-1}. 358 breath samples from New Jersey residents gave a median concentration of 2.2 μg m^{-3} with a maximum of 220 μg m^{-3}. Workers in paint spray booths were exposed to an average level of 27 μg l^{-1}, more than 10 times that of the residents (Howard, 1991).

The xylenes are rapidly decomposed in air by reaction with OH radicals. It is important to exclude them from groundwater. They were detected in landfill leachate at Hamilton, Ontario at 12–30 μg l^{-1} (*ortho*) and 12–191 μg l^{-1} (*para*), and rainwater levels at Los Angeles were 0.011 μg l^{-1} (*meta* + *para*). In the UK, surface levels are generally below 0.01 μg l^{-1}.

The ACGIH has set a TLV (TWA) for the mixed isomers at 100 ppm (434 mg m^{-3}). In the UK the STEL is 150 ppm (650 mg m^{-3}).

References

Crookes, M. J., Dobson, S. and Howe, P. D. (1993) *Environmental Hazard Assessment: Xylenes*, Building Research Station, Watford.

Howard, P. H. (ed.) (1991) *Handbook of Environmental Fate and Exposure Data for Organic Chemicals, Vol. 2, Solvents*, Lewis, Chelsea, MI.

Riihimaki, V. and Hanninen, O. (1990) in Snyder, R. (ed.) *Ethel Browning's Toxicity and Metabolism of Industrial Solvents, Vol. 1, Hydrocarbons*, Elsevier, Amsterdam.

YTTERBIUM CAS No. 7440-64-4

Yb

Ytterbium, a member of the lanthanide or rare earth elements, has an atomic number of 70, an atomic weight of 173.04, a specific gravity of 6.97 (α-form) at 20°, a melting point of 824° and a boiling point of 1194°. Natural ytterbium is a mixture of seven stable isotopes, with atomic masses ranging from 168–176: ^{168}Yb (0.13%), ^{170}Yb (3.05%), ^{171}Yb (14.3%), ^{172}Yb (21.9%), ^{173}Yb (16.1%), ^{174}Yb (31.8%) and ^{176}Yb (12.7%) A bright, silvery and soft metal, it forms compounds in the +2 and +3 oxidation states.

Average concentration in the earth's crust is ∼ 3 ppm. Ytterbium occurs along with other rare earths in a number of rare minerals. It is commercially obtained from *monazite sand* (~0.03% Yb), *euxenite* and *xenotime*. The pure metal is obtained by reducing the trichloride with potassium. Ytterbium levels in human organs are very low.

Uses. Reducing agent of some organic functional groups. Stress gauges. ^{169}Yb ($t_{1/2} = 32$ days) is used in medical diagnostics.

No biological role. Ytterbium and its salts are of low toxicity but as for all lanthanides they should be handled carefully. Ytterbium compounds are eye and skin irritants. They may act as an anticoagulant. The metal dust is a fire and explosion hazard.

Further reading
Gschneidner, K. A., Jr, and LeRoy, K. A. (eds) (1988) *Handbook on the Physics and Chemistry of the Rare Earths*, North-Holland, Amsterdam.
King, R. B. (ed.) (1994) *Encyclopedia of Inorganic Chemistry*, Wiley, Chichester.
Kirk–Othmer ECT, 4th edn, Wiley, New York, 1995, Vol. 14, pp. 1091–1115.
Thompson, R. (ed.) (1995) *Industrial Inorganic Chemicals. Production and Uses*, Royal Society of Chemistry, Cambridge.

Ytterbium (III) chloride CAS No. 10035-01-5

YbCl$_3$

Uses. Preparation of ytterbium compounds and complexes.

Toxicity. Poisonous if injected or ingested into human body. Irritant to mucous membranes and upper respiratory tract. LD$_{50}$ oral mouse 4836 mg kg^{-1}; intraperitoneal mouse 300 mg kg^{-1}.

Ytterbium (III) nitrate pentahydrate CAS No. 35725-34-9

Yb(NO$_3$)$_3$.5H$_2$O

Uses. Preparation of ytterbium compounds and complexes.

Toxicity. Moderately toxic. LD$_{50}$ oral rat 1623 mg kg^{-1}; oral mouse 126 mg kg^{-1}. Affects male reproduction system.

YTTRIUM CAS No. 7440-65-5

Y

Yttrium, a member of Group 3 of the Periodic Table, has an atomic number of 39, an atomic weight of 88.91, a specific gravity of 4.47 at 20°, a melting point of 1527° and a boiling point of 3338°. It is a silver–grey metal and only the +3 oxidation state is stable in solution. It has only one natural isotope, ^{89}Y, but radionuclides ^{90}Y and ^{91}Y are products of atomic explosions.

The abundance in the earth's crust is 28 ppm, 32nd in order of relative abundance. Yttrium is found together with other lanthanides and occurs in minerals such as *monazite, bastnasite* and *xenotime* (YPO$_4$). Typical concentrations in the environment include the following: sea water 0.01–0.3 ppb, soils 2.5–250 ppm; in healthy humans, bones 70 ppb, liver 10 ppb, blood 6 ppb.

Uses. Red phosphor in colour televisions and fluorescent tubes. Superconductors. Ceramics. Electronics. Yttrium alloys in nuclear chemistry. Laser (YAG = yttrium aluminium garnet). ^{90}Y ($t_{1/2} = 64$ h) is used in radio(immuno)therapy.

Not essential for humans or animals. Soluble salts poorly absorbed in the stomach, $\sim 0.05\%$. Y^{3+} has the ability to replace Ca^{2+}. Most severe effects for experimental animals occur after inhalation of yttrium dusts or from injection of yttrium compounds.

Occupational exposure limit: OSHA TLV (TWA) $1\,mg\,Y\,m^{-3}$. Yttrium dust is flammable.

Further reading

Gschneidner, K. A., Jr, and LeRoy, K. A. (eds) (1988) *Handbook on the Physics and Chemistry of the Rare Earths*, North-Holland, Amsterdam.

Gerhardson, L., Wester, P. O., Nordberg, G. F. and Brune, D. (1984) Chromium, cobalt and lanthanum in lung, liver and kidney tissues from deceased smelter workers, *Sci. Total Environ.*, **37**, 233–246.

Kirk–Othmer ECT, 4th edn, Wiley, New York, 1995, Vol. 14, pp. 3595–3618.

Venugopal, B. and Luckey, T. D. (1975) *Metal Toxicity in Mammals*, Plenum, New York, Vol. 2.

Yttrium (III) chloride hexahydrate CAS No. 10025-94-2

$YCl_3.6H_2O$

Uses. Preparation of the pure metal.

Toxicity. Affects liver and spleen. LD_{50} intraperitoneal mouse $158\,mg\,kg^{-1}$, rat $45\,mg\,kg^{-1}$.

Yttrium (III) nitrate pentahydrate CAS No. 57584-28-8

$Y(NO_3)_3.5H_2O$

Uses. Preparation of other yttrium compounds.

Toxicity. Eye and skin irritant. LD_{50} intraperitoneal mouse $1710\,mg\,kg^{-1}$; LD_{50} intraperitoneal rabbit $515\,mg\,kg^{-1}$; LD_{50} oral rat $71\,g\,kg^{-1}$.

Yttrium (III) oxide CAS No. 1314-36-9

Y_2O_3

Uses. Phosphors for colour TV. Yttrium–iron garnets for microwave filters. Stabilizer for high-temperature services materials.

Toxicity. LD_{50} intraperitoneal rat $230\,mg\,kg^{-1}$, mouse $430\,mg\,kg^{-1}$.

ZINC CAS No. 7440-66-6

Zn

Zinc, a member of Group 12 of the Periodic Table, has an atomic number of 30, an atomic weight of 65.39, a specific gravity of 7.13 at $20°$, a melting point of $419.58°$ and a boiling point of $907°$. It is a bluish white metal composed of five stable isotopes: ^{64}Zn (48.6%), ^{66}Zn (27.9%), ^{67}Zn (4.1%), ^{68}Zn (18.8%) and ^{70}Zn (0.6%). ^{65}Zn (half-life 243.8 days) and ^{69m}Zn (half-life 13.8 h) are used in biological experiments. The stable oxidation state of zinc is $+2$.

Widely distributed on all continents, 80 ppm ($80\,mg\,kg^{-1}$), 24th in order of abundance. Principal ores are *sphalerite* (zinc blende, cubic ZnS), *wurtzite* (hexagonal ZnS), *smithsonite* ($ZnCO_3$) and *hemimorphite* [$Zn_4(OH)_2Si_2O_7.H_2O$]. Also found in cadmium, copper, lead, arsenic, selenium, antimony, silver and gold ores. Emissions from zinc smelters can give rise to serious pollution in the local area with deposition of zinc affecting growth in trees and vegetation. Zinc is present in coal. Typical distributions in the environment are: soil $20\,mg\,kg^{-1}$, sea water $0.003–0.6\,\mu g\,l^{-1}$, drinking water $<0.2\,mg\,l^{-1}$, air $0.05–4.0\,\mu g\,m^{-3}$, sewage sludge $500–20\,000\,mg\,kg^{-1}$ dry weight, plants $15–100\,mg\,kg^{-1}$ dry weight, meat products, vegetables and cereals $15–60\,mg\,kg^{-1}$ wet weight, fish $\sim 15\,mg\,kg^{-1}$ wet weight, bones of mammals $75–170\,ppm$, human milk $3\,mg\,l^{-1}$. Some sea creatures bioaccumulate zinc: molluscs (up to $50\,ppm$), crustaceans ($7–100\,ppm$) and oysters ($100–2000\,ppm$). Zinc intake for adult humans is $7–15\,mg\,day^{-1}$; the daily requirement for zinc is $3–10\,mg$ depending on age. Total zinc in the human body (adult, $70\,kg$) is around $2300\,mg$ (30% in bones, 60% in muscles). Concentrations in various organs in the human body are estimated to be ($mg\,kg^{-1}$ wet weight) retina 130, nails 17–304, bone 53–117, kidney 25–85, thyroid 24–37, liver 8–16, skin 6–19, brain 13–39. Whole blood contains $4.8–9.3\,mg\,l^{-1}$.

Uses. Batteries. Dental alloy. Brazes. Brass and other zinc alloys. Protective coating for iron and steel. Zinc dust is used as a pigment and reducing agent.

Zinc as an essential trace element can impair life functions of plants, animals and humans, by either deficiency or surplus. It is a component of more than 200 enzymes and proteins which are necessary in important metabolic processes. It is a constituent of human cells, and several enzymes depend on it as a co-factor. Zinc is vital for the healthy working of many of the body's systems. It helps with the healing of wounds and is particularly important for healthy skin. Confirmed zinc-containing enzymes include alcohol dehydrogenase, superoxide dismutase, RNA and DNA polymerase, pyruvate carboxylase, carboxy- and aminopeptidases and alkaline phosphatase. The toxicity of zinc is species dependent; in plants it is low and is observed only in soils with zinc levels greater than $300–400\,mg\,kg^{-1}$ the symptom being growth retardation. In contrast, some fish species are much more sensitive to zinc concentration and have LC_{50} values in the range $1–10\,mg\,l^{-1}$ due to interference with the operation of gills, affecting respiration and blood circulation. Zinc and many of its compounds are very toxic to animals and humans and can cause acute/chronic poisoning, with a wide range of clinical symptoms. High dosage of zinc salts can result in serious damage to the upper alimentary tract and severe shock. Inhalation of zinc dust causes zinc fume fever – loss of appetite, thirst, fatigue, severe chills, blood content of sugar rises, usually lasts 2–3 days. Competition for sites between Zn^{2+} and the other divalent metals such as calcium, lead, cadmium and copper is believed to be the main reason for zinc toxicity.

Deficiency of zinc in plants, animals and humans arises from inadequate zinc intake, which can usually be overcome by the addition of zinc salts as therapy. A common symptom for all species is stunted growth. Clinical symptoms for humans include delayed wound healing, neurological disorders, anorexia, night blindness, taste and smell malfunction and delay of sexual maturity.

In Japan, an outbreak of a severe disease affecting the skeletal–muscular system occurred among people growing rice in fields heavily polluted by zinc sulphide. Accidental ingestion of $3\,oz$ ($85\,g$) of liquid zinc chloride resulted in pharyngitis and oesophagitis, hypocalcaemia, lethargy and confusion. The effects of excessive zinc inhalation (metal fume fever) include metallic taste, irritability, insomnia, sweating, flu-like symptoms, cough and shortness of breath, emphysema.

Occupational exposure limits.   OEL TWA $0.1 \, mg \, Zn \, m^{-3}$.   Zinc dust is flammable and can cause explosions.

Further reading

ECETOX (1991) *Technical Report 30(5)*, Euopean Chemical Industry Ecology and Toxicology Centre, Brussels.

Flinder, C. J. (1986) Zinc, in Friberg, L., Nordberg, G. F. and Vouk, V. B. (eds), *Handbook on the Toxicology of Metals*, 2nd edn, Elsevier, Amsterdam, Vol. II, pp. 664–679.

King, R. B. (ed.) (1994) *Encyclopedia of Inorganic Chemistry*, Wiley, Chichester, Vol. 8, pp. 4393–4450.

Prasad, A. S. (1993) Essentiality and toxicity of zinc, *Scand. J. Work Environ. Health* **19**, 134–136.

SI 1991 No. 472 The Environmental Protection (Prescribed Processes and Substances) Regulations 1991, HMSO, London, 1991.

Walsh, C. T., Stansted, H. H., Prosad, A. S., Newberne, P. M. and Fraker, P. J. (1994) Zinc health effects and research priorities for the 1990's, *Environ. Health. Perspect. Rev.*, **102**, 5–46.

Zinc (II) chloride CAS No. 7646-85-7

$ZnCl_2$

Uses.   Soldering and welding flux. Dry batteries. Cotton processing. Fire proofing.

Toxicity.   Toxic and corrosive. Severe irrritant, cauterizing effect on skin and mucous membranes. Fumes attack the lungs. Possible mutagen. LD_{50} oral rat $350 \, mg \, kg^{-1}$. LD_{Lo} oral human $50 \, mg \, kg^{-1}$.

Zinc (II) chromate CAS No. 13530-65-9

$ZnCrO_4$

Uses.   Pigment. Corrosion inhibitor in paints.

Toxicity.   Human carcinogen.

Zinc (II) diethyldithiocarbamate CAS No. 14324-55-1

$Zn[SC(S)N(C_2H_5)_2]_2$

Uses.   Natural rubber preservative. Rubber vulcanization accelerator.

Toxicity. LD_{50} oral rat $570 \, mg \, kg^{-1}$.

Zinc (II) oxide CAS No. 1314-13-2

ZnO

Uses.   Pigment in the rubber industry. Glass. Enamels. Plastics. Lubricants. Chemicals. Pharmaceutical agent for burns and ointments.

Toxicity.   Dust may cause metal fume fever. LD_{Lo} oral human $500 \, mg \, kg^{-1}$. LD_{50} oral mouse $7950 \, mg \, kg^{-1}$.

Zinc (II) phosphide CAS No. 1314-84-7

Zn_3P_2

Uses. Rodenticide.

Toxicity. Toxic. Reacts with moist air or water to produce highly toxic phosphine. LD_{50} oral rat $570\,mg\,m^{-3}$.

Zinc (II) stearate CAS No. 557-05-1

$Zn[CH_3(CH_2)_{16}CO_2]_2$

Uses. Soap. Lubricants. Waterproofing of textiles, paper and concrete.

Toxicity. LD_{Lo} intratracheal rat $250\,mg\,kg^{-1}$.

Zinc (II) sulphate heptahydrate CAS No. 7446-20-0

$ZnSO_4.7H_2O$

Uses. Rayon manufacture. Supplement for humans, animals and plants for zinc deficiency.

Toxicity. LD_{Lo} oral human $106\,mg\,kg^{-1}$. LD_{50} oral rat $2150\,mg\,kg^{-1}$. Ingestion of 1–2 g caused nausea and vomiting.

Zinc (II) sulphide CAS No. 1314-98-3

ZnS

Uses. White pigment. Luminous dials. Phosphors in TV and X-ray machines. Semiconductors.

Toxicity. N.a. Avoid acids which produce hydrogen sulphide.

ZIRCONIUM CAS No. 7440-67-7

Zr

Zirconium, a member of Group 4 of the Periodic Table, has an atomic number of 40, an atomic weight of 91.22, a specific gravity of 6.51 at 20°, a melting point of 1853° and a boiling point of 4409°. It is a greyish white, lustrous metal, very slightly radioactive. It has five stable isotopes in nature: ^{90}Zr (51.45%), ^{91}Zr (11.2%), ^{92}Zr (17.15%), ^{94}Zr (17.38%) and ^{96}Zr (2.8%, $t_{1/2} > 3.6 \times 10^{17}$ years). Two oxidation states occur, +2 and +4, the latter being the most stable. The pure metal is obtained by reduction of the tetrachloride with magnesium (Kroll process).

The abundance of zirconium in the earth's crust is $150–300\,mg\,kg^{-1}$, 20th in order of relative abundance. Important minerals are *zircon*, $ZrSiO_4$ and *baddeleyite*, ZrO_2. Obtained commercially as a by-product of titanium production. Typical concentrations in the environment include the following: fresh water 0.002–0.02 ppm, sea water 0.02–0.5 ppb, plants 0.3–2 ppm, dry weight; present in human blood $<0.1\,\mu g\,g^{-1}$; the average daily dose for adults is $\sim0.4\,mg$. The adult (70 kg) body has been estimated to contain $200–300\,mg$ of zirconium. Found in blood, brain, kidney, liver and lung.

Uses.   Aerospace industry. Chemicals. Surgical instruments. Nuclear reactor technology. Superconductive magnets. Zirconium compounds are used in weather-proofing textiles, dyes and pigments, tanning leather, glass and ceramics. Precious stone, zircon.

Not essential for humans or animals. Low toxicity for insoluble zirconium compounds. Zirconium salts are much more toxic after intraperitoneal dosage. Exposure to zirconium(IV) chloride, $ZrCl_4$, over 60 days at a concentration of $6\,mg\,m^{-3}$ resulted in decreased haemoglobin in workers. Allergic skin reactions occurred in individuals who were regular users of lotions and deodorants containing zirconium salts. Occupational hazard of workforce involved in zirconium and its compounds production include cardiac pain, reduced haemoglobin levels, pulmonary granulomas and, specifically, zirconium-containing aerosols, lung granulomas and sodium zirconium lactate skin granulomas. Cases of pneumoconiosis in workers involved in the production of zirconium(IV) oxide, ZrO_2, and zirconium silicate, $ZrSiO_4$.

Occupational exposure limit.   OSHA PEL (TWA) total dust $15\,mg\,m^{-3}$; respirable fraction $5\,mg\,m^{-3}$. STEL $10\,mg\,m^{-3}$. Zirconium dust is flammable.

Toxicity. LD_{50} intravenous rabbit $150\,mg\,kg^{-1}$.

Further reading

King, R. B. (ed.) (1994) *Encyclopedia of Inorganic Chemistry*, Wiley, Chichester, Vol. 8, pp. 4475–4488.

Venugopal, B. and Luckey, T. D. (1975) *Metal Toxicity in Mammals*, Plenum, New York, Vol. 2.

Zirconium(IV) chloride CAS No. 10026-11-6

$ZrCl_4$

Uses.   Catalyst. Synthesis of organozirconium compounds. Electrolyte.

Toxicity.   Toxic and corrosive. Inhalation may be fatal as a result of spasm, inflammation and oedema of the larynx and bronchi, chemical pneumonitis and pulmonary oedema. LD_{50} oral rat $1688\,mg\,kg^{-1}$.

Zirconium(IV) fluoride CAS No. 7783-64-4

ZrF_4

Uses.   Component of molten salts for nuclear reactors.

Toxicity.   Corrosive. Causes burns. Harmful if inhaled, ingested or through skin contact.

Zirconium(IV) hydride CAS No. 7704-99-6

ZrH_4

Uses.   Catalyst. Nuclear reactors. Pyrotechnics. Reducing agent.

Toxicity.   Harmful by all routes. Irritant.

Zirconium(IV) oxide CAS No. 1314-23-4

ZrO_2

Uses. Gem material. Glass and ceramics industries.

Toxicity. May be harmful by inhalation, ingestion or absorption through the skin. Irritant to mucous membranes and upper respiratory tract.

Zirconium(IV) oxychloride CAS No. 7699-43-6

$ZrOCl_2$

Uses. Antiperspirant. Clay stabilizer.

Toxicity. N.a.

Zirconium(IV) nitrate CAS No. 13746-89-9

$Zr(NO_3)_4$

Uses. Preservative.

Toxicity. Low toxicity. Strong oxidizing agent.

Zirconium(IV) sulphate hydrate CAS No. 34806-73-0

$Zr(SO_4)_2.xH_2O$

Uses. Catalyst. Tanning of leather.

Toxicity. Harmful by all routes. Inhalation can cause spasm, inflammation, chemical pneumonitis and pulmonary oedema. LD_{50} oral rat $3500 \, mg \, kg^{-1}$.

Appendix A

Some Medical Terms Used in the Text

ACIDOSIS a serious condition in which the blood pH falls due to kidney disease or diabetes

ALBUMINURIA increased excretion of albumin in the urine

ANAEMIA decrease in the concentration of haemoglobin and in the number of red blood cells

APLASTIC ANAEMIA a serious and usually lethal blood disorder, in which all cellular elements of bone and marrow are reduced in number due to failure of blood cell precursors to reproduce (see also Pernicious anaemia)

ASCITES accumulation of fluid in the peritoneal cavity

ATAXIA inability to coordinate voluntary muscle movement

BILIRUBINAEMIA the presence of bilirubin in the blood, a normal condition

BRADYCARDIA slow heart beat – fewer than 55 beats per minute

BRADYKINESIA abnormal slowness of voluntary movement

CARCINOGEN a cancer-inducing substance or action. Direct-reacting carcinogens do not require metabolic activation, e.g. nitrogen mustards, dimethyl sulphate. Procarcinogens are converted into carcinogens by metabolism in the organism, e.g. nitrosamines, polycyclic aromatic hydrocarbons

Carcinogen Classes

ACGIH

A1 confirmed human carcinogen

A2 suspected human carcinogen

A3 animal carcinogen at relatively high doses

A4 no adequate data – not classifiable in humans or animals

A5 no evidence of human carcinogenesis based on reliable studies

IARC

Group 1 carcinogenic in humans, even if only sufficient in animals with similar mechanism to humans

2A probably carcinogenic in humans

2B possibly carcinogenic in humans

3 Insufficient data for humans – not classifiable

4 Probably not carcinogenic in humans

CENTRAL NERVOUS SYSTEM (CNS) the brain and its continuation down the spinal chord – entirely encased in bone.

CHLOROSIS a form of anaemia due to iron deficiency; or loss of pigmentation in green plants (see also Anaemia)

CHOLINOLYTIC counteracting the action of acetylcholine

CHOROIDITIS inflammation of the choroid (foetal membrane)

COMA a state of deep unconsciousness from which the person cannot be aroused by stimulation. It can result from head injury, lack of oxygen, poisoning, liver and kidney failure

CONJUNCTIVITAL IRRITATION irritation of the transport membrane attached around the cornea

CORNEA the outer and principal lens of the eye through which the coloured iris can be seen

CYANOSIS a blueness of the skin arising from insufficient oxygen in the blood

DISTAL situated at a point beyond or remote from a reference point

DIURETIC DRUGS those which cause an increased output of urine

DYSPHAGIA painful difficulty in swallowing due to disorders of the oesophagus (gullet)

DYSPNEA difficulty in breathing

DYSPROTEINAEMIA any disease associated with abnormal serum proteins

EMESIS vomiting

EMPHYSEMA abnormal distension of tissues by gases, dilation and destruction of the alveoli

ENTERITIS inflammation of the mucosa of the small intestine

ERYTHROPOIESIS formation of erythrocytes

FIBROSIS excessive growth of fibrous connective tissue

GASTROINTESTINAL relating to the stomach and intestines

GINGIVITIS inflammation of the gums

GONADOTROPIC acting to stimulate the gonads

GRANULOMATA tumour-like tissue or nodules of granulation tissue

HAEMOLYSIS loss of haemoglobin from the red blood cells

HEPATIC relating to the liver

HYPERCALCAEMIA elevated concentration of calcium in the blood or blood serum

HYPERCALCIURIA increased excretion of calcium in urine

HYPERPARATHYROIDISM abnormally increased activity of the parathyroid glands

HYPERPERISTALSIS excessively active intestinal peristalsis (successive waves of contraction)

HYPERTENSION abnormally high blood pressure; potentially dangerous as it can lead to arterial damage and adverse effects on kidneys, brain and the eyes

HYPOCALCAEMIA reduced concentration of calcium in the blood or blood serum

HYPOKALAEMIA abnormally low potassium concentration in the blood

HYPOTENSION low blood pressure, which may result from surgical shock or from standing upright for long periods.

HYPOTHERMIA abnormal lowering of the body temperature

INTRAPERITONEAL being or occurring within the peritoneal cavity

INTRAVENOUS into a vein

ISCHAEMIC HEART DISORDER inadequate blood flow to the heart

LARYNX the 'Adam's apple' or voicebox, situated at the upper end of the windpipe

MENINGISM a group of symptoms and signs suggesting meningitis

MITOSIS the division of the nucleus of a cell to produce two identical daughter nuclei

MUCOSA mucus secreting cells

MUCOUS MEMBRANE lines most body cavities and hollow internal organs – mouth, nose, eyelids, intestine and vagina. It contains many cells which secrete mucous so as to moisten and lubricate the surface

MUTAGEN an agent capable of changing the structure of DNA without killing the affected cell; they include ionizing radiation, UV, X-rays, γ-rays, cosmic rays and many chemicals. The consequent mutations may be perpetuated in the cell's descendants

MYOCARDIUM central muscle layer of the wall of the heart

NAUSEA the unpleasant feeling of sickness which may precede vomiting

NECROSIS structural changes which follow the death of body tissue

NEOPLASTIC new abnormal uncontrolled growth of a tissue

NEURASTHAEMIA a condition characterized by fatigue, weakness, multiple aches and pains and insomnia

NEURON a functional unit of the nervous system. A single cell with very long fibre-like extensions carrying nervous impulses and interconnecting with each other at synapses.

NEUROPATHY any neuron disorder inducing pathological changes in the peripheral nervous system

NEUROSIS any long-term mental or behavioural disorder in which contact with reality is retained and in which the condition is recognized as abnormal by the sufferer

OEDEMA pathological accumulation of fluid in the tissue spaces

OESOPHAGITIS inflammation of the oesophagus

OLIGURIA pathologically diminishes excretion of urine

OSTEOMALACIA weakening of the bones, usually due to lack of calcium incorporation

PALPITATION abnormal awareness of the action of the heart due to rapid or irregular action

PARATHESIA numbness of the fingers, hands and feet

PERIPHERAL NERVOUS SYSTEM (PNS) the entire complex that extends from the CNS to supply muscles, skeleton, organs and glands

PERITONEUM the double-edged membrane that lines the inner wall of the abdomen and which covers abdominal organs

PHARYNGITIS inflammation of the pharynx

PHARYNX the common passage to the gullet and the windpipe – a muscular tube lined with mucous membrane

PHLEBOTHROMBOSIS presence of a clot in a vein

PHYTOTOXIC toxic to plants

PLUMBOPORPHYRIA increased blood concentration/urine excretion of porphyrins or porphyrin precursors such as δ-aminolevulinic acid

PNEUMOCONIOSIS lung disease resulting from the deposition of inhaled particles in the lungs

PNEUMONITIS inflammation of the lung

POLYCYTHAEMIA pathological increase in the number of red cells, thrombocytes and granulocytes in the blood

PULMONARY OEDEMA the accumulation of fluid in the lungs leading to severe breathlessness; treatment is with diuretic drugs

RENAL relating to the kidneys

RETINITIS inflammation or degeneration of the retina

RHINITIS inflammation of the mucous membranes of the nose

SEROUS relating to, containing or producing serum

SERUM a clear watery fluid, especially that which moistens the surface of serous membranes

TERATOGEN an agent or factor, such as radiation, that is capable of inducing developmental malformation of the foetus

VERTIGO the illusion that the environment or one's body is rotating. It may arise from travel sickness, alcohol intake, anxiety, drugs or abnormally deep or rapid breathing

Further reading
The principal source of the above was:
Youngson, R. M. (1992) *Collins Dictionary of Medicine*, Harper–Collins, Glasgow.

For a more detailed interpretation see:
Stedman's Medical Dictionary, 26th edn, Williams and Wilkins, Baltimore, 1995.

Further reading

PRINCIPAL SOURCES

American Conference of Governmental Industrial Hygienists (1991), *Documentation of TLVs and Biological Exposure Indices*, 5th edn, ACGIH, Cincinnati, OH.

Chemical Safety Data Sheets (1991–95) Royal Society of Chemistry, Cambridge, vols 1–6.

Commission for the Investigation of Health Hazards of Chemical Compounds (1995) *Maximale Arbeitsplatz-Konzentration (MAK Values)* Deutsche Forschungs gemeinschaft (DFG), VCH, New York.

ECETOX (1991) *Technical Report 30(4)*, European Chemical Industry Technology and Toxicology Centre, Brussels.

Environmental Health Criteria (1976–97) *Monographs 1–167*. UNEP, ILO and WHO, Geneva.

Health and Safety Executive (1995) *EH 40/95*, HSE Books, HMSO, London.

Howard, P.H. (ed.) (1989–93) *Handbook of Environmental Fate and Exposure Data for Organic Chemicals*, Lewis, Chelsea, MI., Vols 1–4.

International Association for Research on Cancer (1972–97) *Monographs 1–62*, IARC, Lyon.

King, R.B. (Ed, 1994) *Encyclopedia of Inorganic Chemistry*, Wiley, (Chichester).

Kirk–Othmer, Encyclopedia of Chemical Technology, 4th edn, Wiley, New York, 1991–96.

Lewis, R.J. (1993) *Hazardous Chemicals Desk Reference*, Van Nostrand Reinhold, New York.

Lewis, R.J. (1994) *Rapid Guide to Hazardous Chemicals in the Workplace*, 3rd edn, Van Nostrand Reinhold, New York.

Lewis, R.J. (ed.) 1996, *Sax's Dangerous Properties of Industrial Materials*, 9th edn, Van Nostrand Reinhold, New York.

The Pesticide Manual (1994), 10th edn, Crop Protection Council/Royal Society of Chemistry, Cambridge.

Richardson, M.L. and Gangolli, S. (1994) *Dictionary of Substances and Their Effects*, Royal Society of Chemistry, Cambridge.

S.I. 1991 No. 472 The Environmental Protection (Prescribed Processes and Substances) Regulations 1991, HMSO, London, 1991

Sittig, M. (1994) *Worldwide Limits for Toxic and Hazardous Chemicals in Air, Water and Soil*, Noyes, New York.

ToxFAQs (1991–96) Agency for Toxic Substances and Disease Registry (ATSDR) Atlanta, GA.

INORGANIC CHEMICALS

Ashby, P.J. and Craig, P.J. (1996) Organoelement compounds in the environment, in Harrison, R.M. (ed.), *Pollution: Causes Effects and Control*, 3rd edn, Royal Society of Chemistry, Cambridge, pp. 309–342.

Cox, P.A. (1995) *Inorganic Chemistry in the Environment, The Elements on Earth*, Oxford University Press, Oxford.

Eisenbud, M. (1987) *Environmental Radioactivity from Natural, Industrial and Military Sources*, 3rd edn, Academic Press, New York.

Fergusson, J.E. (1990) *The Heavy Elements: Chemistry, Environmental Impact and Health Effects*, Pergamon Press, Oxford.

Freiburg, L., Nordberg, G.F. and Voux, V.B. (eds) (1986) *Handbook on the Toxicology of Metals*, 2nd edn, Elsevier, Amsterdam, Vols 1 and 2.

Greenwood, N.N. and Earnshaw, A. (1984) *Chemistry of the Elements*, Pergamon Press, Oxford.

King, R.B. (ed.) (1994) *Encyclopedia of Inorganic Chemistry*, Wiley, Chichester.

Merian, E. (ed.) (1991) *Metals and Their Compounds in the Environment*, VCH, Weinheim.

Seiler, H. G., Sigel, H. and Sigel, A. (eds) (1988) *Handbook on the Toxicity of Inorganic Compounds*, Marcel Dekker, New York.

Thompson, R. (1995) *Industrial Inorganic Chemicals, Production and Uses*, Royal Society of Chemistry, Cambridge.

ORGANIC CHEMICALS

Ackermann, P., Jagerstad, M. and Ohlsson, T. (1995) *Foods and Packaging Materials*, Royal Society of Chemistry, Cambridge.

Ash, M. and Ash, I. (1996) *Handbook of Paints and Coatings*, Gower, Aldershot, Vols 1 and 2.

Carraher, C.E. (1996) *Polymer Chemistry*, 4th edn, Marcel Dekker, New York.

Verschueren, K. (1996) *Handbook of Environmental Data on Organic Chemicals*, 3rd edn, Van Nostrand Reinhold, New York.

Whim, B.P. and Johnson, P.G. (1996) *Directory of Solvents*, Blackie, London.

Witcoff, H.A. and Reuben, B.G. (1996) *Industrial Organic Chemicals*, Wiley, New York.

ANALYTICAL METHODS

Fifield, F.W. and Haines, P.J. (1995) *Environmental Analytical Chemistry*, Blackie, London.

National Institute of Occupational Safety and Health (1984) *Manual of Analytical Methods*, NIOSH, Cincinnati, OH, Vols 1–3.

Occupational Safety and Health Administration (1985) *Analytical Methods Manual*, OSHA, Salt Lake City, UT, Vols 1–3.

Skoog, D.A. and Leary, J.J. (1992) *Principles of Instrumental Analysis*, 4th edn, Harcourt, Brace, Jovanich, New York.

USEPA (1993) *Manual of Chemical Methods*, 2nd edn, AOAC International, Arlington, VA.

Wagner, R. E., Kotas, W. and Yogis, G.A. (1992) *Environmental Analytical Methods*, 2nd edn, Genium, Schenectady, NY.

Watson, C. (1994) *Official and Standardised Methods of Analysis*, Royal Society of Chemistry, Cambridge.

GENERAL TEXTBOOKS

Alloway, B.J. and Ayres, D.C. (1997) *Chemical Principles of Environmental Pollution*, 2nd edn, Blackie, London.

Coultate, P. (1996) *Food, the Chemistry of its Components*, 3rd edn, Royal Society of Chemistry, Cambridge.

Dinham, B. (1993) *The Pesticide Hazard*, The Pesticide Trust, ZED Books, London.

Emsley, J. (1994) *The Consumer's Good Chemical Guide*, Freeman, San Francisco.

Haigh, N. (1991) *Manual of Environmental Policy in the EC and Britain*, Longman, Harlow.

Jones, J.M. (1992) *Food Safety*, Eagan, St Paul, MN.

Palma, R.J. and Espenscheid, M. (1994) *The Complete Guide to Household Chemicals*, Prometheus, New York.

Porteous, A. (1994) *Dictionary of Environmental Science and Technology*, Wiley, Chichester.

Selinger, B. (1986) *Chemistry in the Market Place*, Australian National University Press, Canberra.